Molecular Manufacturing

ELECTRONICS AND BIOTECHNOLOGY ADVANCED (EL.B.A.) FORUM SERIES

Volume 1 FROM NEURAL NETWORKS AND BIOMOLECULAR ENGINEERING TO BIOELECTRONICS
Edited by Claudio Nicolini

Volume 2 MOLECULAR MANUFACTURING
Edited by Claudio Nicolini

Molecular Manufacturing

Edited by

Claudio Nicolini

Institute of Biophysics
University of Genoa
Genoa, Italy

Assistant Editor
Sergei Vakula

EL.B.A. Foundation
Portoferraio, Italy

Plenum Press • New York and London

Library of Congress Cataloging-in-Publication Data

Molecular manufacturing / edited by Claudio Nicolini ; assistant editor, Sergei Vakula.
p. cm. -- (Electronics and biotechnology advanced forum series ; v. 2)
"Proceedings of the 1994 International Workshops on Electronics and Biotechnology Advanced, held March 10-12, 1994, and September 5-15, 1994, on the Isle of Elba, Italy"--T.p. verso.
Includes bibliographical references and index.
ISBN 0-306-45284-7
1. Nanotechnology--Congresses. I. Nicolini, Claudio A. II. Vakula, Sergei. III. International Workshops on Electronics and Biotechnology Advanced (1994 : Isle of Elba, Italy) IV. Series.
T174.7.M65 1996
660'.63--dc20 96-5249
CIP

Proceedings of the 1994 International Workshops on Electronics and Biotechnology Advanced, held March 10 – 12, 1994, and September 5 – 15, 1994, on the Isle of Elba, Italy

ISBN 0-306-45284-7

A Division of Plenum Publishing Corporation
233 Spring Street, New York, N. Y. 10013

10 9 8 7 6 5 4 3 2 1

Printed in the United States of America

PREFACE

The present volume is a continuation of the EL.B.A. Forum Series, which was initiated in the spring of 1995 with the first volume, entitled *From Neural Network and Biomolecular Engineering to Bioelectronics*, in which a brief outline of modern bioelectronics given as "*the use of biological materials and biological architectures for information processing and sensing systems and devices down to molecular level.*"

The present volume highlights the aspects of advanced biotechnology and electronics originating from molecular manufacturing, which has been emerging as an independent branch of research.

This volume appears in a crucial moment, when significant progress has already been made in this strategic field and when technologies derived from it are recognized as critical for the welfare of our society.

In addition, acknowledging to the Italian Ministry of University and Scientific and Technological Research for launching the National Research Program "Technologies for Bioelectronics" in 1992 and for continuation of support of this advanced multidisciplinary research, we would like to acknowledge the support of the National Research Council of Italy through the "Molecular Manufacturing" CNR Strategic Project since 1994.

The significant unique role of Technobiochip in the organization of the EL.B.A. Forums and in bringing to light the enormous industrial potential of bioelectronics is duly acknowledged, as well as its attraction and support of top level scientists to the series of EL.B.A. Forums of which this volume is part.

I would like to express my gratitude particularly to Dr. Sergey Vakula of the EL.B.A. Foundation, acknowledging also the precious assistance of Dr. Paolo Occhialini of the Scientific and Technological Park of the Elba Island, and Dr. Isabella Zolfino of Technobiochip.

Claudio Nicolini
Member of the National Science and Technology Council
President of the EL.B.A. Foundation

CONTENTS

Molecular Manufacturing

1

FROM PROTEIN NANOTECHNOLOGY TO PROTEIN AUTOMATA

Year Zero

Claudio Nicolini

Institute of Biophysics, University of Genoa
Via Giotto, 2, 16153 Genoa Sestri Ponente - ITALY
EL.B.A. Foundation
Marciana Marina (LI), Italy

BACKGROUND

Recent works on protein monolayers (Nicolini *et al.* 1993, 1995a,b,c; Shen *et al.* 1993; Facci *et al.* 1994a, b) appear to have for their exceptional thermal stability "the kind of implications that might capture the immediate interest of a wide audience", as testified by the enthusiastic News and Views published in Nature time ago (Hampp, 1993). Typically, (Langmuir-Blodgett) LB films of most organic compounds are formed in a Langmuir trough by spreading the solutions at the liquid-gas interface and transferring by a vertical lift technique onto solid substrate. In our laboratories protein monolayers are instead usually prepared with horizontal lift (Langmuir-Schaefer technique properly modified) and deposited on solid substrates in the range of surface pressure between 20 and 35 mN/m.

In order to outline the unique structural and functional properties and the numerous potential applications of this protein nanotechnology, we intend in this overview to briefly summarize all the recent structural and functional assays both in solution and in thin film by a combination of circular dichroism, surface potential measurements, atomic force microscopy, scanning tunnelling microscopy, transmission electron microscopy, differential scanning calorimetry, NMR, infrared and EPR spectroscopy, X-ray diffractometry, absorption spectra, nanogravimetry, ellipsometry, CCD high resolution image analysis and electric (DC) measurements.

PROTEIN SYSTEM

Nearly all chosen proteins have been well characterized in solution by X-ray crystallography and their three-dimensional structure is known at atomic resolution (Table I). Up to 13 different proteins, both water-soluble and trans-membrane, have been utilized

Molecular Manufacturing, edited by C. Nicolini
Plenum Press, New York, 1996

Table 1. X-Ray Crystallographic Data of Proteins from B.N.L. Data Bank

PROTEIN	SOURCE	HEIGHT(Å) front view	WIDTH (Å) top view	DEPTH(Å) side view	SHAPE	REF
Glutathione S-Transferase	Rat Liver	54	32 + 32 *	54	sphere	X. Ji et al ., 1992
Alcohol Dehydrogenase	Horse Liver	70	54	48	ellipsoid	F. Colonna-Cesari et al., 1986 H. Eklund et al., 1984
Alkaline Phosphatase	Escherichia coli	101	62	68	ellipsoid	E.E. Kim et al., 1991
Thioredoxin	Escherichia Coli	33	34	30	sphere	S.K. Katti et al., 1990
Azurin	Pseudomonas aeruginosa	42	31	30	tetrahedron	E.T. Adman et al., 1981
Cytochrome C551	Pseudomonas aeruginosa	28	32	29	sphere	Y. Matsuura et al., 1982
Cytochrome C	Bonito Heart	38	33	32	sphere	N. Tanaka et al. 1975
Cytochrome P450 SCC	Bovine	73	71	66	tetrahedron	S. Vijayakumar et al., 1992
Antibodies	Fab fragment and 2 Fac fragment	125	40	90	Y shape	A.T. Brunger et al., 1991
Photosynthetic Reaction Center	Rhodobacter sphaeroides	90	28 + 28 *	68	ellipsoid	T.O. Yeates et al., 1988
Photosynthetic Reaction Center	Rhodopseudomonas viridis	125	55	60	ellipsoid	J. Deisenhofer et al., in press
Bacteriorhodopsin	Halobacterium halobium	48	43	45	parallelepiped	K.C. Chou et al., 1992

* two subunit.

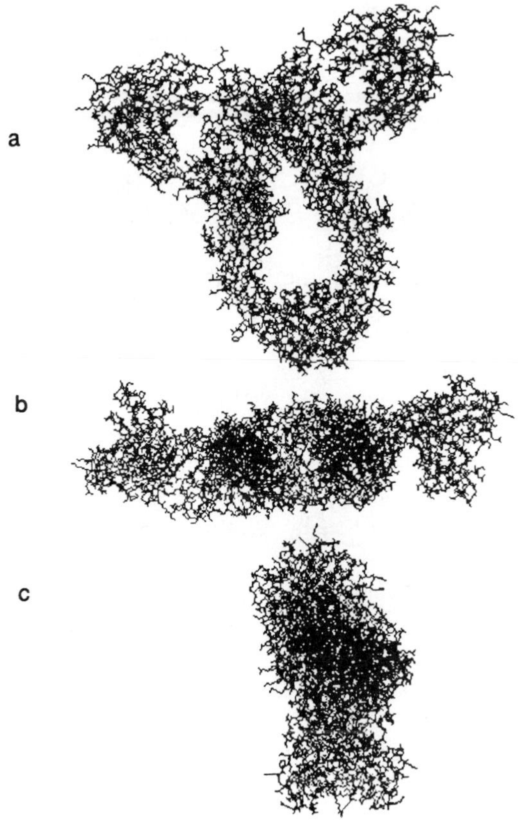

Figure 1. PDB image of IgG, a) front view, b) top view, c) side view.

and extensively characterized by a wide variety of biophysical probes in our efforts to develop new protein-based nanotechnology.

Water Soluble

As model systems for soluble proteins we have explored three enzymes, namely glutathione-S-transferase, alcohol dehydrogenase and alkaline phosphatase, four metallo-proteins, namely azurin, cytochromes c, c551 and P450scc, IgG antibodies (Fig. 1) and thioredoxin, an ubiquitous protein with many functions in thiol-dependent redox reactions.

Trans-Membrane Proteins

As model system for membrane-bound proteins we have chosen photosynthetic reaction centers from *Rhodobacter sphaeroides* (Fig. 2) and from *Rhodopseudomonas viridis*, bovine rhodopsin and bacteriorhodopsin (BR).

In the system of *Halobacterium halobium*, bacteriorhodopsin is found as a 2D crystalline lattice in purple membranes. The purple membrane is usually 1000 nm in diameter and 50 Å in thickness and makes part of the overall cell membrane of Halobacteria. Bacteriorhodopsin is the main protein in the biological function of these bacteria. BR converts light energy into chemical energy by transferring protons from the inside to the outside of cell membrane, thus functioning as a light-driven proton pump. As a result, ATP is regenerated from ADP by an enzyme called ATP-synthase.

LBM THIN FILMS: ORDERED PROTEIN MONOLAYERS AND MULTILAYERS BY A NEWLY MODIFIED LANGMUIR-SCHAEFER TECHNOLOGY

As proteins are not ordinary amphiphilic molecules suitable for LB technique, several points must be taken into account in the formation of thin protein films, hereby called “LBM” to underline the several chemical (reverse micelles, derivatization) and biotechnological (site-specific mutagenesis, subphase) modifications which needed to be introduced in the standard Langmuir-Blodgett technique to preserve the native protein structure and function at the air-water interface. All proteins were transferred by Langmuir-Schaefer technique (horizontal lift) as it is the most suitable to provide homogeneous reproducible coverage of substrates with such type of monolayers. There is a big class of proteins, namely trans-membrane proteins, for which spreading over water surface does not affect the protein structure.

Two possible cases of “spreading solutions” can be distinguished. In the first case the spreading solution is a solution of separate proteins with hydrophobic area surrounded with surfactant molecules (as in the case of RC). In this case, after spreading we have not only protein molecules at the air/water interface but also the surfactant molecules. Taking into account that surfactant molecules have rather short hydrocarbon chains (ca. 12 carbon atoms) not allowing to reach high values of surface pressure, it is possible to remove them from the surface by applying compression-expansion cycles for several times.

In the second case the spreading solution is a solution of membrane fragments. The deposition onto substrates in this case is rather difficult as the fragments are mainly hydrophilic and only small hydrophobic parts are present. It results in the fact that most of the protein molecules penetrate the volume of the subphase, instead of participating in the monolayer formation. To prevent this behaviour it is necessary to use concentrated salt solution as subphase. But a new problem arises, namely, during the deposition salt is also transferred with the film onto the substrate. To remove it, washing of the sample is required after deposition of each layer.

To produce films by classical LB technique the solution of amphiphilic molecules consisting of hydrophilic groups interacting strongly with water and long hydrocarbon chains is spread at the interface. After evaporation of the solvent the monolayer is compressed with barrier up to a definite surface pressure which is measured with the balance and is maintained at the constant level during the deposition.

The solid substrate is moved successively down and up through the air-water interface and successive monolayers are deposited onto the substrate. One important detail is that the molecules of the monolayer deposited last are practically always arranged with inert hydrophobic tails in the direction of the air medium after pulling out the substrate from aqueous subphase. Immersion of the substrate with the film after completion of the process into some solution for adsorption of the dissolved compound, e.g. some soluble protein, onto hydrophilic groups of surfactant molecules will not give any result because active hydrophilic groups are protected with inert hydrophobic layer. Design of usual LB instruments with one compartment usually enables to deposit films consisting of monolayers of one type. However, if two separate troughs or one trough with two separate compartments are used one can deposit the structure of alternating bilayers of different molecules.

This latter design gives the possibility to deposit films of alternating monolayers, but always the last layer is arranged with hydrophobic inert layer outside, so that dissolved molecules will not interact with hydrophilic groups of amphiphilic molecules if the deposited film is placed in some solution.

A new alternative approach is to "close" the substrate with the monolayer deposited by dipping down the latter with some plate situated very near to the substrate.

If then the substrate is removed together with the closing plate from the water subphase, water will be held between the plate and the substrate by capillary forces. Thus the monolayer is protected by the layer of water and the hydrophilic surface becomes the external boundary of the film. Then the system consisting of the substrate and the plate can be transferred to any other compartment, allowing the successive deposition of new active monomolecular layers and the adsorption of any dissolved compound.

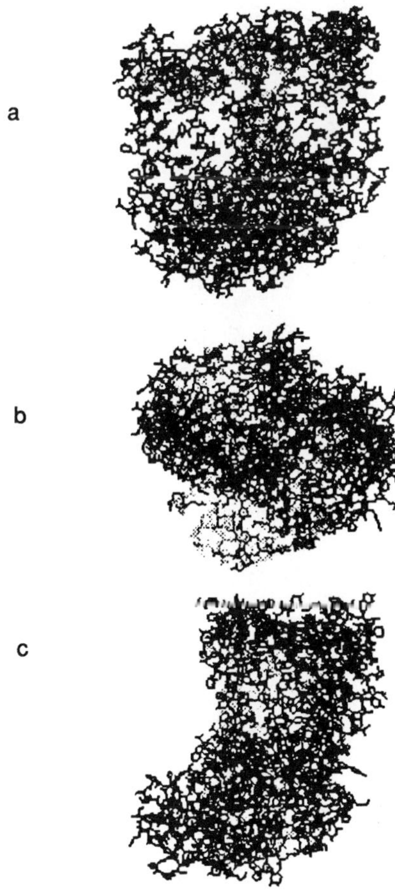

Figure 2. PDB image of Photosynthetic Reaction Centers from *Rhodobacter sphaeroides*, a) front view, b) top view, c) side view.

IS THE SOLUTION STRUCTURE AND FUNCTION PRESERVED AT THE AIR-WATER INTERFACE?

The issue of the protein monolayer formation at the air/water interface is critical for the preservation of native protein structure and function. The structure of most proteins is such that the hydrophobic nucleus is inside the globule and the hydrophobic and partially polar interactions maintain the protein structure. The value of forces is comparable with that of the surface tension. Thus, during spreading of protein monolayers over the water surface there is always a risk to denature the protein. There are some examples in literature (see cytochrome c) where such behaviour really takes place.

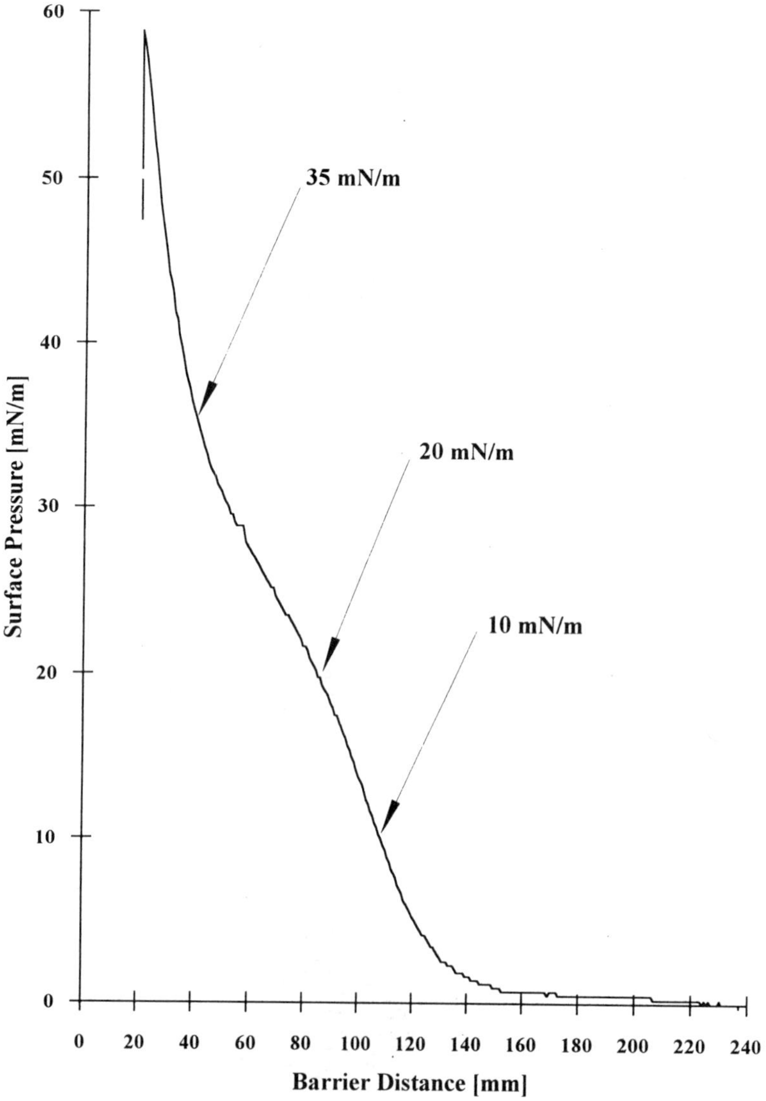

Figure 3. Isotherm of GST. The arrow indicates the deposition surface pressure. The surface pressure of deposition for subsequent structural analysis is indicated.

Table 2. LB film weight and thickness versus number of layers as determined by (1) nanogravimetry in 10^{-10} g/100 nm^2 and (2) ellipsometry in Angström at the saturating surface pressure

PROTEIN	GST	AP	Thioredoxin	P450scc	IgG	Rhodopsin	Reaction Centres (R.S.)
one layer	(1) 2.80 (2) 21±3	(2) 70±4	(1) 1.88 (2) 25æ5	(1) 3.2 (2) 35	(1) 1.3 (1a) 3 (2) 90±10 (2a) 35±10	(1) 7.54 (2) 40	(2) 48
two layers	(1) 4.98 (2) 39±3	(2) 130æ4	(1) 3.47 (2) 57±5	(1) 6.7 (2) 77	(1) 2.7 (1) 6.2 (2)175±15 (2a) 80±10	(1) 11.18 (2) 72	(2) 66
three layers	(1) 7.92 (2) 48±4	(2) 205æ4	(1) 4.53 (2) 90±6	(1) 10(2) 127	(1) 4 (1a)8.9 (2)290±20 (2a)120±15	(1) 15.23 (2) 90	(2) 97.5
four layers	(1) 1088 (2) 56±4	(2) 245æ4	(1) 6.06 (2) 115±6	(1) 13.2 (2) 177	(1) 5.5 (1a)11.8 (2)420±20 (2a)150±15	(1) 22.07 (2) 125	(2) 164
six layers	(1) 16.63 (2) 79±5	(2) 400æ4	(1) 9.88 (2) 170±10	(1) 19.6 (2) 271	(1) 8 (1a) 17.5 (2) 630±30 (2a) 250±20	(1) 33.87 (2) 179	-
eight layers	(1) 21.26 (2) 107±6	-	(1) 13.06 (2) 231æ12	(1) 26 (2) 367		(1) 46.04 (2) 264	(2) 32.5
ten layers	(1) 26.13 (2) 135±7	-	(1) 16.54 (2) 285æ12	(1) 32.4 (2) 457		(1) 58.84 (2) 385	(2) 30
twenty layers				(1) 64.4 (2) 907			
Average monolayer	(1) 2.62 (2) 14	(2) 66	(1) 1.6 (2) 27	(1) 3.2 (2) 44.8	(1) 1.6 (1a) 4.2 (2) 40 (2a) 40	(1) 5.7 (2) 42	(2) 29.2
Surface pressure (mN/m)	(1) 20 (2) 20	(2) 45			(1) 25 (1a) 40 (2) 40 (2a) 15	(2) 45	

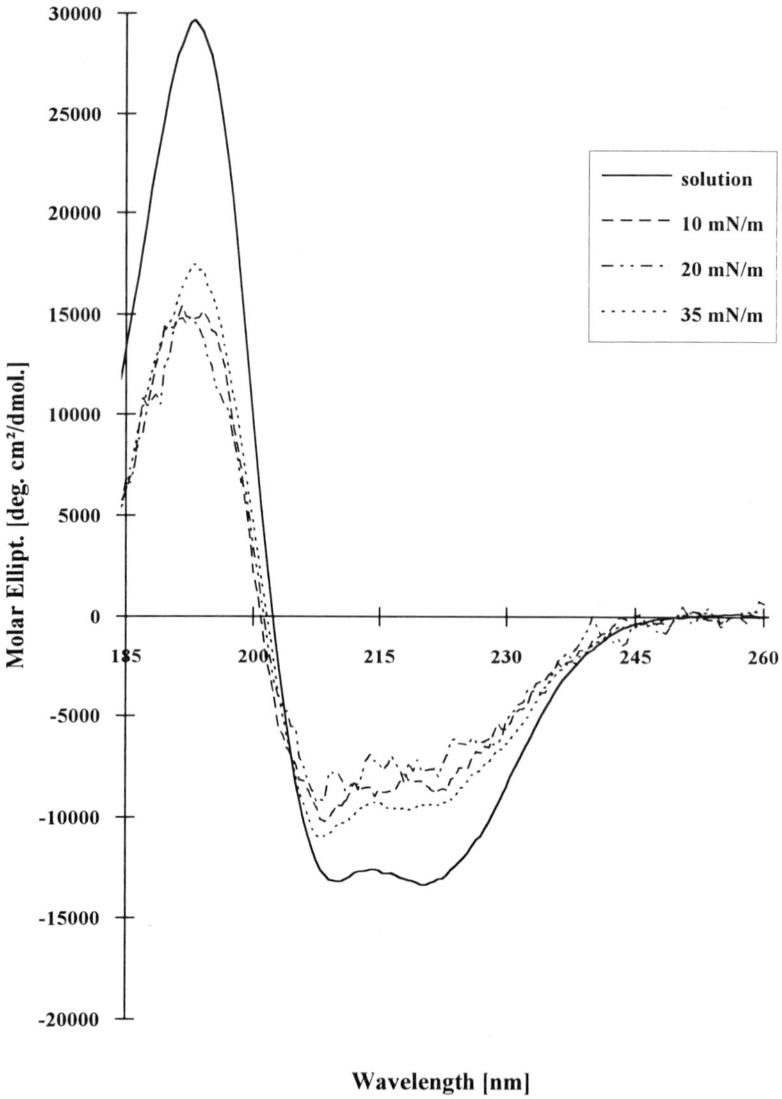

Figure 4. CD spectra of GST in solution (solid line) and in films at three different pressures (dashed lines). The spectra are recorded between 185 and 260 nm.

Nevertheless, several proteins being deposited onto solid substrates by LB method demonstrate preservation of structural and functional properties, confirmed by numerous techniques in different groups. There are two possible explanations of such non-denaturing behaviour, namely, either proteins do not denature at the surface because of their inherent structure and binding energy, or the proteins are denatured at the interface, but by forming a layer, which decreases the value of the surface tension, leave the remaining proteins being subsequently adsorbed on this layer in their native structure. What is happening in the reality appears still an open question. From the applied point of view the answer to this question is irrelevant, because in both cases well organized 2D films with reproducible structure and properties can be obtained. On the other hand, it would be important in order to understand

Figure 5. STM image of a monolayer of GST deposited by LBM technique on plate of HOPG. Image size 750x900 Å^2, V = 0.2 Volt. (carried out on an STM by Asse-Z, Padua).

the real process of film formation to experimentally probe the existence or absence of a denatured layer

The surface concentration of proteins cannot be estimated from compression isotherm analysis (Figure 3), but can be measured on the solid substrate by nanogravimetric technique (Table II). The sub-phase is typically a buffer solution of phosphate 10^{-3} M pH 7.2 (or other salt of varying concentration depending on the type of protein). The total volume of the subphase is ~80 ml, with a protein solution in phosphate 10^{-3} M pH 7.2 deposited at a concentration of about 2.5 x 10^{-6} M.

Several layers can be quickly deposited successively with LS technique without compression and their thickness determined by ellipsometry.

In order to assess the effect of film formation on the native protein structure CD spectra have been recorded in solution and in LS films at different surface pressure (see Fig. 4 for GST enzyme): by this and other structural and functional probes (e.g., STM, Fig. 5) optimal conditions can be determined for the LS film assembly of GST enzymes.

Similarly for each individual protein under consideration. In the case of IgG antibodies the surface pressure strictly determines their orientation and subsequent monolayers thickness (Figure 6), in perfect agreement with the geometric features of the very same antibodies independently determined by X-ray crystallography (Table III). STM analysis of IgG, while impossible in LS film due to the large monolayer thickness (about 120 Å), strikingly confirms in solution the above conclusion (Figure 7).

Table 3. Geometric features in Angstrom of LB protein films at initial (I: 10-15 mN/m) and at saturating (S: 35-40 mN/m) surface pressure

PROTEIN	GST (horse)	AP (calf intestine)	Thioredoxin	P450scc	IgG	Rhodopsin (bovine)	Reaction Centres (*)spheroid (+)viridis	Cyt c
Ellipsometry (Thickness)	9 (I) 30 (S)	29±3(I) 70±5(S)		56±2 Å rec 33±3 Å wt	40 (I) 120 (S)	25±3 (I) 42±4 (S)	*49±4 Å	No film w/o 34 A° w/rev mic
Interference (Thickness)								15X15Å w/o (subphase)
X-ray low angle (Thickness)			865 Å	54 Å re O wt 72 rec Scat Vol52 wt "	108 Å (S) 50 A° (I)	59 Å (35mN/m)		34 Å (w/micelle) 10 A w/o micelle
Nanogravimetry (Area/molecule)	2800700 Å (		1000 Å	4500 (I) 2907 (S)	8000(I) 3500 (S)		4000 (I) 2500 (S)	
STM (length x width)	30X32 Å	O			125x90(I) O (S)		*30x64 +50x60	O
TEM (length x width)				13 x 76 rec.8 x 15 wt O				O
Crystallography Data-PDB-BNL (height x width x depth)	54x64**x54	70x54x48	33X34X30	73x71x60	125X90X40	Nodata (bovine) bacteial 48x43X45	*90x56**x68 +125x55x60	38X33X32
Molecularweight (kDaltons)	59**©	140**	11.7	65	160	39	*114** +163	14

**Resulting from two subunits of equal width
O No molecular resolution achieved
© In the PDB data bank only data on GST from rat liver are available

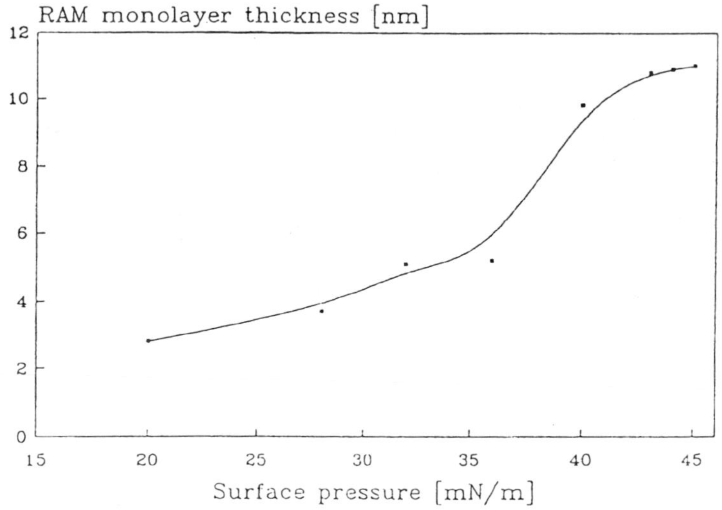

Figure 6. Dependence of IgG LBM film thickness on the surface pressure of deposition. Ellipsometric measurements.

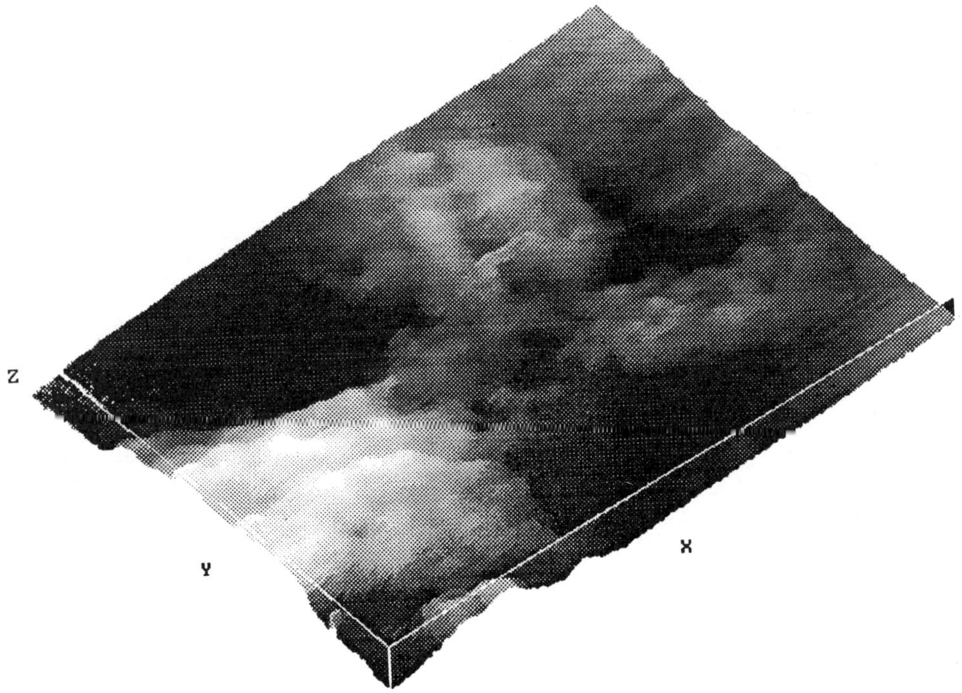

Figure 7. STM image of IgG antibody molecule deposited on graphite (carried out on an STM by Asse-Z, Padua).

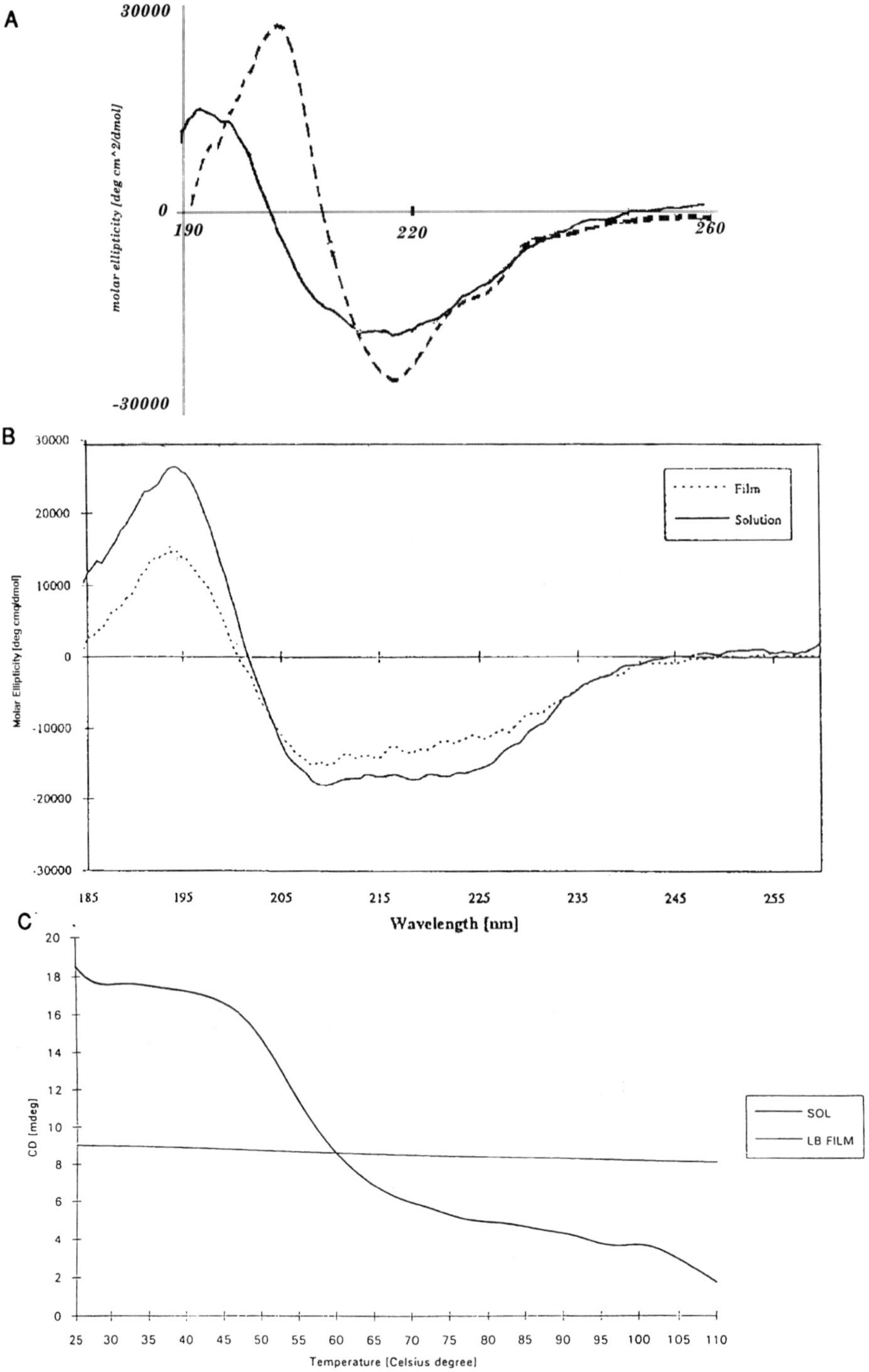

Figure 8. CD spectra of IgG molecule (A), photosynthetic reaction centers from *Rhodobacter sphaeroides* (B) and Cytochrome P450 (C) in solution (.....) versus LBM films (____).

Figure 9. STM image of photosynthetic RC monolayer at room temperature (25°C).

Dealing with water soluble proteins it is possible to distinguish several different situations by determining their secondary structure in solution and in LBM film, as apparent from circular dichroism spectra in the 190-250 nm range (Figure 8).

There are stable proteins that are not destroyed at the interface. One of the examples of such proteins are antibodies. Numerous covalent S-S bonds prevent denaturation of these molecules at the interface, as the interaction forces are much stronger with respect to surface tension. In this case the usual Langmuir technique can be applied for the formation of such layers. Other examples are the Photosynthetic Reaction Centers and the P450scc cytochromes, which structure (Figure 9) and function (Figure 10) remain invariant upon monolayer formation.

There are proteins, which undergo partial denaturation at the air/water interface. Glutathione-S-transferase (GST) and cytochrome P450scc are two examples of such proteins, whereby the GST molecule is divided into two subunits at the water surface. One can apply usual LB technique also in this case, but it is necessary to take into account that both structure and function (Figure 11) of the GST in the thin layer will be somehow decreased, even if largely preserved. Similarly for most other proteins, being metalloproteins, enzymes or antibodies, as it will be shown in subsequent paragraphs.

Finally, there are proteins that are absolutely unstable at the water surface. Cytochrome c is one example of such proteins, as conclusively shown by infrared spectroscopy (Figure 12). Indeed the bonds which stabilise cytochrome c secondary structure have energy per bond of about 4 kcal/mol. Considering that the area occupied by one cytochrome c is about 1000 Å^2, with the area per one bond stabilising secondary structure being equal to 10 Å^2, the density of hydrogen bonds responsible for the tertiary structure is approximately 10 times less, i.e. area per one bond is equal to 100 Å^2. The maximum change of surface tension value due to “optimal reorganisation” of the molecule (all hydrophilic groups are in water, all hydrophobic groups are in air) is 72-28=44 mN/m=44 mJ/m^2 (but in reality still considerably less because such “ideal reorganisation” is impossible). Assuming the above values for the area occupied by one bond responsible for the secondary and the tertiary structure, the corresponding energy per bond becomes about 0.6 kcal/mol and 6 kcal/mol respectively.

Three different techniques have been usefully integrated for the identification of the secondary structure of cytochrome c monolayer at the air-water interface: attenuated total reflection-Fourier transform infrared (ATR-FTIR) spectroscopy, interferometry and FT-IR spectrometry. The spectra of native cytochrome c were obtained using KBr pellet preparation. Heat-denaturation in solution and LB film formation at room temperature appear to produce similar spectral alterations, which in turn can be associated to similar alteration in the native secondary structure of the protein. Transmission spectra were measured in a Perkin-Elmer 2000 spectrometer using a SPECAC grazing angle accessory under the

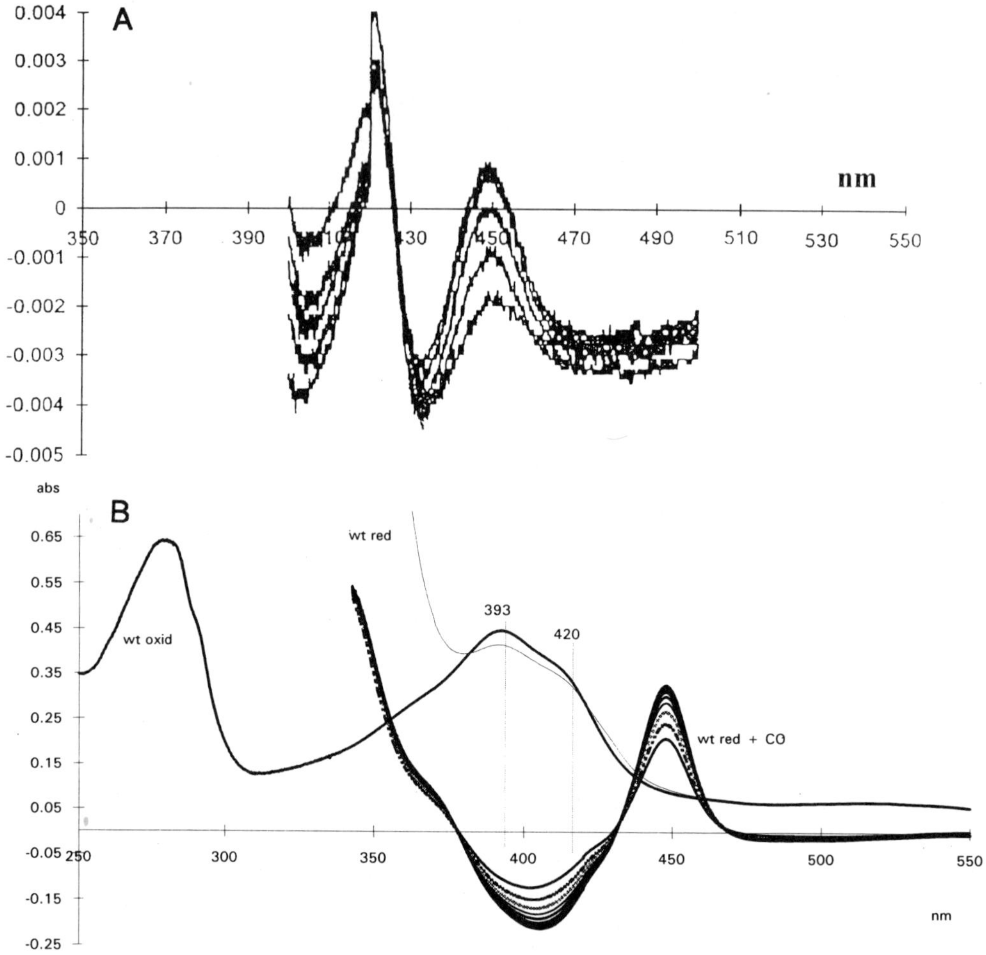

Figure 10. CO-reduced difference spectra of the P450scc wild-type in the LB film (A) and in solution (B).

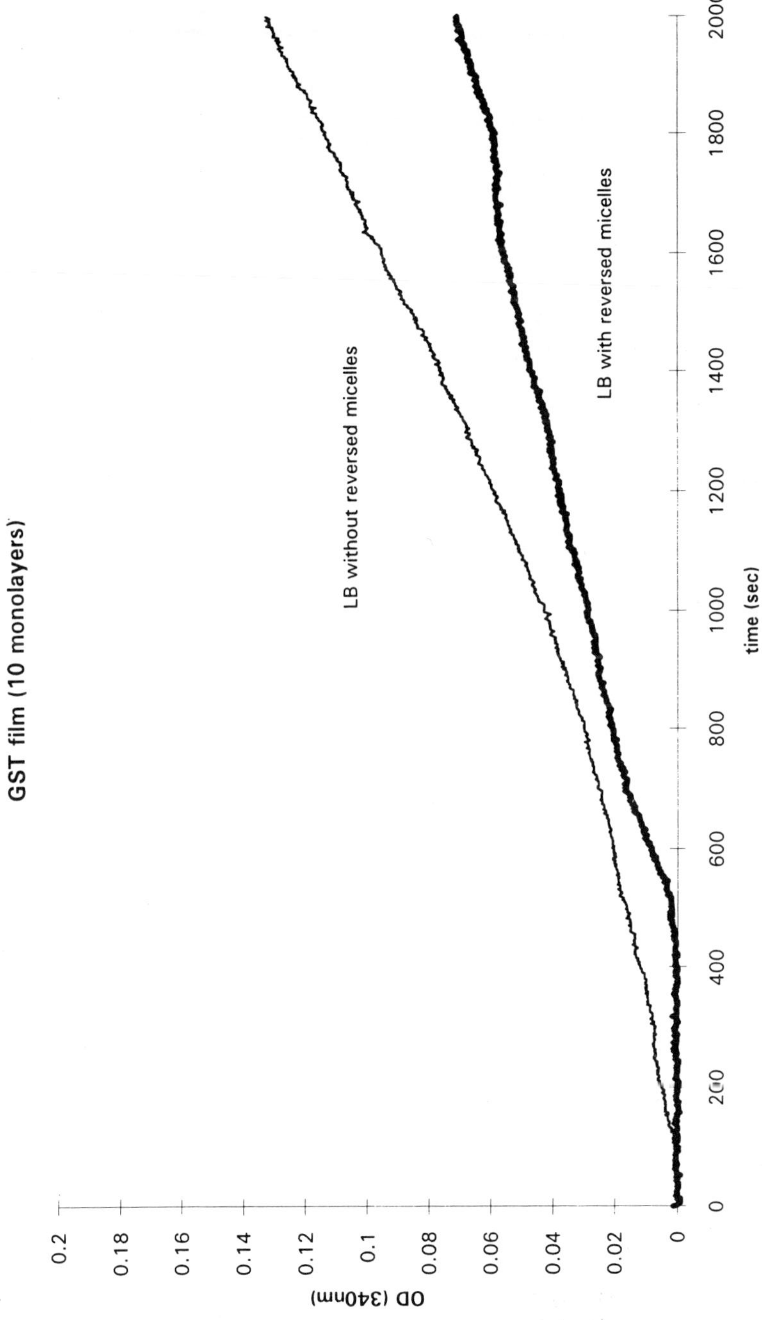

Figure 11. Activity of GST enzyme in LBM film, with and without reverse micelles (see later).GST film (10 monolayers).

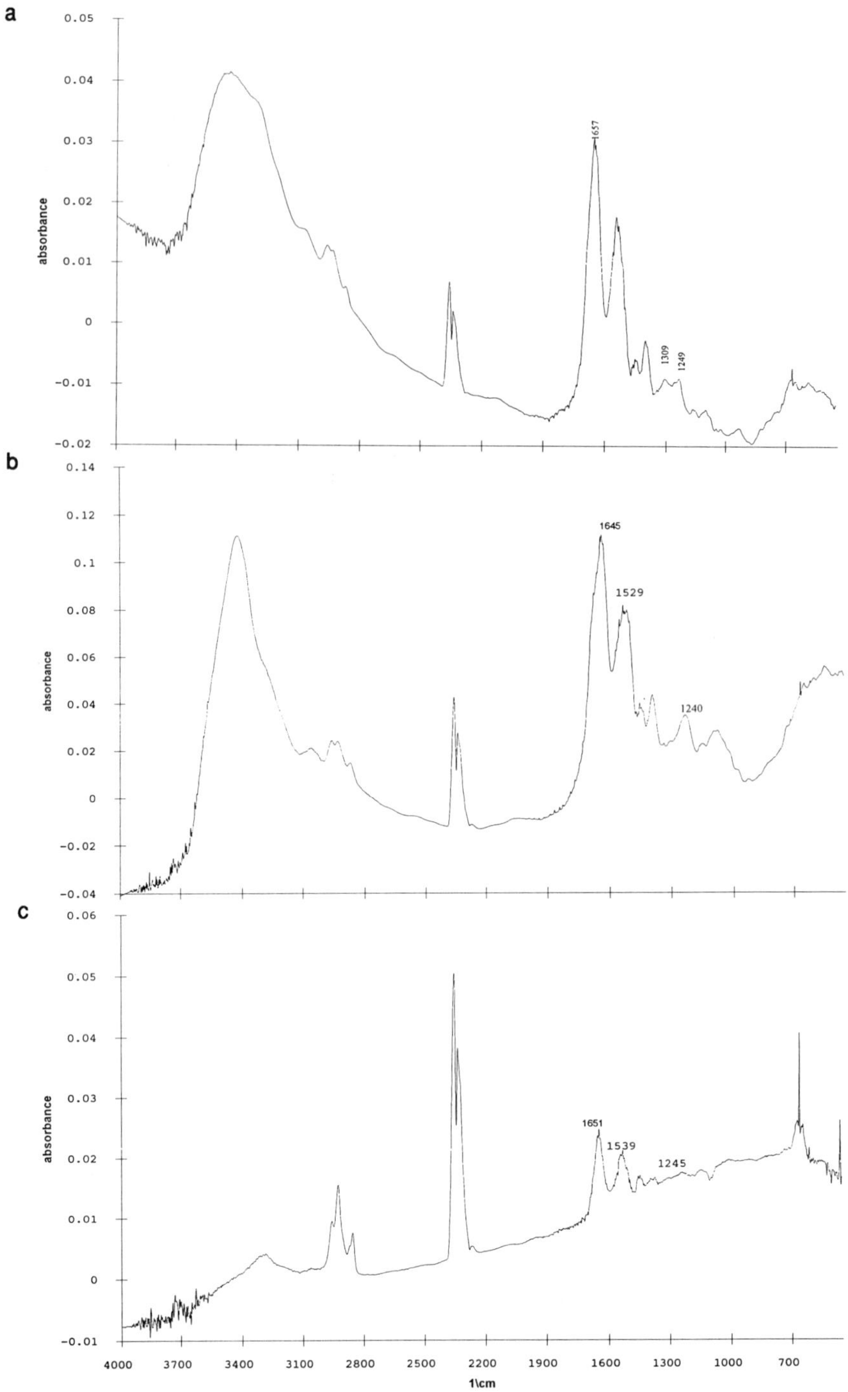

Figure 12. Infrared spectra of native cytochrome c in solution (A) and in LBM film (B). For reference the heat- denatured cytochrome c in given in panel (C) The panel C shows the spectra of denatured protein. It must be noted that the frequency of the amide I band shifted at 1645 cm^{-1}. According to the literature this correspond to a transition from an α-helix to a β-sheet configuration (Bramanti *et al.* 1995).

condition of angle of incident of 82°, 4 cm^{-1} resolution and 30 scans. Generally the bands at 1650, 1630, 1670, cm^{-1} were respectively assigned to α-helix, β-sheet and random coil structures; that at 1309, 1240 cm^{-1} to α-helix and β-sheet.

Following denaturation there is a clear shift to higher frequency at 1665 cm^{-1} and in the region of the Amide III band there appears a striking difference, namely the band at 1309 cm^{-1} disappears and that at 1240 cm^{-1} increases significantly; following film formation similar effects are apparent, pointing particularly in the peak region around 1665 cm^{-1} to the cytochrome c denaturation.

The interference method confirms the above conclusion allowing to assess film thickness, even in this case where X-Ray diffraction patterns do not show any Bragg reflections and reproducible data cannot be obtained by ellipsometry because of significant inhomogeneity.

Spreading the cytochrome solution on the water surface and waiting for some time, one can see the increase of the surface pressure even without compression, which implies enlargement and uncoiling of polypeptide chains. In order to make possible the deposition of such proteins it is necessary to diminish the surface tension. Two different approaches can be considered. First, to attach the protein to head groups of preformed monolayer of lipids and to analyse the corresponding films by X-ray scattering (Figure 14a). An X-ray pattern can exhibit Bragg diffraction peaks and Kiessig oscillations. In case of Bragg diffraction we have the interference of X-ray beam, reflected from each repetitive unit (molecular planes)

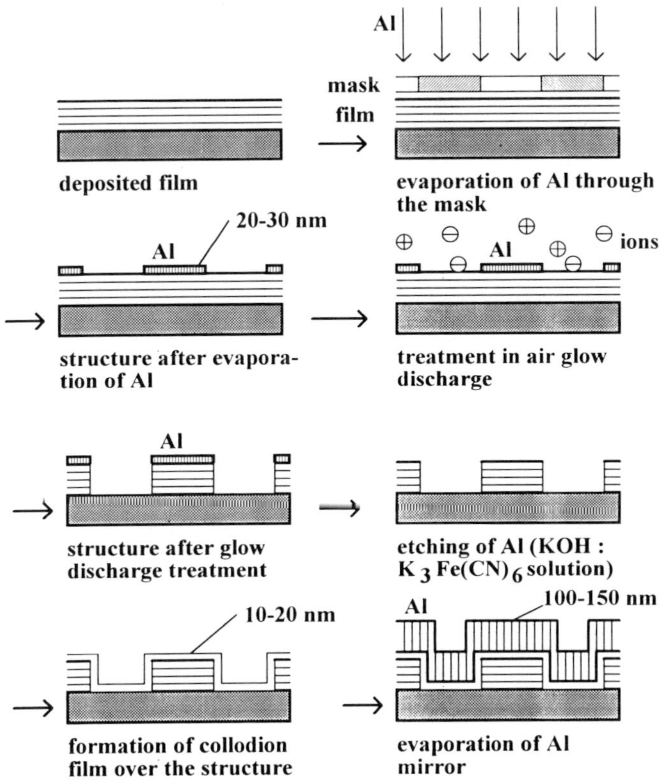

Figure 13. Procedure for interference measurements. Measurements were carried out with interference Mirau attachment for optical Carl Zeiss microscope Axioscop.

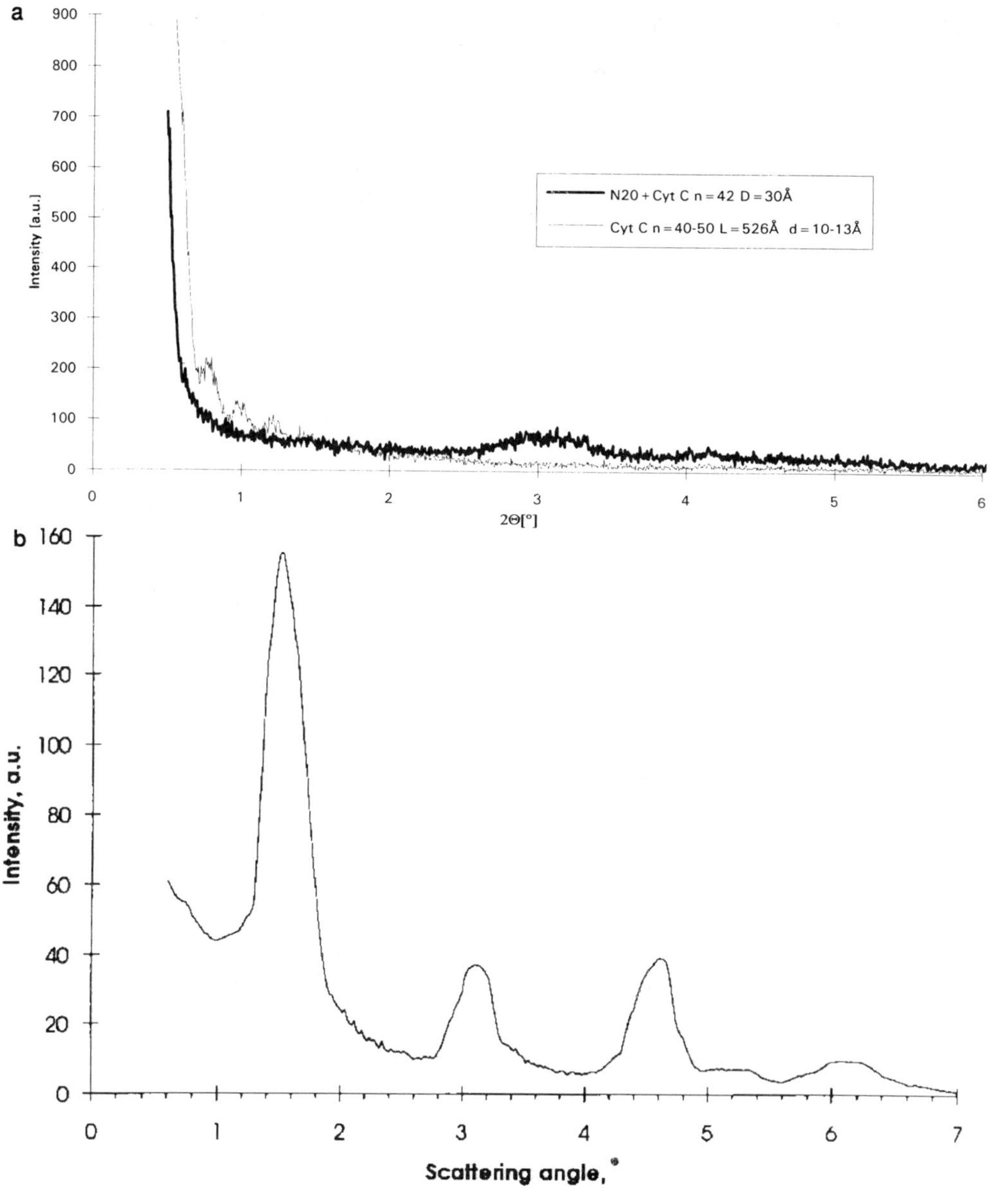

Figure 14. X-Ray pattern of a film of cytochrome c deposited using a) preformed film of lipids and b) reverse micelles of sodium diisooctylsulphosuccinate.

and several peaks are observed, the angular position of which is determined by Bragg equation:

$$2\,D\sin\Theta = n\lambda$$

where D is the spacing (thickness of the repetitive unit), Θ is the incident angle of the X-ray beam, λ is the wavelength of the X-ray and n is the order of the reflection. Thus, when we have Bragg reflections we know exactly the sizes of the repetitive unit, and when we know the composition of the unit we can estimate the sizes of the elements. For example, for cytochrome c the repetitive unit of 30 Å contains one layer of lipid and one layer of protein. By knowing that one CH_2 group is 1.25 Å long, since we have 16 groups the thickness of lipid layer is 20 Å. Thus, for the cytochrome c we have less than 10 Å.

At the same time the X-ray pattern of Langmuir-Schaefer film of pure cytochrome c does not contain Bragg reflections, so there appears no lateral ordering or periodicity in the film. Nevertheless, there are Kiessig oscillations, which indicate homogeneity of the film and allow to calculate the total thickness of the film. The thickness was found to be 526 Å, which corresponds to the layer thickness a little bit more than 10 Å. In case of Kiessig oscillations, we deal with the interference of X-ray beams, reflected from upper and lower boundaries of the layer, and we can estimate only the total thickness according to the following formula:

$$L = \frac{\lambda}{2(sin\Theta_1 - sin\Theta_2)}$$

where L is the total thickness of the film, λ is the wavelength, $\sin\Theta$ is sinus of angular position at two adjacent oscillations.

The second attempted approach to overcome protein denaturation at the interface was to use reversed micellar approach (Erokhin et al., 1994). This seems to be more promising, as it does not use any adsorption of protein molecules to the monolayer and, thus, gives more possibilities to control the film formation process. Under these latter conditions the native structure of cytochrome c appears indeed preserved as indicated by the X-ray pattern (Figure 14b).

With the only exception of cytochrome c, we can exclude in general the possibility of the presence of a denatured protein in the LBM film on the water-surface interface and on the solid substrate; the thickness dependence from the number of layers as determined by means of nanogravimetry, ellipsometry and interferometry represents indeed an unambiguous experimental proof of the absence of any denatured protein in the layer.

Furthermore, typically prior to deposition on the activated substrate, the films are not only dried, but washed in the water flow rather vigorously, which is very important in order to exclude the eventual presence of denatured sublayer. As for the two-layer model utilized in computing thickness from ellipsometric data, the first, lower layer, cannot be the denatured sublayer since the determination of its parameters was undertaken before the deposition: this layer characterizes indeed only the substrate. Our experimentation conclusively shows that a single non-denatured protein monolayer is formed in most instances at the air-water interface, which remains deposited on the activated substrate after washing. Namely should the water surface carry a bilayer of proteins, with the upper layer being denatured and the second (lower) layer containing the native protein adsorbed onto the upper layer, we could not obtain the presence of the native protein in the deposited layer. In this case all covalent bonds would be covered by denatured protein and native protein would be washed out by the water yet. Indeed, in the emblematic case of GST film discussed earlier (Fig. 11), our procedure of deposition provides the presence of the active native GST molecules on the substrate when is covalently bound with the surface. Everything that is not bound is washed from it by the water jet.

Recombinant Versus Wild Type

The properties of the two P450scc cytochromes - wild-type versus recombinant - are systematically characterized in structural and functional terms, in order to probe if confor-

Table 4. Functional and structural melting temperatures (°C) of proteins in solution and in LB films

PROTEIN	GST	AP	Thioredoxin	P450scc	IgG	Rhodopsin	Reaction Centres
Solution secondary and higher order structure	60 C	60 C	60°C	~60°C	60°C	75°C	55°C
LB secondary structure	150°C		190°C	>100°C	>200°C	>200°C	~200°C
LB tertiary-quaternary structure							110°C
Solution functional activity	60°C	60 C				~75°C	~60°C
LB functional activity	>150°C	>150 C			190C	~200°C	110°C
Solution dehydration							50°C
LB dehydration							100°C

mational variations are associated with the two different processing pathways used for protein maturation (cotranslational and post-translational). Initially sequencing was carried out using pulsed-liquid phase sequencer Mod. 477A (Applied Biosystems) in order to determine the primary structure of both proteins. As recombinant type with two signal sequences (secretory and mitochondrial) and wild type ("wt") with mitochondrial sequence were examined according to the same routine, the obtained results show identity of the primary structure of the N-terminus of both proteins. The first amino acid was determined to be Ile, which agrees with theoretical suggestion about the position of the digestion site of the signal protease. These data give evidence that the recombinant protein is the mature one and goes trough the entire processing pathway. The fact that the recombinant construction has two functioning digestion sites for signal proteases seems to be very promising in view of biotechnological applications (Paschkevitsch *et al.* 1996).

The identity of their primary structure, in the presence of significant structural (Figure 15) and functional alterations, is consistent with the hypothesis that processing pathway has a role in determining the function and the structure of proteins.

From compression isotherms at the air-water interface both recombinant and wild type cytochromes P450 seem to be rather similar. Both isotherms differ strongly from that of cytochrome c, where both shape of the curve and low value of the surface pressure indicate protein denaturation.

X-ray study of LB films revealed Kiessig fringes for the recombinant but not for the wild type, indicating higher homogeneity. At the same time ellipsometric and X-ray study gave a monolayer thickness equal to 54 Å for the recombinant and 30 Å for the wild type. Comfortingly the size expected for this protein from its 3D crystallographic structure present in the BNL data bank is about 60 Å.

In summary, the structural difference between cytochrome P450scc "wild type" and recombinant is caused by the difference of signal structure of these proteins. The "wild type" cytochrome P450scc possesses a mitochondrial signal which addresses it by the mitochondrial pathway of processing. The low level of folding allows to conserve this protein during processing into cytosols.

On the other hand, the secretory signal of recombinant cytochrome that was appended as a result of the cloning, addresses the protein to the outside of the cell through the secretory pathway. In this case the protein must be adapted to coated vesicle transport mechanism which means higher level of folding.

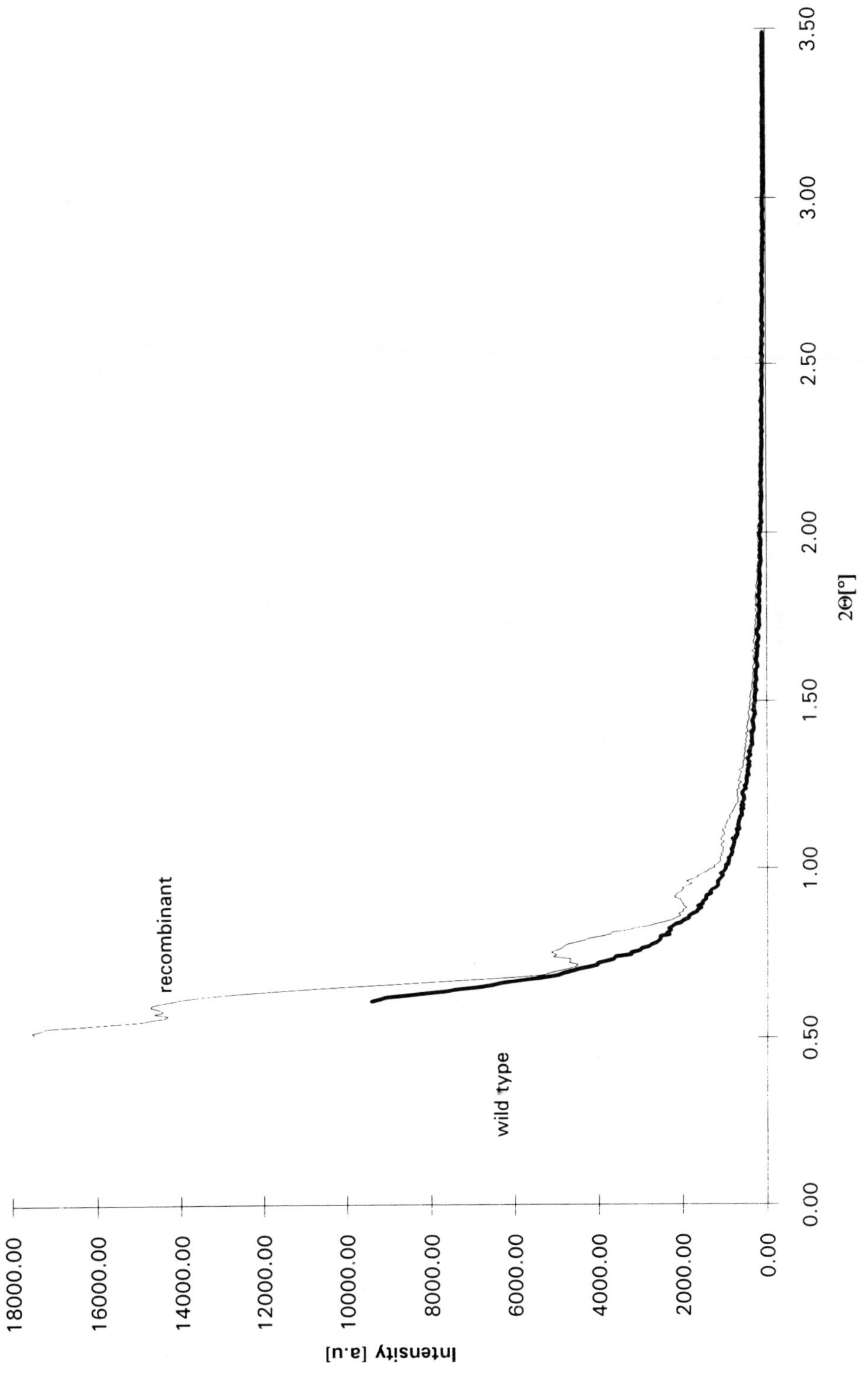

Figure 15. Low-angle X-ray scattering of yeast recombinant versus wild-type P450scc.

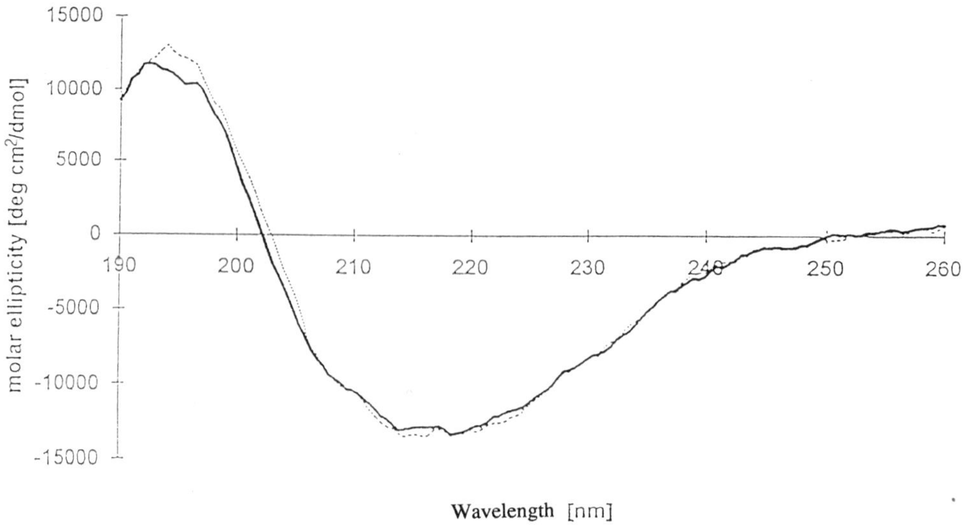

Figure 16. CD spectra of IgG LB film (20 monolayers) before (dotted line) and after (solid line) thermal treatment at 150°C for 10 min.

THERMAL STABILITY OF PROTEIN STRUCTURE AND FUNCTION IN SOLUTION VERSUS LBM FILM

One common feature of all protein films is the high thermal stability of the protein structure, both secondary and higher, and function as shown in Table IV; practically all the proteins used in the study (RC, P450, antibodies etc.) demonstrated the preservation of the secondary structure till 200°C when organized into the film.

The understanding and the attainment of protein thermal stability is a very important goal both from basic and applied point of view as it requires a deep knowledge of all the parameters involved in the structural stabilization and since it is of fundamental importance in all those cases where proteins are used as functional elements for biodevices and bioreactors.

Indeed, in recent time proteins, mainly those involved in the electron transport, in molecular recognition and in catalytic processes, have undergone increasing utilization in many technological applications, from biosensors and biotransistors to biocatalysis. However, several of these applications require operating temperatures higher than 100°C, a condition under which most of the proteins are known to denature, including those isolated from extreme thermophiles.

The usual approach to this problem is protein engineering which, in addition to its inherent limitations despite overemphasised claims, requires enormous efforts and needs a specific approach for each individual protein under study. Therefore, it was very amazing when Nicolini *et al.* in March 1993, by forming highly packed protein films by LS technique, were able to find out a general procedure for stabilising up to 200°C protein structure which instead was lost at 55°C in solution and at 150°C in dried "smears" (a kind of self-assembled films). The procedure turned out to be of general validity and was proved to be effective on both water soluble (antibodies, cytochromes) and membrane proteins (Reaction Centres from *Rhodobacter sphaeroides*), stabilising not only the structure but also the function to a certain extent.

Comfortingly enough, these findings were confirmed by X-ray study with synchrotron radiation only 5 months later by Shen *et al.* (1993) for dry self-assembled films of bacteriorhodopsin and recently for dried photosynthetic membrane film by Miyake *et al.* (1994).

It is worth noting that thermal behaviour of dry self-assembled smears of membrane proteins such as Photosynthetic Reaction Centres (RC) and water soluble ones as, e.g. antibodies, displays a pattern which is intermediate between that of solution and of LB film (Dubrovsky *et al.* 1995a,b; Facci *et al.* 1994a; Erokhin *et al.* 1995a).

The first question we addressed was about the generality of the preservation of protein secondary structure in LS films under thermal treatment previously reported for membrane-proteins (see above). Fig. 16 depicts circular dichroism spectra of antibody LB film before and after thermal treatment, while Table IV summarises similar findings on other water-soluble and membrane-proteins. Up to now heat-proof protein structure has been firmly established for LB films of Photosynthetic Reaction Centers (Nicolini *et al.* 1993; Facci *et al.* 1994b), glutathione-S-transferase (Antolini *et al.* 1995a; Paddeu *et al.* 1995a), alkaline phosphatase (Petrigliano *et al.* 1995), bacterial and bovine rhodopsins (Maxia *et al.* 1995), urease (Paddeu *et al.* 1995b), thioredoxins (Dubrovsky *et al.* 1995), P450 (Nicolini *et al.* 1995a), C551 and azurin metalloproteins (data not published yet). The circular dichroism data confirm that also in all above cases, protein secondary structure is not affected by temperature.

Protein Function

The second issue was about the preservation of functional properties of proteins in LB film after heating. Fig. 17 shows that functional activity for RC in LB film is retained up to a temperature about 40°C higher than in the case of dry smears. In case of RC in solution the functional thermal stability is drastically lower. Even if the disactivation temperature does not match the thermal denaturation threshold, it represents a strong improvement of protein thermal stability.

Similarly functional activity appears strikingly preserved up to 150-200 °C in LB films of a wide variety of membrane and water-soluble proteins, as shown in Table IV and in Figure 18 (for IgG antibodies). Only in RC protein films the light-induced charge separation appears preserved at a lower temperature, namely 110°C (Erokhin *et al.* 1995a), as shown earlier. Fig.18 represents the functional activity of antibody LB films. In this case, due to the structure of these molecules, the functional activity was preserved at least after heating up to 150°C and above.

Tertiary and Higher Order Structure

The aim of this overview is to draw general conclusions about thermal stabilisation in both structural and functional terms of different proteins organized in regular 2D structures by LB technique (Erokhin *et al.* 1995a; Nicolini *et al.* 1993, 1995a, b, c) Considering that most of the proteins utilized are water-soluble and all stem from mesophilic organisms, this is an astonishing result, quite generally applied and with far reaching implications. The temperature dependencies were always performed by heating the samples at a fixed temperature in a commercial oven and performing measurements at room temperature after the samples have attained thermal equilibrium.

The amazing increase of immunological activity of antibody LB film, evident from the value of the saturation level and reaction rate (Figure 18), was attributed to the improvement of the film order. The improvement of the film order turned out to be a general property and was demonstrated both for RC and antibody LB films. Furthermore the whole

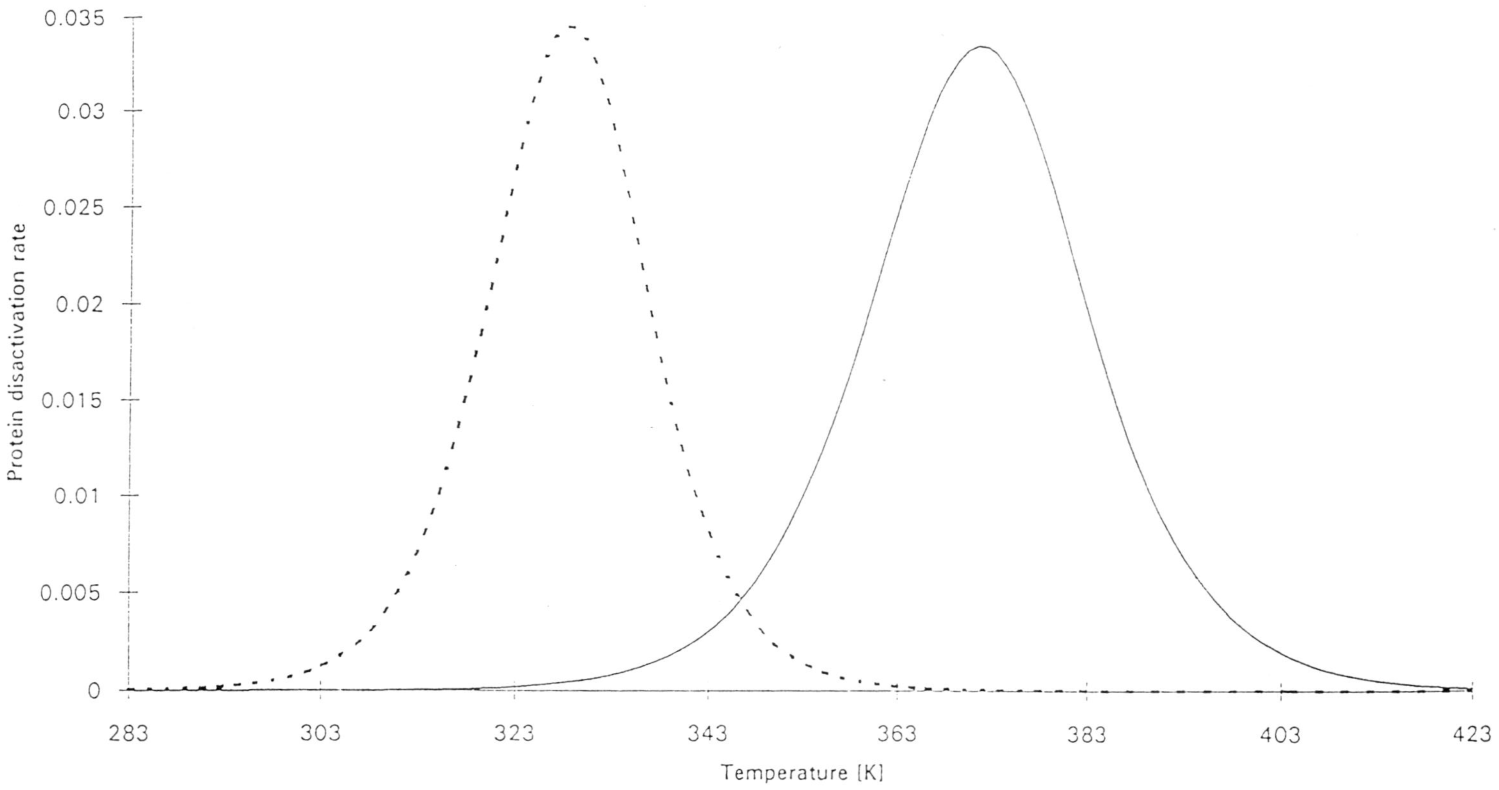

Figure 17. Deactivation rate of photosynthetic reaction centre from *Rhodobacter sphaeroides* as a function of temperature assayed by optical absorbance for dried smears (dashed line) and by surface potential measurement for LB monolayer (solid line).

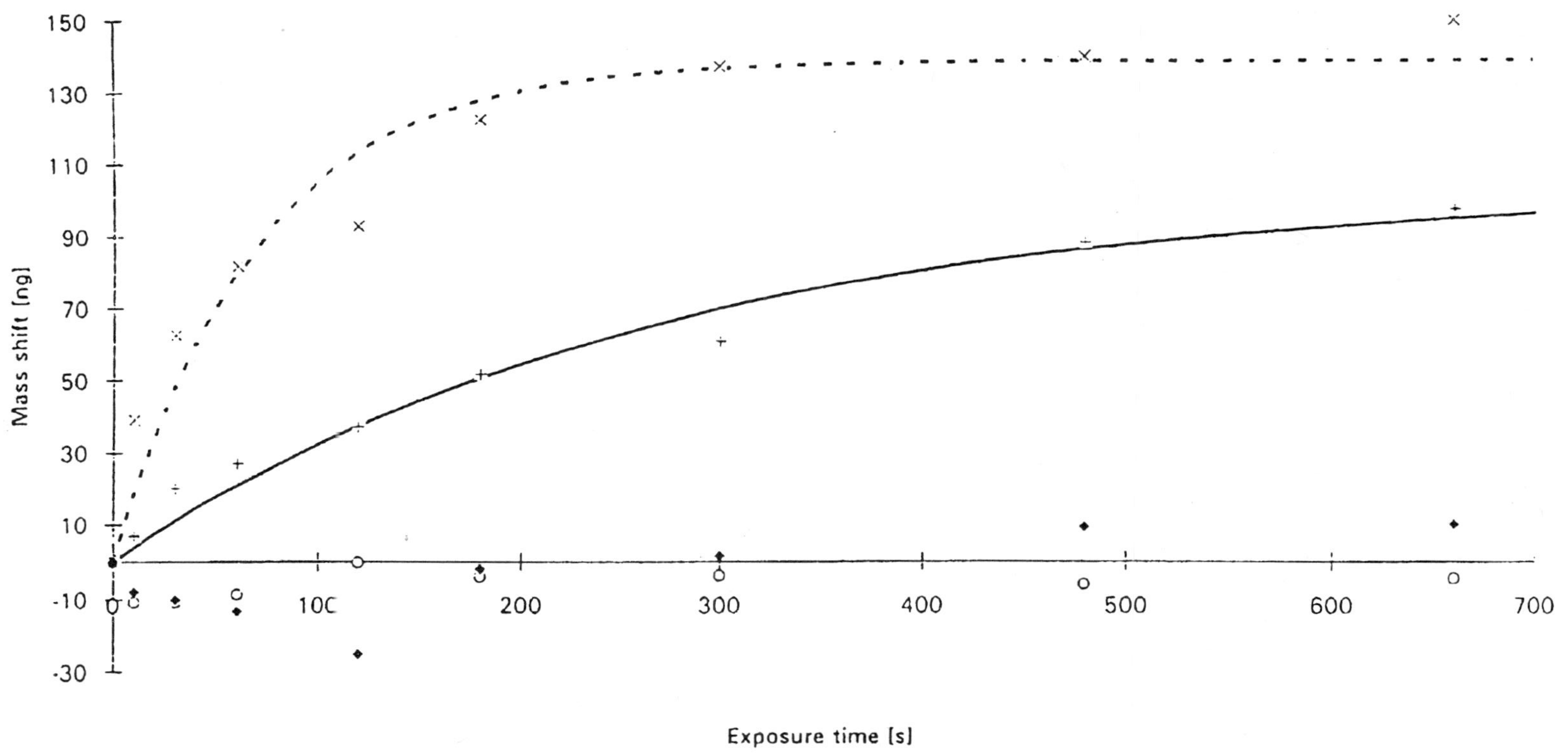

Figure 18. The amount of antigen attached to the LB IgG layer deposited onto the quartz resonator surface: + - experimental data on not heated film; - - best fit of (+); x - experimental data on films, previously treated at 150° C for 10 minutes; - - - - best fit of (x); o, × - control experiments.

sequence of structural alterations induced in RC protein LB film with increasing temperature points to a progressive loss of functional activity, tertiary-quaternary structure, and only at quite higher temperature of secondary structure.

Summarizing these results, we can safely conclude that thin protein film formation by the Langmuir-Blodgett technique provides to both hydrophilic and trans-membrane proteins the conformational constraints suitable for preventing thermal denaturation within a wide temperature range (up to 150°C and higher), thereby making all classes of proteins "heat-proof" in both structural and functional terms. Besides, the LB technique confers to the protein film the appropriate initial conditions, namely the decreased conformational entropy sufficient to allow protein recrystallization by heating and cooling down the film to room temperature.

Lamellar structure of the films was studied by X-ray diffraction (Figure 19). Protein films do not demonstrate high lamellar ordering since the molecules are much more complicated with respect to usual amphiphilic molecules. Nevertheless, it was very interesting that heating improved lamellar ordering. In particular, for antibody films the Bragg reflection appeared in the X-ray pattern after heating the sample till 150°C.

From the other point of view, in-plane structure of the film was also improved after heating. STM of RC film revealed strong improvement of the order after the heating. All the previous conclusions were also checked by surface potential measurements at the macroscopic level.

Table V reports surface potential data for both samples, showing the increased anisotropy in protein films after heating.

Enzymatic Activity

The formation of LB film without reverse micelles is successful with three enzymes: alkaline phosphatase from calf intestine (AP), glutathione-S-transferase and urease. The study of AP required extensive experimentation in order to optimize film formation.

The activity with the micelles remains high without extensive detachment, being about 5% of the solution value for GST film deposited without micelles during 30 minutes.

The surface density of deposited enzymes was for urease 0.79 + 0.19 pmol cm^{-2}, 1.05 + 0.21 pmol cm^{-2}, respectively for non-silanized and silanized quartz, for GST 5.82 10^{-2} ±

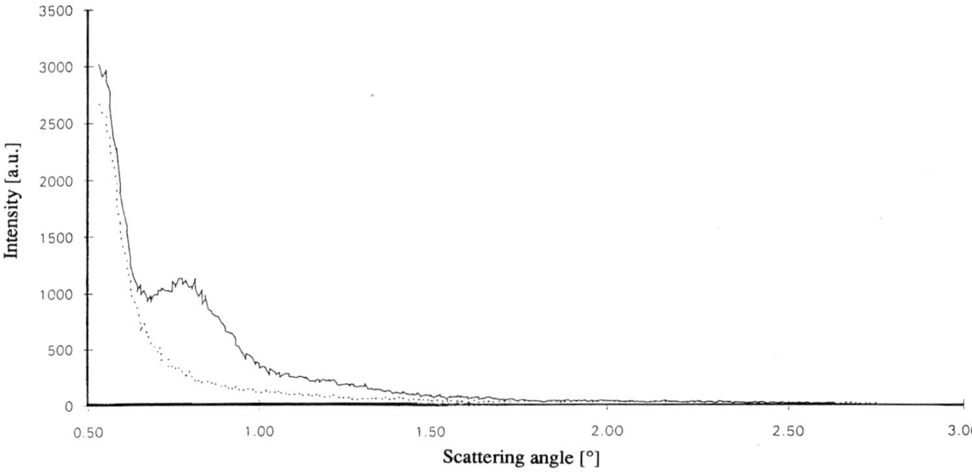

Figure 19. X-Ray pattern of antibody film before and after heating.

Table 5. Surface potential of 1 monolayer of RC and antibodies onto SnO_2 substrate showing the increase in film anisotropy due to thermal treatment. RC monolayer was measured in the dark. Measurements were performed with a home-made surface potential gauge suitable to measure surface potential on solid substrate with a sensitivity of less than 1 mV (Facci et al., 1994b)

	RC	Antibody
before heating	25 mV	-84.5 mV
after heating	192 mV	-370 mV

1.93 10^{-2} pmol mm^2, 4.88 $10^{-2} \pm 1.22\ 10^{-2}$ pmol mm^2, respectively for non-silanized and silanized quartz.

The enzymatic activity appears unaltered at three increasing temperatures, confirming that even the protein function is heat-proof in LB films (Figure 20).

AP detached from silicon nitride and optical glass surfaces during the activity assay. Indeed, with consecutive assays the catalytic activity decreased and the reaction persisted after removal of the film from the cuvette due to dissolved enzyme. Anyway, the functional assay is a rough measure of the amount of active enzyme lifted in each case. The activity increased with the number of monolayers deposited at fixed surface pressure far from the saturation value. With the same number of monolayers the activity increased with surface pressure, until it reached the saturating value. Comparable results were obtained with subphases containing NaCl in the concentration of 3M and $(NH_4)_2SO_4$ in the same concentration (not shown).

AP was bound to Hybond-N^+ membranes by direct spreading and by Langmuir-Schaefer deposition technique. Before use, 9x30 mm pieces of membrane were attached to glass slides to ensure a flat surface for the horizontal lift.

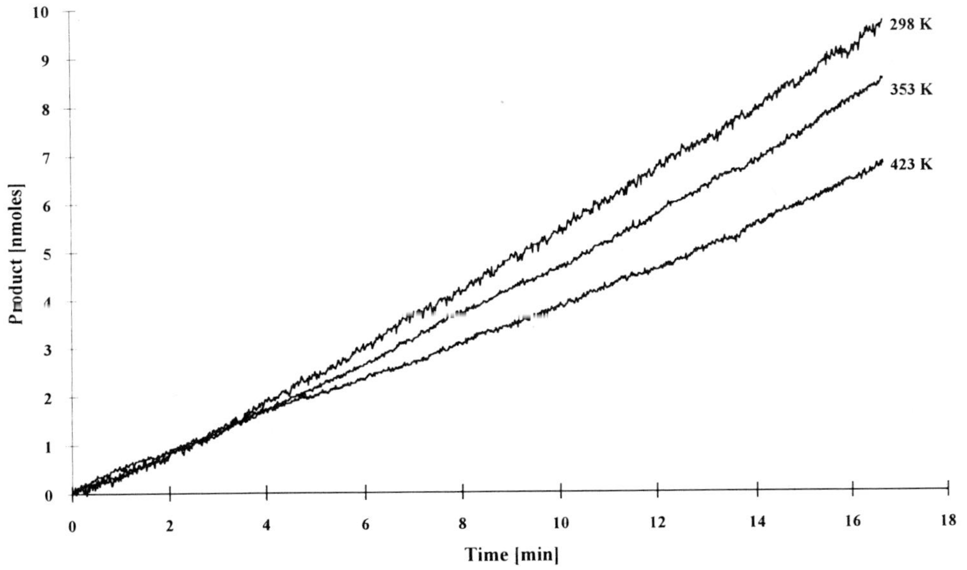

Figure 20. Kinetic curve of the enzymatic reaction of one GST monolayer deposited onto silicon substrate utilizing the LB technique at three different temperatures. The produced GDNB is determined versus time at room temperature for 2.5 mM GSH phosphate buffer pH 6.5.

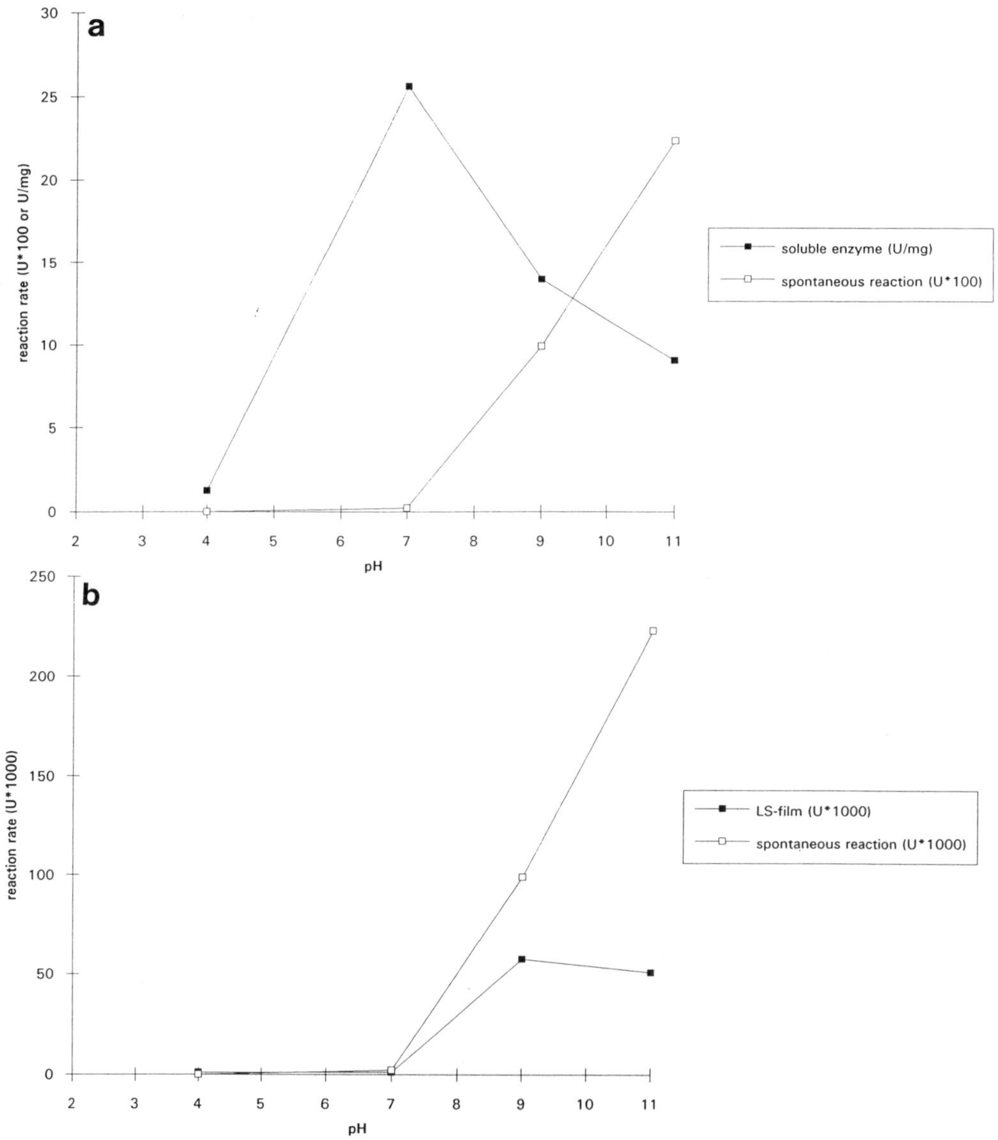

Figure 21. pH dependence for the soluble enzyme (above) and for the film (below).

The activity assays were performed by direct monitoring of product formation by spectrophotometry with magnetic stirring inside the cuvette. Comfortingly no detachment of enzyme was observed during the assay of membrane-bound LB films. Indeed, consecutive assays did not show a decrease of activity and no reaction was observed after removal of the membrane from the reaction volume. Drying of the membrane at room temperature in a desiccator resulted in a sensible loss of activity for the spread but not for the LB-deposited enzyme.

The spontaneous reaction increases monotonously with the pH. Evaluation of the catalyzed reaction is difficult at high pH values due to background reaction. Therefore the pH dependence for the LS-film has to be taken in consideration with caution. Fig. 21 shows the pH dependence for the soluble enzyme and for the film, respectively. The rate of spontaneous reaction is plotted for comparison.

Immunological Activity

In order to optimise immunological activity the orientation of the IgG molecules in Langmuir films needs to be immobilized on a solid substrate. This technique is based on measuring the binding of the secondary antibodies specifically recognising Fab and Fc fragments of immobilised IgG molecules (Dubrovsky *et al.* 1994, 1995a). Specific binding of anti-Fab and anti-Fc rabbit anti-mouse (RAM) polyclonal antibodies to the film of mouse IgG shows strong dependence on surface deposition pressure of the film, thus indicating changes of molecular orientation occurring in the film. At high deposition pressure the binding parameters of anti-Fab RAM antibodies are higher than that of anti-Fc; this means that IgG molecules of the film are preferentially oriented with their Fab fragments outside. The ratio of molecules oriented in this way to the opposing molecules is estimated to be 3.125. The technique can be considered rather general and suitable for similar study of any protein two-dimensional arrays since it is possible to obtain antibodies towards different parts or fragments of any protein.

Immunological reaction ability of IgG Langmuir-Blodgett monolayers organized by means of a Protein A sublayer was studied by gravimetric technique. In order to discriminate the effects of molecular orientation and to preserve the native structure of IgG molecules using the protein sublayer, the kinetics of specific and non-specific binding at immunoglobuline monolayer deposited onto different substrates, such as silanized aluminium surface and a surface covered by ovalbumin and Protein A sublayers were compared. It was shown that the prevailing effect is molecular orientation. Sublayer of Protein A appeared not only to increase the specific sensitivity of IgG monolayer but to decrease the non-specific binding as well. For this structure the sensitivity of the monolayer of rabbit antimouse IgG towards mouse IgG the value of 10 pmol/l was achieved. The sensitivity in case of bare metal surface and ovalbumin sublayer is 10 and 100 times less respectively.

Protein A and ovalbumin sublayers were obtained by Langmuir-Schaeffer technique. The formation of monolayers on the water-air interface and the deposition onto the solid substrate were studied by means of compression isotherms, gravimetric and ellipsometric techniques. It was shown that the solubility of Protein A depends on the surface pressure, increasing sharply after certain value of the pressure. It was also shown that Protein A can be deposited onto the substrate in the form of dense 2D monolayers (Dubrovsky *et al.* 1995b).

The structure of immobilized Langmuir-Blodgett films of IgG and their immunological activity were studied by means of ellipsometry. The dependence of film thickness on surface pressure of deposition gives evidence of the tilting of molecules with an increase of the pressure. Below pressure 30 mN/m the thickness of the film is approximately 4 nm which coincides with the smallest dimension of the IgG molecules. Between 30 and 40 mN/m the thickness increases sharply achieving the value of about 10 nm which is equal to the largest molecular dimension. Further increase of pressure does not come out in the growth of the thickness. This means that the films are transferred from the water-air interface in the form of 2-D ordered monomolecular layer. The dependence of the immunological activity on the pressure of deposition was shown to have a descending pattern. We propose different mechanisms which explain the decrease of the immunological activity at the IgG molecules in the film with the increase of the surface density such as the blocking of the active sites and the decrease of the conformation mobility of the Fab fragments.

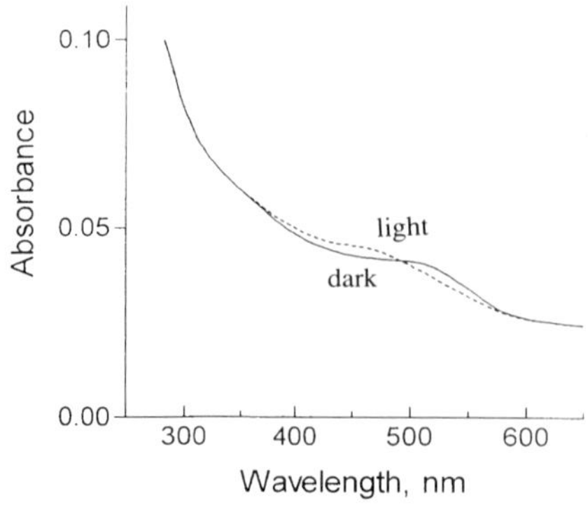

Figure 22. The adsorbance spectrum of rhodopsin in LB film within a range between 250 and 650 nm in the dark (solid line) and in the light (dotted line) (adapted from Pepe et al.).

Electron Transfer and Redox

Electron transfer reactions are extremely important in biology, particularly in bioenergetic reaction pathways. In the early steps of photosynthesis, high efficiency solar-energy conversion is achieved with a complex of protein-bound electron donors and acceptors. Control of charge separation and recombination rates is required to ensure that productive forward electron-transfer reactions within this complex occur rapidly, while wasteful back reactions occur orders of magnitude more slowly. Initial light-driven transfer steps are complete within 3 ps, and charge separation is subsequently stabilised for tens of milliseconds with a quantum efficiency near 100%. A comparable selective acceleration of electron transfer reactions has not achieved in any artificial system.

Our goal is to develop with the heat-proof metalloprotein film previously summarized (RC, P450 ..) a computational design capability for electron-transfer in protein array. A minimal theoretical description is needed to model adequately the fundamental mechanisms of protein-mediated electron tunnelling. Although the description does not include every detail of the protein electronic structure, recent models make concrete, testable predictions about primary, secondary, tertiary, and quaternary structural effects on electron-transfer rates. Measurements of electron transfer rates in ruthenium-modified (ruthenated) proteins test the reliability of the method. It has been found that inclusion of protein structural features neglected in structure less barrier models is essential for understanding the observed transfer rates in these systems. Further efforts in this direction are essential.

As model system for molecular recognition and redox reaction thioredoxin was utilized which is a protein that occurs in all living systems and is capable of catalysing thiol-disulphide exchange reaction and protein disulphide isomerization.

Another interesting protein, quite promising for optoelectronic application, is rhodopsin, which adsorbance spectrum in LB film is shown in Figure 22 within a range between 250 and 650 nm. From this spectrum it is possible to evidence the conservation of characteristic adsorbance at 500 nm also after the exposure of the desiccated film to white light.

If the sample under study is repeatedly subjected to flashes of green light, its adsorbance characteristics are shifted with a maximum around 480 nm. This phenomena shows the formation of metarhodopsin, an intermediate of rhodopsin, that in film results stable while in solution it has a life-time of msec, hydrolysing itself rapidly.

Dealing with electron transfer properties of trans-membrane proteins, an interesting point is the physical reason which allows RC LB film to be imaged by STM microscopy, the most probable mechanism seemingly being connected with the presence of the hydration shell around the sample. In fact, studies (Guckenberger, *et al.* 1989; Amrein *et al.* 1989; Kell *et al.* 1992) devoted to the understanding of the mechanism of image formation in biological

Table 6. Relative amount of water in samples of Reaction Centers from *Rhodobacter sphaeroides*

Sample	Relative water amount (%)
RC dry smear	70
RC LB	9

samples have shown the existence of a critical value of the relative humidity (threshold) below which it is impossible to get resolution on RC samples by STM microscopy. This value varies according to different conditions (e.g. the amount of ions in the sample environment) but generally speaking is in the range 30-50% (relative humidity). This evidence points out the role of ionic currents in the image formation process.

The data listed below on the amount of water desorbed from our samples do not contradict these last consideration since they are independent upon the relative humidity of the environment (the data are acquired as difference in mass between the sample fresh and after thermal treatment but after coming back to standard conditions of T, P relative humidity, etc).

LBM TECHNOLOGY VERSUS SELF-ASSEMBLY

The fabrication of protein 2D arrays is yet closer to an art than to a science. The level of difficulty is in the choice of pH, temperature, type of salt and its concentration, the incubation time and above all, the way to present the protein at the interface. Many methods have been proposed in the literature, all of them proven able to yield 2D protein arrays in different circumstances (Table VII). These methods can be reconducted to either Langmuir-Blodgett (this overview) or to Self- Assembly, based on subphase lifting (Yoshimura *et al.*, 1994) and mercury surface (Yamaki *et al.* 1993, Nagayama 1994a, b).

Subphase lifting was very efficient with apoferritin protein. It is possible to make crystals that reach the size of 5 micron. The quality of crystals is very good, and the transfer from the air-water interface is relatively easy. The amount of protein to inject near the interface is small, which means a very low risk of denaturation. The subphase must be denser than the solution, which means utilization of glucose concentration of about 2%. However, while very successful with apoferritin, other proteins proved to be much more difficult to

Table 7. List of the methods utilized for protein 2D crystals so far

Methods	Motion		2D particles arrays	
	Internal	*External*	*Internal*	*External*
Traditional way	Brownian	——	NIL	≤ 10 mm
Dense subphase	Brownian	——	NIL	≤ 5 mm
LB binding	Brownian	——	Not tested	≤ 100 mm
LB compression	Brownian	$\forall F = 0$	≤ 5mm	≤ 1 mm
Mercury trough	Surface Tension Capillary Brownian	Random water flow direction	Not tested	≤ 0.1 mm
Water evaporation	Brownian	$\forall F \neq 0$	≤ 10000 mm	≤ 0.1 mm

where $\forall F$ is the horizontal gradient of the forces.
if $\forall F = 0$: no directional surface flow
if $\forall F \neq 0$: directional surface flow

make 2D crystals. Even holoferritin proves quite difficult. In our study a simple method for making two-dimensional (2D) protein arrays using a new spreading and self-assembly technique (Nagayama *et al.* 1993, 1994) was utilized. A solution of the iron protein was injected into an aqueous subphase that had a higher density and surface tension than protein solution. The buoyancy of the protein solution made it rise to the surface where it spread quickly and smoothly. For the stage of rising and spreading, the buoyancy and surface tension were critical factors. The proteins that reached the air-water interface unfolded instantaneously and formed a continuous film. These films were transferred onto holey carbon films for transmission electron microscopy analysis. The holes of the carbon film were found to be covered with a smooth, amorphous film of unfolded protein onto which intact ferritin was adsorbed. For the stage of 2D array formation, the presence of cadmium ion in the subphase was critical factor. The 2D array formation occurred in the thin layer of protein solution caught between the unfolded protein film of the surface and the subphase below.

Mercury surface proved to be successful with many proteins. The best one is again holoferritin, that seems to resist spreading over mercury, and finally succeed to make 2D crystals. The problem of this method is the high protein concentration (200 mg/ml) that is required to produce crystals. This high concentration is unsuitable for a large range of protein, as those discussed in this paper. Nevertheless, there is a low but real production of 2D crystals even at low concentration. Perfectly flat substrates are regarded as essential for supporting the growth of 2D crystals. Therefore, freshly cleaved surface of mica or surface of mercury have been used for crystallization. Although several proteins have successfully been crystallized on mercury, this technique has the disadvantage that sophisticated equipment is required to ensure a clean surface. Recently there has been considerable success in using lipid-water interfaces for 2D crystallization. The design of lipid polar head groups that interact with proteins is the key point of this method. Creating a suitable charge density is sufficient for some systems, while others require that specific ligands be attached to the lipid polar head groups.

In addition to recent work on 2D crystallization, the surface chemistry of protein spreading and protein monolayer formation at the air-water interface has been the subject of extensive investigations. To achieve complete spreading of the proteins, concentrated salt solutions that have a higher surface tension than water (e.g., 3 M NH_4SO_4) have been used as the subphase.

Alternatively, alcohols such as iso-propyl alcohol which decrease the surface tension were added to the protein solution to promote spreading. It was reported that unfolded protein films were formed at the air-water interface when protein solutions were spread on an aqueous subphase. The smallest occupied area of the unfolded protein film was measured and found to be about 0.78 mm^2 per mg protein (0.15 nm^2/amino acid residue). At this area density, the air-water interface is presumably covered completely with a monolayer of the unfolded protein film. When larger amount of protein than former condition was spread, the behaviour of the π-A isotherm (surface pressure versus occupied area per molecule) changed from the behaviour that is normally observed for monolayer coverage. Here the enzymatic activity of spread protein (pepsin, trypsin, acetylcholinesterase, and catalase) was conserved. Many molecular models were proposed to explain this phenomena, e.g. adsorption of intact proteins on the unfolded protein film, or that intact protein was incorporated in the unfolded protein film. There have been many discussions regarding the structure of these film, but these have not led to a unanimously accepted conclusion. This discussion reflects the lack of information about the real structure of such films, which can only be provided by direct imaging techniques such as microscopy.

Langmuir-Blodgett technique finally combined with heating at 150°C has proven to be capable of making 2D quasi-crystals of all types of proteins, both water soluble and trans-membrane as shown in the previous pages. In fact, the protein film is rapidly com-

Figure 23. a) TEM image of a 2D crystal of cytochrome P450scc obtained by self-assembly based on subphase lifting, b) FFT of the image.

pressed after spreading at the air-water interface. Up to this point, no crystallization occurs. But when heated to high temperature, the film becomes crystalline and acquires a surprising resistance to denaturation at high temperature. In this section we intend to carry out a close comparison between the above two technologies in terms of heat-proof, long range storage and array homogeneity and size, limiting however ourselves to only the first part of our modified LB technology, which is the transfer of one layer, without heating.

Heat Proof

Contrary to what shown earlier for LB protein films, self-assembled 2D protein crystals do not display a similarly striking thermal stability, melting their native secondary and higher order structures at a temperature only few degrees higher than the corresponding values in solution.

Long Range Storage

Similarly LB protein films display long range stability, which lacks in 2D crystal protein arrays.

GST multilayers deposited by the Langmuir-Schaefer technique at the surface pressure of 20 mN/m and exposed to phosphate buffer 0.1M pH 6.5 with 1mM EDTA at room temperature reveal long range stability in functional and structural terms (see Figure 24).

Whenever a 30-monolayer GST film was deposited onto an optical quartz surface, within the range of the experimental error the film absorbance was constant over a week of exposure (Figure 24) and has been stored dry for over five months without any significant structural and functional alterations. Similarly, whenever a 10-monolayer GST film was deposited on a silicon nitride surface the film thickness (about 160 Å) determined by ellipsometry was unchanged over 75 min exposure and no measurable activity was detected in the buffer, even under shear conditions typical of biosensor applications. Substantial difference in stability was apparent between one-monolayer samples deposited at 20 and 35 mN/m. At 20 mN/m the activity was constant over ten consecutive repetitions of the assay. GST activity assays were performed according to the Sigma protocol by using 1-chloro-2,4-dinitrobenzene (CDNB) as a substrate and enzyme detachment was monitored by ellipsometry, absorbance spectrophotometry and activity assay.

Similar stability was apparent in AP LB films stored on membranes versus time (Figure 25) and versus increasing pH.

Array Homogeneity and Size

The characterization of a protein monolayer at the air-water interface requires an instrumentation capable to obtain a resolution at the nanometer scale. Among the instruments well known to reach routinely these limits is the transmission electron microscope. Transmission Electron Microscopy images (Figure 23) were acquired on the P450scc arrays obtained by self-assembly or by LB technology and then digitalized by CCD camera (Photometrics). 512x512 pixels (corresponding to 550x550 Å) images were filtered by means of an Enhanced Spatial Filter with a 20x20 window dimension in order to increase spatial resolution of the patterns and to decrease the influence of tilted plane in the image.

The inspection of the Fourier Transform of self-assembled recombinant and wild-type P450 scc shows that it is possible to identify ellipsoidal aggregations randomly and periodically distributed. In LB films there are preferential orientations in the spatial distribution of single elements: 45° (high frequency) and 135° (low frequency) directions for the recombinant, 30° (high frequency) and 120° (low frequency) directions for the wild-type. In the Fourier Transform images superposed spots on orthogonal directions represent periodicities produced by Spatial Filter. High frequency signal corresponds to short distances while low frequency to longer ones. No periodicity due to bidimensional crystal ordering can be recognized by means of this LB study. However, in the regions were the film was deposited, it was possible to observe highly ordered and symmetric zones, exhibiting even long range (hundreds of nm^2) 2D order.

In summary, while heat-proof and long range storage appear present in the LB protein films but lacking in 2D self-assembled arrays, the contrary appears true for array homogeneity and size. The LB arraying of proteins never exceeded few microns large at the best, and moreover, the lattices appear at random on the surface with occasional holes. It is probable that if arraying and crystallization could be done and controlled over macroscopic surfaces, the whole film would become heat resistant and crystalline at the same time. In general, arraying of particles is still not a fully controllable process. The prevailing philosophy is to let the natural forces to make this arrangement through self-organisation. Our approach is instead to take the control over a wide range of translational motions, while letting the natural forces act at nanometer scale.

Long-range translational displacement of molecules over macroscopic surfaces is championed by the Langmuir-Blodgett method. A huge body of experimental data demonstrates that this method is extremely suitable to orient amphiphilic molecules at the air-water interface. It is also capable of making a protein film at the air water interface, and at a certain point, to make 2D array. As revealed with TEM observations, the main problem is that these particles do not close-pack easily over macroscopic scale. Resulting from this situation is the creation of local vacancies.

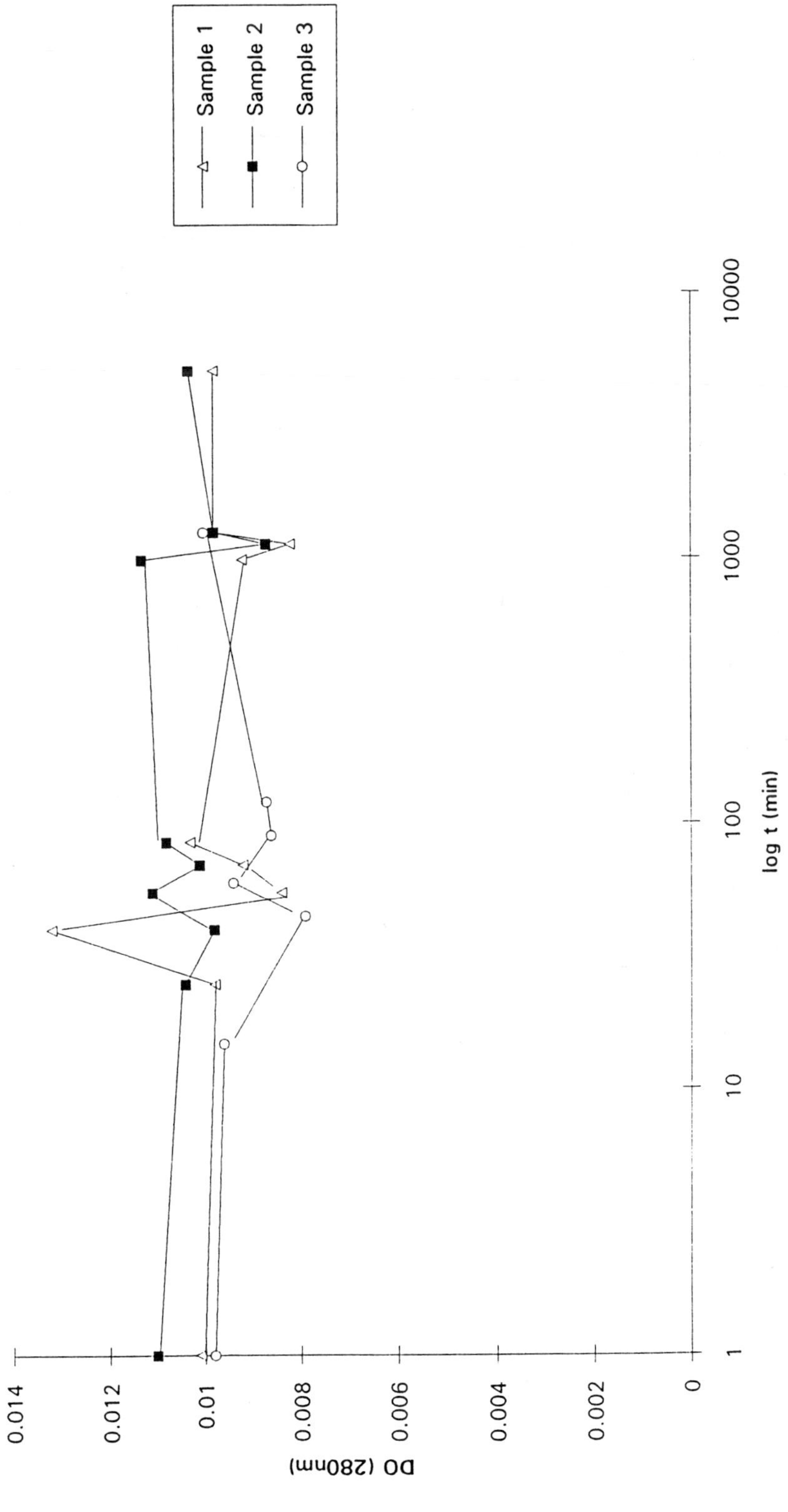

Figure 24. GST detachment from LS-film (30 monolayers at 20 mN/m) deposited onto quartz substrates.

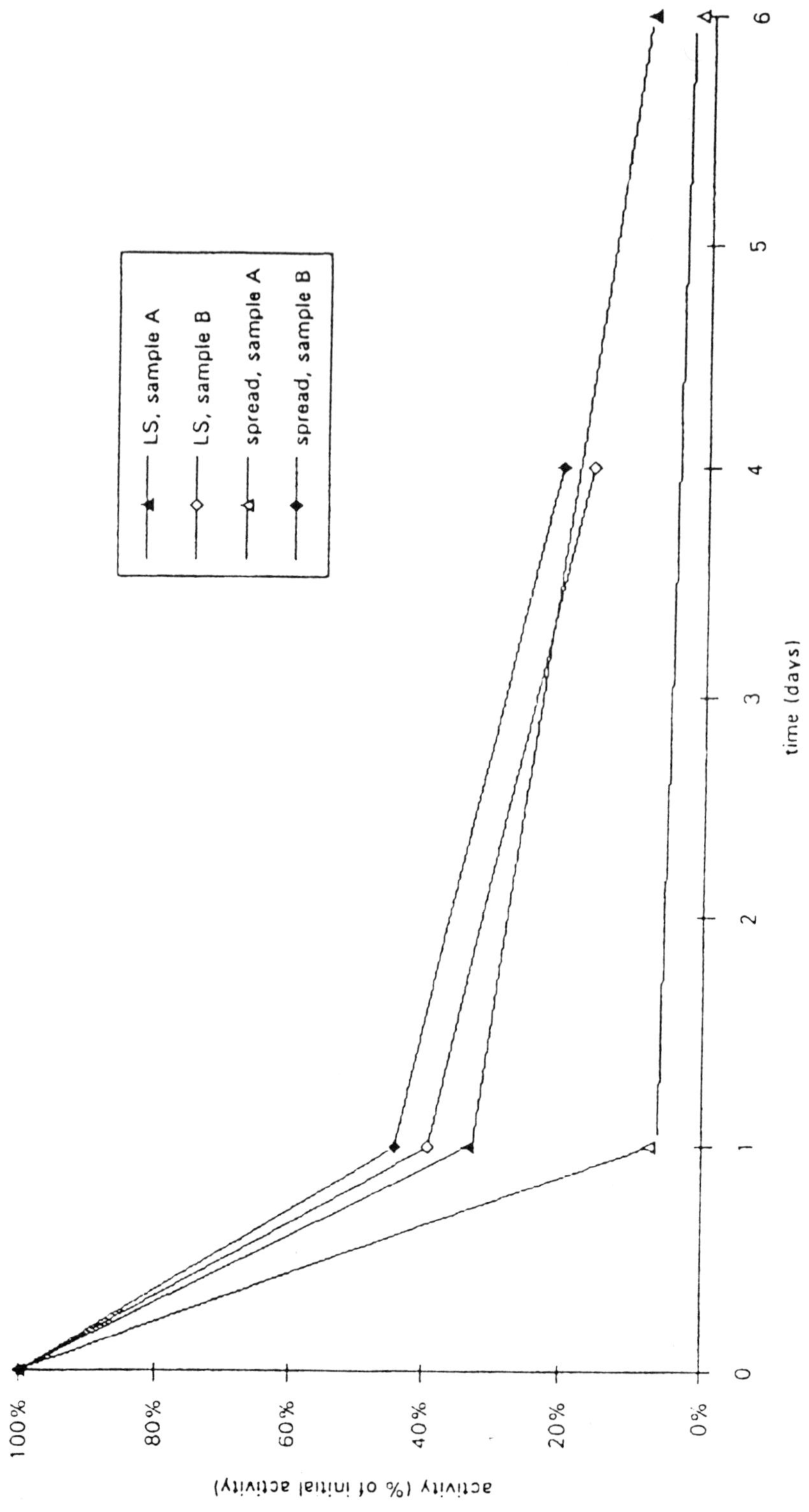

Figure 25. AP enzymatic activity in LB film immobilised on membranes stored in the refrigerator.

Another long-range processing was the usage of a directional flow driven by evaporation. A glass slide was pulled out of water which created a meniscus ending with a triple contact line: liquid, solid and gas. Due to evaporation, the particles moved up this line and finally stopped. They were hexagonally close-packed and immobilised due to the lack of water after evaporation. This packing with a directional flow, driven by evaporation, demonstrated capabilities to make submillimeter size of polycrystalline layer of 800 nm latexes. However, instabilities near the contact line make this technique very sensitive. This technique works in fact in "suction" mode and not in compression. In other experiments this directional flow was also capable of packing latexes with diameters from many microns down to 55 nm, but no proteins were successfully packed this way.

One of the reasons concerns the protein structure. It is well known that proteins denature at the air-water interface. Thus, the necessary evaporation process will deplete the protecting water layer surrounding the protein and finally will provoke its denaturation. So this process is intrinsically harmful for the protein unless it is protected against denaturation. Another reason is the contact angle at the triple line. This angle is related to the rupture of the water thin film over the solid substrate. The thickness of rupture is in the order of 30 nm. This is why no particles below 55 nm proved to be suitable for the evaporation technique. Nevertheless, this method has given one clue for the packing of latex particles. It is to bring them one after the other, in order to avoid a "jam" of particles. Then, although attractive in principle, each of two methods (self-assembly versus LB) has its own advantages and limitations. The best of each approach, if combined in one set-up, should be the most suitable one to successfully make large 2D arrays of fine particles or proteins in a controllable way.

Towards this end a new method for continuous production of uniform particles 2D array at macroscopic scale has been recently implemented, whereby the particles - either latex spheres or proteins - are kept in solution or suspension; in this case a hydrophilic roller is first used to transfer particles from one side to the other side with the particles being in a thin liquid film around the roller; subsequently in the other side, a second hydrophobic roller will compress the floating particles on the third surface. A third roller is finally introduced to make the transfer on solid substrate, making multilayer of particles 2D arrays on its own surface.

APPLICATIONS

Fabrication of protein arrays has many fundamental and industrial goals. For the industry, uniform coating of glass surface with latexes should reduce light scattering, while providing special optical effects. As a matter of fact, beautiful colous were produced using 55 nm and 122 nm latexes. Consequently, optics, as well as cosmetic industries, are potentially interested in such perfect large scale uniform coating.

On the other hand, it was demonstrated in this overview that proteins become heat-proof after having been 2D arrayed. This is a breakthrough in the domain of applied bioelectronics and nanobiotechnology in general. Proteins being very specific functional systems working at the molecular level, their usefulness appears indeed concrete also in the domain of biotechnology.

Moreover, it was shown, that heating can even improve the performance of the total layer, namely, heating of antibody layer resulted in the increase of the amount of bound antigen and acceleration of the reaction. The driving force of the improvement was linked to the improved order (Erokhin *et al.* 1995a). Several potential applications have been recently discussed in a comprehensive overview of bioelectronics, defined as the use of biological materials and biological architectures for information processing systems and new devices (Nicolini, 1995b). We hereby summarize only the most recent pertinent advancements with emphasis in emerging nanobiotechnological applications.

Biocatalysis and Bioactuators

The application of biological materials to different branches of modern technology has expanded enormously during the recent period. Some of the applications which already found their implementation are, for example, gene engineering-based production of biologically essential macromolecules, stereospecific biosynthesis with the utilisation of specific enzymes and production of protein-based bioactuators.

An example of heat-proof and stable bioactuators are LB films based on phenol oxidases and laccase which are specific towards bioactive phenols (Tyr, DOPA, dopamine, adrenaline and noradrenaline), while enzymes from fungi and higher plants are active towards a wider spectrum of substrates (mono- and polyphenols).

Main application of protein based nanotechnology is indeed biocatalysis in the various areas related to biotechnology, of which laccase represents only an already successful example still in progress.

Laccase (p-diphenol: oxygen oxidoreductase) is an ubiquitous enzyme which catalyzes

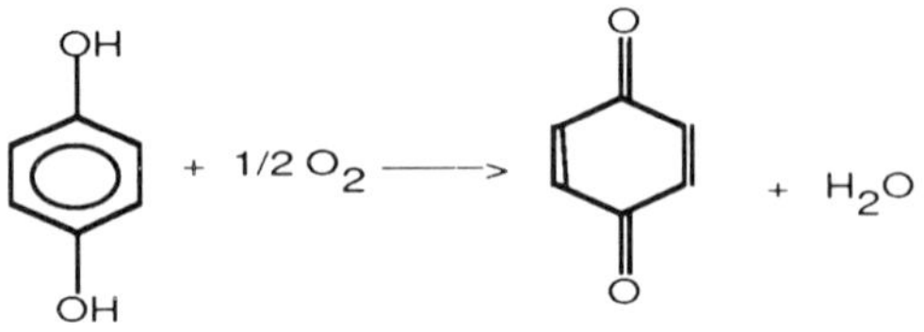

while the enzyme cathecol oxidase (o-diphenol: oxygen oxidoreductase) catalyzes a similar reaction:

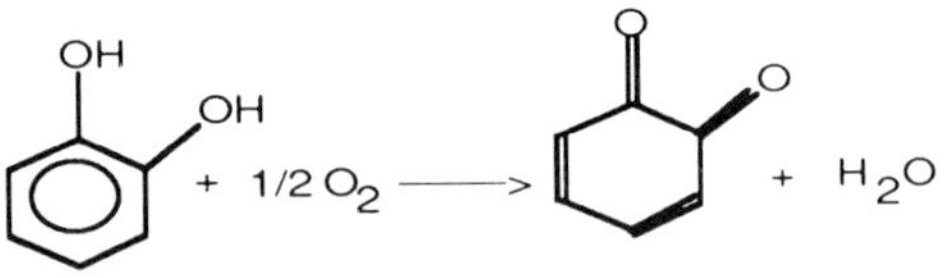

These two enzymes have been used:

1. as bioelements in amperometric biosensors for the selective determination of neurotransmitters;
2. for the construction of bioreactors for the enzymatic degradation of phenolic compounds in waste water.

In these reactors, laccase and catechol oxidase have been immobilised with several well-known procedures.

Recently we have been successful in immobilising them by the following steps:
a) silanizations of glass beads with appropriate reagents;
b) incubation of these beads in a concentrated solution of enzyme.

After this procedure, we placed these beads in a common chromatography column and we have monitored heat-proof enzymatic activity, immobilization lifetime and other industrially interesting parameters by means of standard spectrophotometric assays. Whenever a single protein monolayer is properly formed by Langmuir-Blodgett technique at the

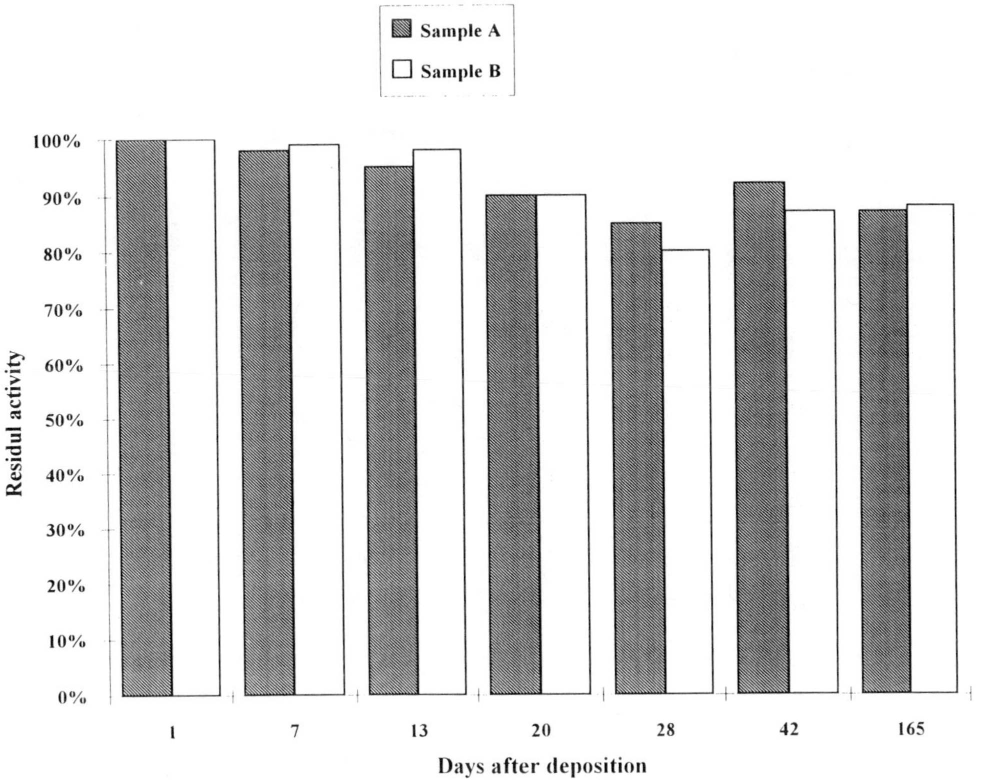

Figure 26. Residual enzymatic activity over time after storage of the sample under dry conditions at 277 K. The averaged decrease of the activity is of about 20% after 5 month. The activity determined at the first day were 1.1 nmoles*min^{-1} and 1.2 nmoles*min^{-1} for the samples A and B, respectively.

saturating surface pressure (see earlier paragraphs), the enzymatic functional activity appears nearly independent of the number of layer (Figure 26). As recently proven for GST (Antolini *et al.* 1995a) and alkaline phosphatase (Petrigliano *et al.* 1995), the implication for the industrial applications of biocatalysis, biosensors and biotechnology in general are quite promising.

Biosensors

Exciting recent developments, based on either a newly designed potentiometric biosensor named PAB (Adami *et al.* 1994, 1995) or a nanogravimetric sensor (Nicolini *et al.* 1995e), are the possibility of quantifying with high resolution specific antigens or molecular compounds by immobilising on the transducer the corresponding LB monolayers of antibodies or enzymes.

Successful examples are the detection of alcohol by alcohol dehydrogenase (Adami *et al.* 1994), urea by urease (Figure 27; Paddeu *et al.* 1995b), pesticides by GST (Antolini *et al.* 1995a), and various antigens by proper monoclonal antibodies (Nicolini *et al.* 1995a).

The development of a completely new generation of extremely sensitive biosensors is presently being explored by utilizing LB technology in an imaginative way, namely

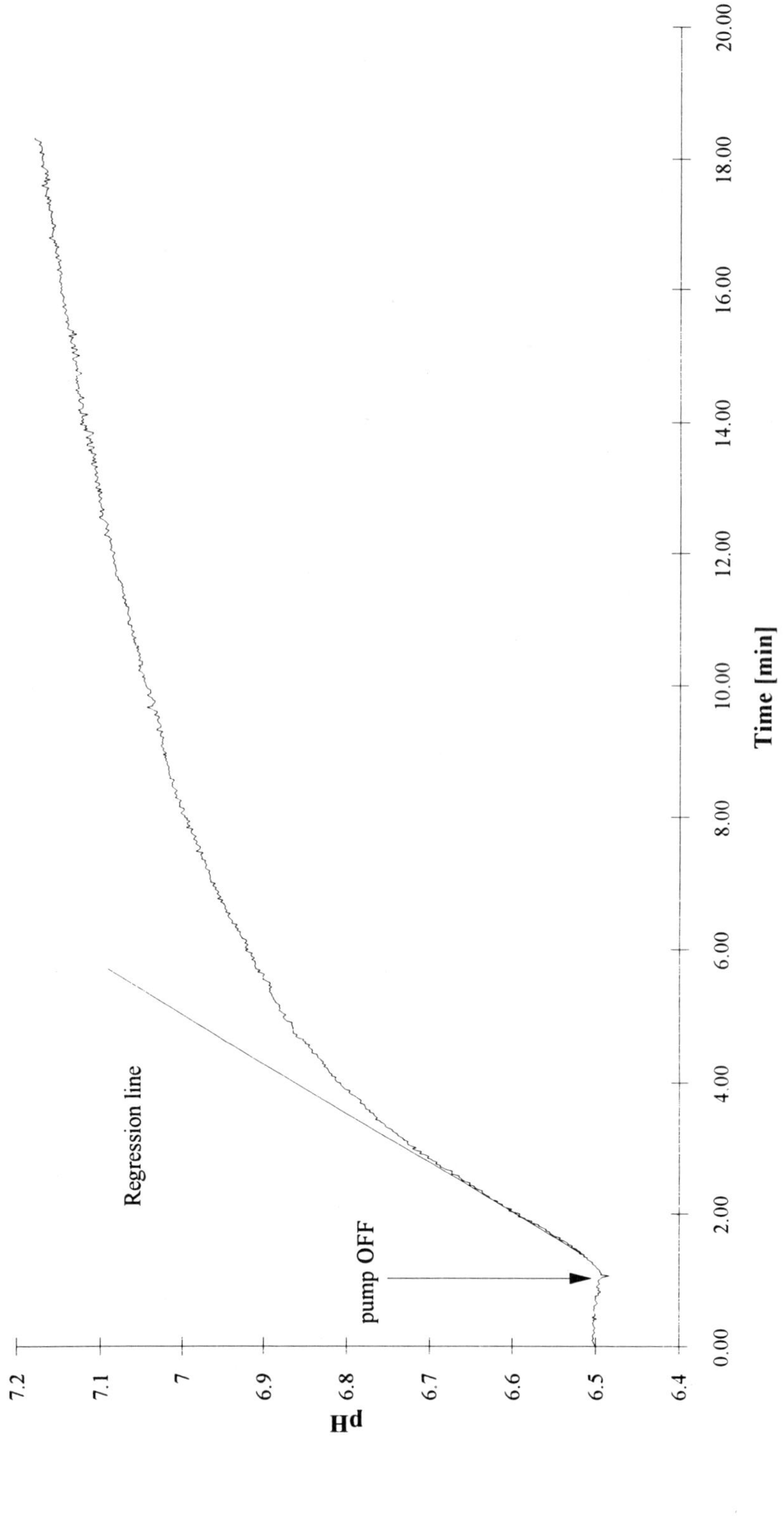

Figure 27. Activity of immobilised urease LB monolayer, as monitored by PAB.

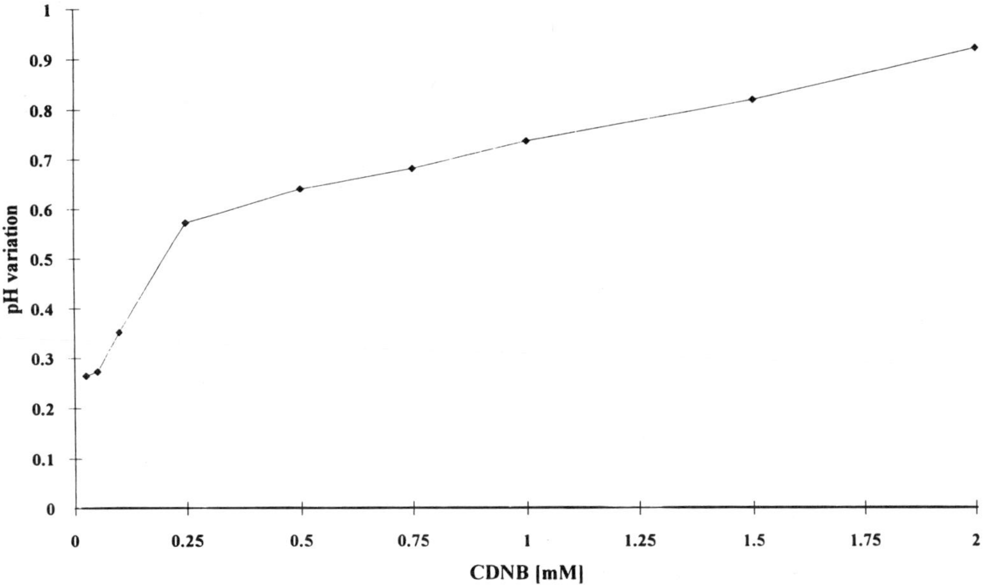

Figure 28. pH variation, referred to infinitive time, versus CDNB concentration. The concentrations of the GST enzyme and of the GSH in PAB chamber are 0.22 M and 2.5 mM respectively. The CDNB concentrations are 0.025 mM, 0.040 mM, 0.25 mM, 0.5 mM, 0.75 mM, 1 mM, 1.5 mM and 2 mM. The pH decreases starting from 6.5.

alternating monolayers of conductive $C_{16}H_{33}$-BEDT-TTF lipids to monolayer of insulating ADH enzymes.

Since NAD^+ molecules cannot diffuse through conducting layers to reach ADH layers inside the system of alternating monolayers, we introduced NAD^+ molecules into the film during the process of film deposition. Then, by treating the sample with the vapour of ethyl alcohol which did penetrate easily through the film, the reaction taking place inside the alternating multilayers caused the release of protons and the modulation of the film conductivity, which in turn could be easily and strikingly monitored by a simple LRC meter to a sensitivity unmatchable by any existing sensor.

Two effects yet not quite understood were discovered for the above multilayers. First of all, irreversible increase of conductivity happens for every sample when they are treated with ethyl alcohol vapour for the first time during 2-3 min. Then, the value of conductivity stabilizes. It seems that there is no systematic dependence on the values of conductance increase, but the increase is usually less if initial conductivity of the sample is high. As a rule, conductance increases approximately 10-20 times and is stabilized at this value. It seems that some irreversible structural reorganization of the film takes place under such treatment. Indeed, when the structure is well organized and initial conductivity is high for this reason, increase of conductivity is not very strong and vice versa. Later we treated the samples in a simple way, i.e. we kept them during 2-3 min over the surface of ethanol and obtained immediately a stabilized value of conductivity.

The second effect is the increase of conductance value at high frequency when the sample is kept in ethanol vapour. For example, let us suppose that initial value of current (for stabilized sample) is equal to I_0. When we put the sample over the surface of ethyl acetate during 30-40 s current reaches the saturated value $I_0+\Delta I$. Then we remove the sample in air medium and within 5-10 s current returns to the value I_0. Values of I_0 and $I_0+\Delta I$ are very well reproducible if we repeat the procedure of treatment immediately for several times.

Biomaterials for Electronics

Semiconductor technology has changed our lives within only a few decades. Sand, the starting material, has been turned into a versatile "high-tech" product. Techniques have been developed for the production of silicon wafers and for the modification of this raw material into the final "high-tech" product - miniature electronic logic functions. Research now focuses on the molecular level, from the micro to the nanostructures.

The rapid increase in our knowledge of the function of biological materials has also directed interest towards this area. The functionality, efficiency, and flexibility of biomaterials is impressive, but they are beyond the capabilities of synthetic chemistry. The development of gene-technological methods has opened the way to controlled modifications of proteins, which in the future will allow the production of newly designed biomaterials. A first example was the modification of the retinal protein bacteriorhodopsin and the generation of a versatile media for optical processing on its basis. This example will be used to describe the principle of the approach and to give an idea of possible future developments. Gene technology as the key technique in this new direction of materials research may be the counterpart to photolithography in semiconductor technology.

Furthermore, since electrical properties are necessary "macroscopic", macroscopic structures must be built-up for obtaining information on biological samples. A possible architecture is presented by Langmuir-Blodgett films.

After formation of such films at air-water interface, eg. of metalloproteins with their coenzymes present in nature to warrant electron transfer, they can be transferred onto a substrate with interdigitated micro-electrodes (see Fig. 29).

This permits us to investigate the current flowing from one micro-electrode to another through LB film between them, utilizing an Operational Integrated Amplifier.

LB films were formed in a LB trough (MDT Corp.); the trough size was 240x100 mm^2, with volume of 300 ml and speed of compression was 0.2 mm/sec. The resistivity of the water subphase was 18.2 MΩxcm.

Twenty layers were transferred onto a sapphire substrate with aluminium electrodes previously cleaned by ethanol and chloroform, by Langmuir-Schaefer methods at 35 nN/m^2 surface pressure for RC and rhodopsin, 25 nN/m^2 for cytochrome P450; the electrode dimensions were equal to 50 μm in width, 40 nm in depth and were interdigitated with spacing of 50 μm.

The current signal (mA) was measured using an operational amplifier in inverting configuration, used as current-voltage converter, and the I-V characteristic was obtained by potential-step method.

The voltage-current characteristic for LB films shows that protein films behave like perfect insulators. Table VIII summarizes the estimates of resistivity of our samples with respect to other conductive materials.

Fig.29 shows voltage-current characteristics on fresh protein LB films; before doping the sample appeared to behave like a conducting one but with very low conductivity.

We doped the same samples both by gaseous iodine and solution containing iron ions; in the second case no results were obtained while in the first one a drastic increase in conductivity of the film was observed. Figure 29 shows the results in these cases, e.g. direct current measurements on LB films of P450 immediately after doping.

However, the increase of conductivity in the films due to doping has proven to be unstable in time; in fact, lower curve in figure 29 reports the results of the same sample shown on upper curves but measured after two days from doping: in this case the conductivity

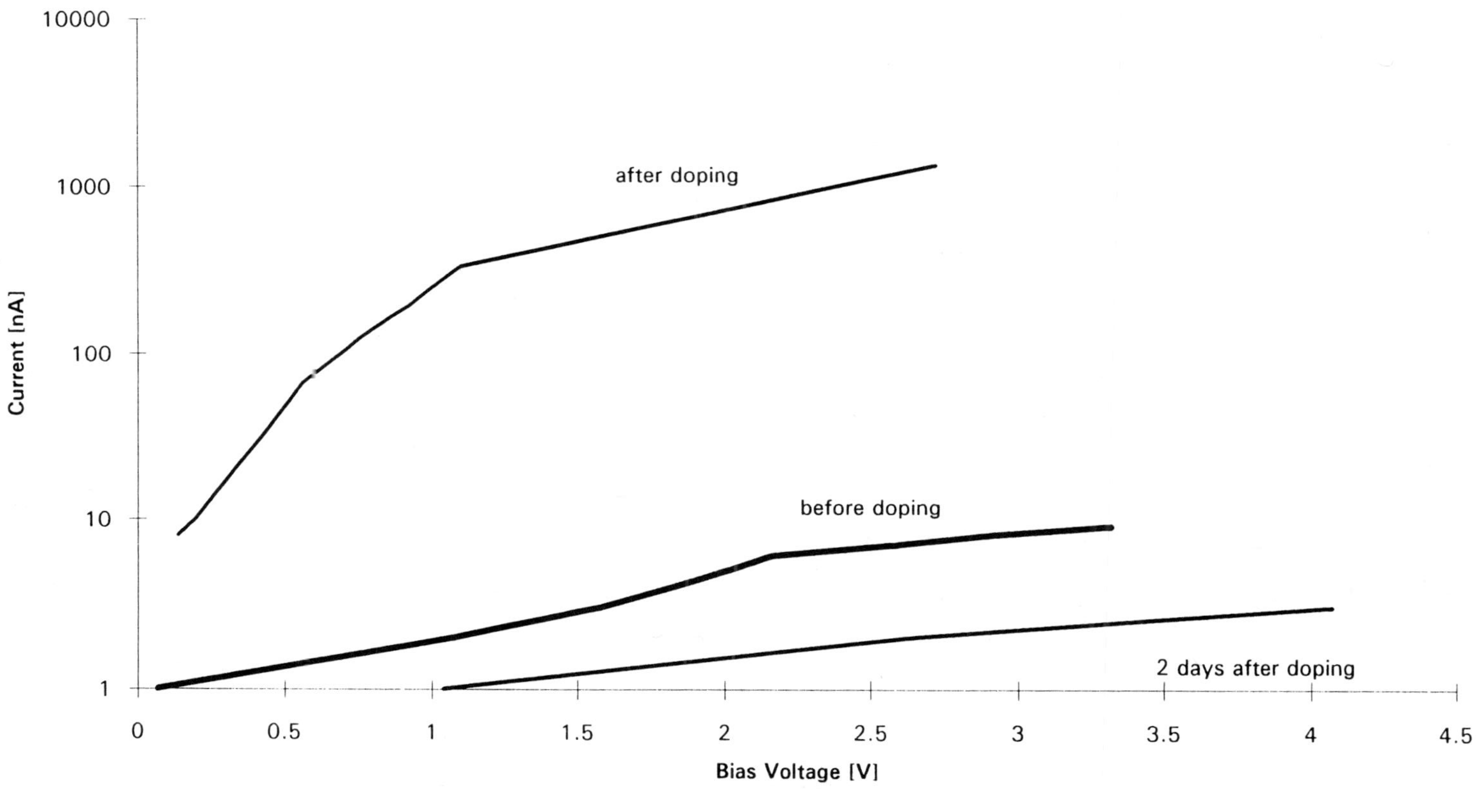

Figure 29. Direct current measurements on a fresh sample of recombinant cytochrome P450scc with co-factor before and after doping. Direct current measurements on these samples doped with iodine but measured after two days from the previous measurements reveals that the conductivity of the LB film was decreased to values found in samples before doping.

Table 8. Resistivity estimated on LB films of proteins with respect to other materials

materials	resistivity [Ωcm]
usual conductors	1-100
conducting LB films	$>10^7$
molecular semiconductors	$>10^{10}$
LB films of proteins	$>10^{14}$

decreased to a level comparable with one without doping (or before doping) and therefore we can suggest that doping by gaseous iodine is not stable in time but the volatility of the doping is such that few days are enough for the disappearance of the doping effect. Routes are presently explored to warrant such stability over time.

RC proteins can conduct an electron in only one direction in presence of a suitable light source (photo conduction) or of an external electric field (conduction). Both these conductive processes are due to the tunnelling of a single electron into the structure of the protein.

Each film consisting of ten RC layers was deposited onto a suitable interdigitated electrode which is made of a sapphire support on which aluminium conductive tracks are fabricated by a photolithography process. The output waves and all significative signals have been acquired by means of an acquisition board installed onto a PC. All signals were visualised by means of an analogue oscilloscope and the input waves have been generated by a function generator, with the desired shape (square or sigmoidal), frequency (500 Hz - 350 kHz), amplitude (CO-IV) and rise-time. Our first measurements were obtained in the dark in order to exclude all the photo conductive processes, by applying a modulated electrical field.

In absence of protein film, the frequency characteristic of the electrode displays cut-off frequencies in close correspondence with those simulated by a circuit simulator (SPICE) installed on a PC utilizing the known value of the R-L-C components.

Preliminary measurements revealed occasionally that depositing ten layers of RC proteins on the electrode, the measured cut-off frequencies and band widths had quite different value depending on their orientation.

Photovoltaic Cells

When a non compensated dipole moment in the molecules exists, there is a possibility to arrange them in regular manner, and to use this dipole moment for the construction of devices exhibiting such phenomena as piezoelectricity and piroelectricity. Both these phenomena are based on the induced changes of the dipole moment, which result in the appearance of voltage across the layer. Even more interesting systems can be assembled with a specific pathway for the electron along the molecule, which can be activated by light. In this case, a neutral molecule becomes a dipole after absorption of light.

One example of such molecule is a photosynthetic reaction centre (RC) where absorption of light transforms a bacteriochlorophyll molecule into an ecited form, and electron tunnel from it continuously to bacteriopheophetin and two quinones, thus causing voltage to appear across the layer (Fig. 30)

The first observation of this photovoltaic phenomenon was reported in 1980 with the following experimental configuration: platinum electrode - RC layer - electrolyte electrode containing hydroquinones. The voltage signal, appeared after the illumination of the structure, was recorded.

In our laboratory the possibility is under investigation to make photovoltaic cells even without liquid electrode. The main difficulty is due to the formation of upper electrode. The most common technique, evaporation, turned out to be not useful, as the monolayer becomes damaged. Even more soft techniques, such as use of mercury drop, results in the destruction of the protein layer. Probably, the problem will be solved when extremely soft techniques of electrode formation will be applied.

One of the best candidates can be conductive LB films since deposition of lipid monolayers should not affect strongly the protein underlayer.

A photoelectric device was prepared using Langmuir-Blodgett (LB) films of photosynthetic reaction centers (RCs). Monolayers of RCs from *Rhodopseudomonas viridis* were deposited on transparent electrodes. The orientation of the RCs on the electrode surface was controlled using various substrates with different surface wettabilities. The degree of alignment was evaluated by measuring the polarities of light-induced electric responses. The orientation was also measured by means of an enzyme-linked immunosorbent assay that uses antibodies to distinguish opposite sites of the RCs. Photocells comprised of RC-LB films showed a steady-state photocurrent. The direction of the current flow was regulated by altering the orientation of the RCs (Miyake *et al.* 1994).

An alternative approach is the construction of photocell using photosynthetic bacterial membranes. The work presents experimental data, which allow to make a conclusion about the possibility of constructing such type of cells using biological materials. To optimise performance we plan to try different types of LB deposition, namely orientation by pressure, electric field and different types of RC: from *Rhodobacter sphaeroides*, *Rhodopseudomonas viridis* and *Chromatium minutissimum*. Next step will be the use of proteins instead of membrane fragments to yield an increase in efficiency of the light conversion.

Optical Filtering and Holography

Optical information storage and processing utilized up to now only synthetic, photochromic and refractive materials as reversible recording media. However, the effect of side reactions on photochemical processes, the complicated and usually expensive crystal growth procedure impeded the realization of several important applications. The use of light-sensitive proteins, such as photosynthetic reaction centers and rhodopsins of various origin has been the subject of intensive studies in recent years. Bacteriorhodopsin, a retinal protein, the main source of which is *Halobacterium salinarium* (formerly *Halobacterium halobium*), is one of the most widely used in the experiments aimed at the application of

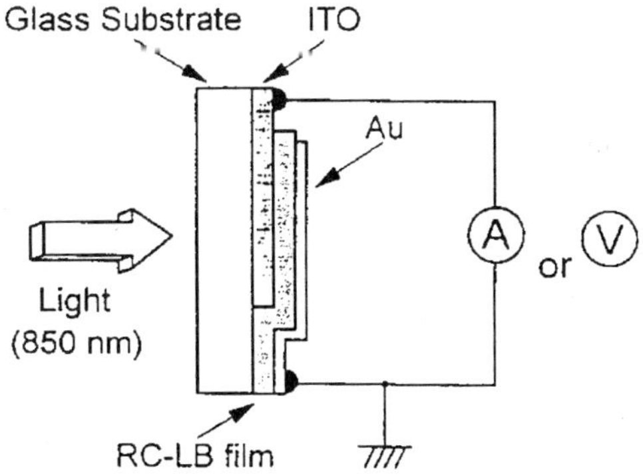

Figure 30. An RC-based photovoltaic cell.

proteins for the creation of alternative dynamic recording material to substitute conventional synthetic media. The extraordinary features, matured during the long natural selection process, include, among others, very high stability, and reversibility of physicochemical processes, in the form of optically homogenous thin films.

The possibility of gene engineering manipulations of bacterial strains that produce the protein permits the creation of mutants with optimal photochromic characteristics for specific applications such as optical filtering, re-writeable media and holography.

Vertebrate rhodopsin is a well-know intrinsic membrane protein which spans the disk membranes of retinal rod outer segment with seven α-helices and has a hydrophobic pocket between the helices, where the chromophore, retinaldehyde, is located and covalently attached to lysine 296.

The absorption of a photon by a rhodopsin molecule triggers the cyclic GMP cascade of phototransduction which constitutes the molecular events of vertebrate vision. Rhodopsin has a maximum of absorbance at about 500 nm which is lost (bleaching) upon a few minutes of illumination in detergent solution. When embedded in dry multilayer of phospholipids, rhodopsin did not bleach upon illumination but light causes the formation of stable metarhodopsin intermediate with absorbance maximum at about 480 nm (Zeisel and Hampp, 1992). However, in this artificial membranes the ratio lipid/rhodopsin was 150:1, while with the Langmuir Blodgett technique a film can be made up mainly of rhodopsin molecules, as shown earlier in this paper.

The advantage to use rhodopsin to make thin films rather than bacteriorhodopsin is that rhodopsin can be extracted by detergent from the membrane without loosing its properties, while isolated molecules of bacteriorhodopsin did no longer behave like a proton pump (Fisher and Oesterhelt, 1979). Bacteriorhodopsin films actually are made with intact purple membranes which contain about ten lipid molecules for one bacteriorhodopsin.

We show that thermal stability is also a property of bovine rhodopsin in LB films which can reach, in fact, a temperature up to 175°C with a negligible loss of secondary structure.

According to the mechanism of functioning of bacteriorhodopsin, the absorption of a photon by a protein molecule leads to a fast (500 fs) *trans-cis* photo-isomerization of retinylidene around the 13, 14 double bond with a quantum yield of 0.64. The subsequent steps involve thermal relaxation of several amino acids and of Schiff base linkage, involved in the proton transport, plus the backbone conformational changes.

The states of bacteriorhodopsin are characterized by the retinylidene conformation and the initial state is the B-state with an all-*trans* configuration. In the dark conditions, there is an equilibrium between the D-state (dark-adapted, with 13-*cis* configuration) and the B-state. The proton adsorption converts the BR molecule into light-adapted B-state (all-*trans* retinal configuration). The subsequent conformational changes bring the protein into the L_{550}-state via the intermediate K_{590}-state by a thermal relaxation process. The subscript indices refer to characteristic wavelengths of the absorption spectra. These features of photochromism are the most interesting ones in the determination of the applicability of BR to optical storage and information processing. However, an important point to be kept in mind is the operational lifetime of such media, evaluated in this case by the number of write/erase cycles that can be effected.

In a holographic film the intensity and phase distribution of an object is recorded. The resulting hologram is a diffractive element which generates the original object wave in amplitude and phase when illuminated with the same reference beam as was employed during recording. The same happens with BR-films, but since they are reversible recording materials, these holograms can be either erased at any time or decay with the time constant of the M-lifetime.

Table 9. Holographic Properties of Bacteriorhodopsin Films

Resolution	$\geq$ 5000 lines/mm
Optical Density	1-20 [OD $_{570}$]
Bleaching	90-100 percent (e.g., $BR_{D96N)}$
Index of Refraction	1.47
Refraction Index Change	10^{-3} - 10^{-2}
Diffraction Efficiency	1-7 percent (typ. 2-3 percent)
Light Sensitivity	1-80 mJ/cm^2 (B-type)
	30 mJ/cm^2 (M-type)
Polarization Recording	possible
Reversibility	3 10^6 cycles
Thickness	10-500 mm (typ. 20 mm)
Speed	msec-sec
Aperture	unlimited

The BR-film is used as short time storage for the reference hologram. With BR_{D96N}-films, a lifetime of the reference hologram of up to several minutes is obtained. After this time, a new reference hologram has to be recorded.

The interference of the holographic image and the directly transmitted light can be described as a processing of two images that correspond to different times but originate from the same location.

Another example of BR-based image processing is holographic pattern recognition. This technique allows correlation of two images that exist at the same time but are spatially separated. Common features of both patterns are detected.

Since this technique is an analogue computing method, signal-to-noise ratio is of central importance. Due to the polarisation recording properties of BR, an effective suppression of noise can be realised.

Bacteriorhodopsin films containing the variant BR_{D96N}, which differs from the wild type BR_{WT} by a single amino acid exchange, Asp 96 $\rightarrow$ Asn, show significantly higher holographic diffraction efficiencies (η) than BR_{WT} films.

The light-induced changes of both the absorption a(λ,I) and the refractive index n(l,I), which influence the diffraction efficiency $\eta(\lambda,I)$, have been investigated, and their spectral dependence has been measured in order to analyze both parameters separately. A maximal change of the refractive index of n Δ= 0.008 at 633 nm related to a modulation of the absorption of 0.55 OD, induced by actinic light of the wavelength 568 nm and an intensity of 20 mW/cm^2, was observed for a BR film containing BR_{D96N} at pH 9.5. From steady-state difference spectra of the absorption, the spectral refractive index change was calculated by the Kramers-Kronig relation, and a good correlation of the theoretically derived and experimentally measured values of the refractive index changes was found. This indicates that the chromophore system of bacteriorhodopsin, which is formed by the retinal molecule, its Schiff base linkage to the protein moiety, and an inner shell of amino acids, behave like an almost undisturbed chromophore with respect to the photo refractive properties at low actinic light intensities, despite the fact that all components of the chromophoric system are covalently linked to the amino acid matrix. Further, it was demonstrated that it is possible to calculate the spectral dependence of the diffraction efficiency $\eta(\lambda,I)$ from the easily accessible absorption changes $\Delta a(\lambda,I)$, and the refractive index change $\Delta n(\lambda,I)$ measured at a single wavelength. This method will be a useful tool for the characterization and optimization of bacteriorhodopsin films.

TOWARDS PROTEIN AUTOMATA AND NANOTECHNOLOGY

In summary, while the design of complex molecular functions is a goal difficult to achieve with the classical approach of chemistry and supramolecular chemistry, nature developed a large manifold of such "nanomaterials". Today, it is not yet possible to construct such complex materials *ab initio*, but we can start out from the molecules that we discover in nature and try to modify their properties and adapt them towards the demands of technical applications. As shown in this paper the tools that are necessary to put these ideas into practice are now available giving birth to a protein-based nanotechnology. Bacteriorhodopsin (and even better octopus rhodopsin) is one of the many possible models for this approach, as motivated by the extraordinary properties of this molecule and because it is one of the few biopolymers where a sufficiently detailed structure-function relation is available. Not only may the photochromism of bacteriorhodopsin be of technical importance, but also its photoelectric properties may be of equal value. However, for optical applications a wide range of classical materials exist which can be used as references for the evaluation of whether the biological materials can compete with them.

The improvements attained in the photochemical properties of BR demonstrate the future importance and effectiveness of gene-technological methods in material science (Hampp and Zeisel, 1994). Modification of the genetic code facilitates the manipulations of organism that can produce these new "high-tech" materials with conventional biotechnological methods. Therefore, the price of such products will be affordable. An intensive screening for functional biopolymers with technically relevant functions and their variation and identification will create a new class of "biomimetic" materials. The experimented studies done with bacteriorhodopsin have shown that this new approach leads to competitive materials in selected areas.

Another possible model are the cytochromes here introduced, because of their multiple pathways for the electron transfer being instead unidirectional in all inorganic material such as silicon and gallium arsenate. A main goal in molecular electronics is the search for molecular assemblies that could be used as circuits which process the information at molecular scale. The physical nature of these structures implies a great number of interacting molecules, so the assembly of a molecular circuit is not feasible by using the conventional techniques for integrated circuits. The picture that every micro structure should interact with its neighbours only, suggests to describe a molecular circuit by means of a logic -functional model computationally equivalent to a cellular automaton essentially to a device that assembles itself.

Instances may be found in organic mono- and multilayers made by Langmuir-Blodgett technique: typical examples are protein or lipid (Berzina *et al.*, 1995) films, whose surface organisation shows domains corresponding to a pseudo-crystalline phase embedded in a fluid lipid phase. An interesting property comes from the analysis of the lateral interactions among the electric dipoles associated to the hydrophilic heads of the lipids in a homogeneous phase. These dipoles have a non zero component of the moment parallel to the film surface, and can be considered as basic elements of a molecular circuit. The surface distribution of such components shows characteristics similar to a cellular automaton: any dipole depends above all on its near dipoles because the interaction decays as the cube of the distance, and the time evolution of its orientation is ruled by the same law for all dipoles.

In 1948, von Neumann embarked on an ambitious project: to show that phenomena as complex as life - the survival, reproduction, and evolution of complex forms of organisation - can be reduced in principle to the dynamics of many identical, very simple primitives capable of interacting and maintaining their identity. First, von Neumann considered the interaction of vortices and particles in suspension in some "primordial soup". Obviously, such a model was intractable, so, following a suggestion by Ulam, he adopted a fully discrete approach: space, time and even the dynamical variables were defined to be discrete.

The resulting cellular-automation theory describes a universe consisting of a homogenous array of cells. Each cell is endowed with a finite number of states and evolves in discrete time according to a uniform local transition rule. the rule can be seen as a function whose arguments are the states at time τ of the neighbouring cells (and possibly the state of the considered cell itself) and whose value is the state of the considered cell at time $i + 1$. the rule is uniform in that it is the same all over the array . Since all the cells "compute" their new state simultaneously, cellular automata are often seen as a paradigm of distributed computation. Dyson (1991) and Bernstein (1992) have given lively accounts of the surprising success of von Neumann's enterprise, a self-reproducing cellular-automaton that anticipated the discovery of the duplicative function of DNA.

Further uses of cellular automata in biology have been numerous. The very name "Life" for the famous cellular automaton that John Convay invented in 1970 still reflects the biology-motivated origin of cellular automata. The similarities between cellular-automation behaviour and that of many physical systems are also quite suggestive. Are these similarities which should be exploited in order to build a protein-based cell automata from heat-proof, homogeneous and highly stable LB film of any of the metalloprotein here described.

ACKNOWLEDGEMENTS

The work has been supported by the Strategic Project "Molecular Manufacturing" of the National Research Council and by the EL.B.A. Foundation (Portoferraio -LI, Italy).

REFERENCES

Adami, M., Piras, L., Lanzi, M., Fanigliulo, A., Vakula, S. and Nicolini, C., 1994, *Sensors and Actuators* B18:178-182.

Adami, M., Sartore, M., Nicolini, C., 1995, *Biosensors and Bioelectronics* 10:155-167.

Adman, E.T., and Jensen, L.H., 1981, *Israel Journal of Chemistry* 21:8-12.

Al-Mohamad, A., Smith, C.W., Al-Saffar, I.S., and Slifkin, M.A., 1990, *Thin Solid Films* 189(1):175-181.

Amman, M., Wilkins, R., Ben-Jacob, E., Maker, P.D., Jaklevic, R.C., 1991, Phys. Rev., B43:1146-1149

Amrein, M., Durr, R., Stasiak, A., Gross, H., and Travaglini, G., 1989, *Science* 243:1708-1711.

Antolini, F., Paddeu, S., and Nicolini, C., 1995a, *Langmuir* 11:2719-2725.

Antolini, F., Trotta, M., and Nicolini, C., 1995b, *This Solid Films* 254:252-256.

Arnold, F.H., and Zhang, J.H.,1994, *Trends in Biotechnology* 12:189-192.

Averin, D.V, and Likharev, K.K, 1986, *J. Low Temp. Phys.* 62:345-373.

Ayrapentiants, S.V., Berzina, T.S., Shikin, S.A., and Troitsky, V.I., 1992, *Thin Solid Films* 21(1-2):261-264.

Barraud, A, Vandevyver, M., 1983, *Thin Solid Film* 99:221-225

Berzina, T.S., Troisky, V.I., Stussi, E., Mulè, M., De Rossi, D., 1993, *Synthetic Metals* 60:111-114.

Berzina, T.S., Troitsky, V.I., Neilands, O.Ya., Sudmale, I.V., and Nicolini, C., 1995a, *Thin Solid Films* 256:186-191.

Berzina, T.S., Troitsky, V.I., Vakula, S., Riccio, A., Morana, A., De Rosa, M., Dante, S., Maccioni, E., Rustichelli, F., Accossato, P., and Nicolini, C., 1995b, *Material Science And Engineering*, C3:13-21.

Berzina, T.S., Troitsky, V.I., Vakula, S., Riccio, A., Gambacorta A., De Rosa, M., Gobbi, L., Rustichelli, F., Erokhin V. and Nicolini, C., 1995d, *Material Science And Engineering*, C3:23-31.

Berzina, T.S., Troitsky, V.I., Vakula, S., Riccio, A., Morana, A., De Rosa, M., Gobbi, L., Rustichelli, F., Erokhin, V. and Nicolini, C., 1995c, *Material Science And Engineering*, C3:33-38.

Bramanti, E., Benedetti, E., Nicolini, C., Berzina, T., Erokhin, V., Benedetti, E., 1995, *Biochim. Biophys. Acta*, submitted.

Brunger, A.T., Leahy, D.J., Hynes, T.R., and Fox, R.O., 1991, *J.Mol.Biol.* 221:239-256.

Bryce, M.R., Petty, M.C., 1995, *Science* 374:771-776.

Butt, H.J., Downing, K.H., and Hansma, P.K., 1990, *Biophysical Journal* 58:1473-1480.
Caffrey, M.S., Cusanovich, M.A., 1994, *Biochim Biophys Acta* 1187:277-288.
Canters, G.W., Gilardi, G., 1993, *FEBS Letters* 325:39-48.
Carrara, E. A., and Nicolini, C., 1992, *Biomedical Modeling And Simulation*, IMACS 183-188.
Carrara, E. A., Anselmino, A., Gavotti, C., Vakula, S. and Nicolini C., 1995, in "From Neural Networks and Protein Engineering to Bioelectronics", EL.B.A. Forum Series, Plenum: N.Y.-London, 1:103-110.
Carrara, E. A., Pagliari, F., and Nicolini, C., 1993a, *Neural Networks* 6:1023-1032.
Carrara, E. A., Pagliari, F., and Nicolini, C., 1993b, *Proc. Ijcnn 93- Nagoya* 3:983-986.
Carrara, S., Facci, P., and Nicolini, C., 1994a, *Rev. Sci. Instrum.* 65(9):2860-2863.
Carrara, S., Gussoni, A., Erokhin, V., Nicolini, C., 1995, *J. Mat. Sci.: Mat. Electronics* 6:79-83.
Chou, K.C., Carlacci, L., Maggiora, G.M., Parodi, L.A., and Schulz, M.W., 1992, *Protein Sci.* 1:810-827.
Clayton, R.K., and Wang, R.T., 1971, Methods Enzymology, Academic American Inc. 23:696-704.
Collier, D.N., Bankaitis, V.A., Weiss, J.B., and Bassford, P.J., 1988, *Cell* 53(2):273-283.
Colonna-Cesari, F., Perahia, D., Karplus, M., Eklund, H., Bränden, C.I., and Tapia, O., 1986, *J.Biol.Chem.* 26:5273-5280.
Crooke, E., Guthrie, B., Lecker, S., Lill, R., and Wickner, W., 1988, *Cell* 54:1003-1011.
D'Auria, S., Raia, C., Rossi, M., Vakula, S., Nicolini, C., 1995, *Arch. Biochem. Biophys.*, submitted
De Chiara, C., Nicastro, G., Mariani, M., and Nicolini, C., 1995, in preparation.
Degtyarenko, K.N., Archakov, A.I., 1993, *FEBS Lett.* 332:1-8.
Deisenhofer, J., 1981, *Biochemistry* 20:2361-2370.
Deisenhofer, J., Epp, O., Sinning, I., Michel, H., 1995, *J. Mol. Biol.* 246:423-457.
Denkov, N.D., Velev, O.D., Kralchevsky, P.A., Ivanov, I.B., Yoshimura, H., Nagayama, K., 1993, *Nature* 361:26-27.
Detlefesen, D.J., Thanabal, V., Pecoraro, V.L., and Wagner, G., 1991, *Biochemistry* 30:9040-9046.
Devoret, M.H, Esteve, D., Urbina, C., 1992, *Nature* 360:547-553.
Dubrovsky, T., Tronin, A., and Nicolini, C., 1995a, *Thin Solid Films* 257:130-133.
Dubrovsky, T., Vakula, S., and Nicolini, C., 1994, *Sensors And Actuators* 22:69-73.
Dubrovsky, T.B., Demcheva, M.V., Savitsky, A.P., Mantrova, E.Yu. and Yaropolov, A.I., 1993, *Biosensors and Bioelectronics* 8:377-385.
Dubrovsky,T., Tronin, A., Dubrovskaya, S., Vakula, S. and Nicolini, C., 1995b, *Sensor And Actuators* 23:1-7.
Dushkin, C.D., Nagayama, K., Miwa, T., Kralchevsky, P.A., 1993, *Langmuir* 9:3695-3701.
Eilers, M., and Schatz, G., 1986, *Nature* 322:228-232.
Eklund, H., Samama, J.P., and Jones, T. A., 1984, *Biochemistry* 23:5982-5996.
Erokhin, V., Facci, P., and Nicolini, C., 1995a, *Biosensors And Bioelectronics* 10, 25-34
Erokhin, V., Facci, P., Carrara, S., and Nicolini, C., 1995b, *Journal of Physics*, D:Appl. Phys., 28:2534-2538.
Erokhin, V., Feigin, L., Ivakin, G., Klechkovskaya, V., 1991, *Makromol. Chem., Macromol. Symp.*, 46:359-363.
Erokhin, V., Vakula, S., and Nicolini, C., 1994, *Thin Solid Films* 238:88-94.
Facci, P., Erokhin, V., and Nicolini, C., 1993, *Thin Solid Films* 230:86-89;
Facci, P., Erokhin, V., and Nicolini, C., 1994b, *Thin Solid Films (LB6)* 243:403-406.
Facci, P., Erokhin, V., Antolini, F., and Nicolini, C., 1994a, *Thin Solid Films* 237:19-21.
Facci, P., Erokhin,V., Tronin, A., and Nicolini, C., 1994c, *J. Phys. Chem.* 98(50):13323-13327.
Fersht, A., and Winter, G., 1992, *Trends in Biochemical Sciences* 17:292-294.
Fisher, U., and Oesterhelt, D., 1979, *Biophys J.* 28:211-230.
Franzoni, L., De Chiara, C., Nicastro, G., and Nicolini C., 1995, in preparation
Gavazzo, P., Paddeu, S., Sartore, M., and Nicolini, C., 1994, *Sensors And Actuators B.* 18-19:368-372.
Germanas, J.P., Di Bilio, A.J., Gray, H.B., Richards, J.H., 1993, *Biochemistry* 32:7698-702.
Glazmann, L.I., Shechter, R.I., 1989, *J. Phys., Condens. Matter* 1:5811-5815.
Goldman, V.J., and Su, B., 1995, *Science* 267:1010-1012.
Gonzales, F.J., 1990, *Trends Pharmacol. Sci.* 13:346-352.
Groshev, A., Ivanov, T., Valtchinov, V., 1991, *Phys.Rev.Lett.* 66:1082-1085.
Guckenberger, R., Kosilinger, C., Gotz, R., Bren, H., Levai, N., and Brainmeister, V., 1988, *Ultramicroscopy* 25:111-121.
Guengerich, F.P., 1992, *FASEB J.* 6:667-668.
Gupta, C.L., Gupta, A., Parashar, T.R., and Jain, V.K., 1993, *Thin Solid Films* 230:45-47.
Hampp, N., 1993, *Nature* 366:12.
Hampp, N., and Zeisel, D., 1994, *IEEE Engineering in Medicine and Biology*, February/March:67-74.
Harison P.M. (ed), 1985, Topics in Molecular and Structural Biology: Metalloproteins. Chemie-Verlag, Weinheim, Germany.
Harlow, E., and Lane, D., 1988, Antibodies: A Laboratory Manual, Cold Spring Harbor Laboratory.

Hennessey, J.P., and Johnson, W.C., 1981, *Biochemistry* 20:1085-1094
Holmans, P.L., Shet, M.S., Martin-Wixtrom, C.A., Fisher, C.W., Estabrook, R.W., 1994, *Arch. Biochem. Biophys.* 312:554-565.
Hua, Y.L., Jiang, D.P., Shu, Z.Y., Petty, M.C., Roberts, G.C., and Ahmad, M.M., 1990, *Thin Solid Films* 192:383-390.
Imai, T., Globerman, H., Gertner, J.H., Kagawa, N., Waterman, M.R., 1993, *J. Biol. Chem.* 268:19681-19689.
Jeng, M.-F., Campbell, A.P., Begley, T., Holmgren, A., Case, D.A., Wright, P.E., and Dyson, H. J., 1994, *Structure* 2:853-868.
Ji, X., Zhang, P., Armstrong, R.N., and Gilliland, G.L., 1992, *Biochemistry* 31:10169-10184.
Katti, S.K., Le Master, D.M., and Eklund, H., 1990, *J.Mol.Biol.* 212:167-184.
Kell, D.B., 1992, *Bioelectrochem and bioenerg.* 27:235-237.
Kim, E.E., and Wyckoff, H.W., 1991, *J.Mol.Biol.* 218:449-464.
Kouwenhoven, L.P., van der Vaart, N.C., Johnson, A.T., Kool, W., Harmans, C.J.P.M., Williamson, J.G., Staring, A.A.M., Foxon, C.T., 1991, *Z.Phys.B, Condensed Matter* 85:367-373.
Koyama, K., Yamaguchi, N., and Miyasaka, T., 1994, *Science* 265:762-765.
Kozarac, Z., Dhathathereyan, A., Möbius, D., *FEBS Lett.*, 1988, 229, 372
Kralchevsky, P.A., and Nagayama, K., 1994, *Langmuir* 10:23-36.
Kralchevsky, P.A., Denkov, N.D., Paunov, V.N., Velev, O.D., Ivanov, I.B., Yoshimura, H., and Nagayama, K., 1994, *J. Physics-Condensed Matter* 6(23):A395-A402.
Lafarge, P., Pothier, H., Williams, E.R., Esteve, D., Urbina, C., Devoret, M.H., 1991, *Z.Phys. B, Condensed Matter* 85:327-332.
Langmuir, I., and Schaefer, I., 1938, *J.Am.Chem.Soc.*60:1351-1360
Lu, Y., Casimiro, D.L., Bren, K.L., Richards, J.H., Gray, H.B., 1993, *Proc. Natl. Acad. Sci USA* 90:11456-11459.
Lvov, Yu.M., Erokhin, V.V., and Zaitsev, S.Y., 1991, *Biol. Mem. (Russ.)* 4(9): 1477-1513
Magri, S., Pepe, I., Traverso, S., and Nicolini, C., 1993, *Proc. Ijcnn 93* 3:2694-2697.
Malatesta, F., Antonini, G., Sarti, P., Vallone, B., and Brunori, M., 1990, *Gazzetta Chimica Italiana* 12:475.
Martin, A. S., and Sambles, J. R., 1993, *Phys. Rev. Lett.* 70:218-221.
Matsuura, Y., Takano, T. and Dickerson, R. E., 1982, *J.Mol.Biol.* 156:389-409.
Mauk, A.G. Electron transfer in Genetically Engineered Proteins. The cytochrome C paradigm. Structure and Bonding, 1991, Springer-Verlag. p132-156.
Maxia, L., Radicchi, G., Pepe, I. M., and Nicolini, C., 1995, *Biophysical J.*, 69:1440-46.
McLean, M.A., Stayton, P.S., Sligar, S.G., 1993, *Anal. Chem.* 65:2676-2678.
Miyake, J., Majima, T., Namba, K., Hara, M., Asada, Y., Sugino, H., Ajiki S., Toyotama, H., 1994, *Mat. Sci. Eng. C1* 2:63-67.
Mogilevski, L.Yu., Dembo, A.T., Svergun, D.I., and Feigin, L.A., 1984, *Crystallography* 29:587-591.
Mullen, K., Ben-Jacob, E., Jaklevic, R.C., Shuss, Z., 1988, *Phys. Rev.* B37:98-105
Nagayama, K, 1994a, *Mater. Sci. Engineer.* C1: 87-94.
Nagayama, K, 1994b, *Membrane* 19:311-316.
Nagayama, K., 1993, *Phase Transitions* 45:185-203.
Nejoh, H., 1991, *Nature* 353:640-642.
Nevernov, I., Kurnikov, I., and Nicolini, C., 1995, *Ultramicroscopy* , 58:269-274.
Nicastro, G., de Chiara, C., Franzoni, L., Tatò, M., Carrara, E.A., and Nicolini C., 1995, *Bull. Magn. Res.* in press.
Nicolini, C., 1995a, in *From Neural Networks and Biomolecular Engineering to Bioelectronics*, EL.B.A. FORUM Series, New York-London:Plenum Publishing Corp. 1:1-36
Nicolini, C., 1995b, *Biosensors And Bioelectronics* 10:105-127.
Nicolini, C., 1995c, in *From Neural Networks and Biomolecular Engineering to Bioelectronics*, EL.B.A. FORUM Series, New York-London:Plenum Publishing Corp. 1:135-166
Nicolini, C., 1995d, Molecular Bioelectronics, World Scientific Publishing Co., Singapore, in press.
Nicolini, C., Adami, M., Antolini, F., Beltram, F., Sartore, M., and Vakula, S., 1992, *Physics World* 5:30-34.
Nicolini, C., Adami, M., Dubrovsky, T., Erokhin, V., Facci, P., Paschkevitsch, P., and Sartore, M., 1995e, *Sensors and Actuators B* 24:121-128.
Nicolini, C., and Rigo, A., 1992, Textbook, "Biofisica e Tecnologie Biomediche", Zanichelli Bologna 1-875
Nicolini, C., de Chiara, C., Facci, P. and Nicastro, G., 1995a, *in preparation*.
Nicolini, C., Dubrovsky, T., De Chiara, C., and Facci, P., 1995b, *Biochim. Biophys. Acta*, in preparation.
Nicolini, C., Erokhin, V., Antolini, F., Catasti, P., Facci, P., 1993, *Biochim. Biophys. Acta* 1158:273-278.
Niemeyer, C.M., Sano T., Smith, C.L., and Cantor C.R., 1994, *Nucl. Acids Res.* 22:5530-5539.
Nygren, P-A., Stahl, S, and Uhlen, M., 1994, *Trends in Biotechnology* 12:184-189.

Pabo, C.O., 1986, *Biochemistry* 25:5992-5998.
Paddeu, S., Antolini, F., Dubrovsky, T., and Nicolini C., 1995a, *Thin Solid Film* , 268:108-113.
Paddeu, S., Fanigliulo, A., Lanzi, M., Dubrovsky, T. and Nicolini, C., 1995b, *Sensor and Actuators* 25:876-882.
Paschkevitsch, P., Sivozhelezov, V., Vakula, S., Sidorovich, V., Eldarov, M., Nozza, F., Nicolini, C., 1996, this volume.
Pepe, I.M., Maxia, L., Nicolini, C., 1995, *Thin Solid Films*, in press.
Peterson, I.R., 1992, in Nanostructures Based on Molecular Materials, eds. W.Göpel and Ch.Ziegler, VCH, Weinheim.
Petrigliano, A., Tronin A., and Nicolini C., 1995, *Thin Solid Films* in press.
Reed, M.A., Randall, J.N., Aggaewal, R.J., Matyi, R.J., Moore, T.M., Wetsel, A.E., 1988, *Phys. Rev. Lett.* 60:535-537.
Riordan, J.F., 1973, *Biochemistry* 12:3915
Roberts, G.G., 1990, Langmuir-Blodgett Films, Plenum Press N.Y.
Sa'ar, A., Feng, J., Gravé, I., Yariv, A., 1992, *J.Appl.Phys.* 72:3598-3601.
Sambrook, J., Fritsch, E.F., Maniatis, T., 1989, Molecular Cloning. A laboratory manual. 2 nd edn. Cold Spring Harbor Laboratory Press.
Sartore, M., Adami, M., Bousse, L., Hafeman, D., Mostarshed, S., and Nicolini, C., 1994, *J. Appl. Physics* 75(8):4000-4008.
Sattler, K., 1991, *Z.Phys. D*, 19:287-292.
Schonemberger, C., van Houten, H., Donkersloot, H.C., 1992, *Europhys. Lett.* 20:249-254.
Shen, Y., Boragno, C., Rolandi, R., and Valbusa, U., 1990, *Thin Solid Films* 187:155-163.
Shen, Y., Safinya, C.R., Liang, K.S., Ruppert, A.F., and Rothschild, K.J., 1993, *Nature* 366:48-50.
Stayton, P.S., Bohn, P.W., Sligar, S.G., 1990, *J.Cell.Biochem., Suppl.*14B:367.
Studier, F.W., Rosenberg, A.H., Dunn, J.J., Dubendorf, J.W., 1990, *Methods Enzymol.* 185:60-89.
Tanaka, N., Yamane, T., Tsukihara, T., Ashida, T., Kakudo, M., 1975, *J.Biochem.(Tokyo)* 77:147-162.
Troitsky, V. I., Berzina, T. S., Katsen, Ya. Ya., Neilands, O. and Nicolini, C., 1995, *Synthetic Metals*, 74:1-6.
Truan, G., Cullin, C., Reisdorf, P., Urban, P., Pompon, D., 1993, *Gene* 125:49-55.
van Bentum, P.J.M., Smokers, R.T.M., van Kempen, H., 1988, *Phys. Rev. Lett.* 60:2543-2546.
Vergani, L., Gavazzo, P., Mascetti, G., Nicolini, C., 1994, *Biochemistry* 33:6578-6585.
Vijayakumar, S., and Salerno, J. C., 1992, *FASEB Journal* 6:132.
Wald, , G., Durrel, J., And St. George, R.C.C., 1950, *Science* 17:179.
Weiner, J.S., Hess, H.F., Robinson, R.B., Hayes, T.R., Sivco, D.L., Cho, A.Y., Ranade, M., 1991, *Appl. Phys. Lett.* 58:2402-2404.
Wilkins, R., Ben-Jacob, E., Jaklevic, R.C., 1989, *Phys. Rev. Lett* 63:801-804.
Yamaki, M., Matsubara, K., Nagayama K., 1993, *Langmuir* 9(11):3154-3158.
Yeates, T.O., Komiya, H., Chirino, A., Rees, D.C., Allen, J.P., and Feher, G., 1988, *Proc. Nat. Acad. Sci. USA.* 85:7993-7997.
Yoshimura, H., Scheybani, T., Baumestier, W., and Nagayama, K., 1994, *Langmuir* 10:3290-3295.
Zasadzinski, J.A., Viswanathan, R., Masden, L., Garnaes, J., Shwartz, D.K., 1994, *Science* 263:1726-1733.
Zeisel, D., and Hampp, N., 1992, *J. Phys. Chem.* 96:7788-7792.
Zhu, D.G., Petty, M.C., 1989, *Thin Solid Films* 176:151

2

FABRICATION OF TWO-DIMENSIONAL CRYSTALLINE ARRAYS OF PROTEIN MOLECULES

Kuniaki Nagayama

Nagayama Protein Array Project
ERATO, JRDC
5-9-1 Tokodai, Tsukuba, 300-26 Japan
Department of Life Sciences, College of Arts and Sciences
The University of Tokyo
Komaba, Meguro-ku, Tokyo, 153 Japan

Life in the universe as we know it is based upon systems of cells: from single-cell bacteria to trillions of complexly interacting cells -an organism. It is sometimes instructive to consider these cellular units as being "factories" containing various molecular machines consuming energy and material while carrying out complex processes and manufacturing new materials (Alberts et al., 1994). Interestingly, many of the components of cells -amino acids to proteins to protein complexes (supramolecules) to cellular organelle -are formed by self-assembly under the instructions of DNA. These cells also produce all of the proteins and supramolecules necessary to form tissue and ultimately organisms that are a million-times larger than the original simple protein.

If humans are to connect the manufacturing carried out at the bottom level of life with their future technology it will be necessary to both understand and harness the self-assembly abilities of protein molecules -mesoscopic hands for human use. Until recently it has been generally believed that proteins are structurally weak, and thus unsuitable to join metals, semiconductors, organic polymers, and inorganic materials that form the backbone of human technology. We, however, believe that proteins are far tougher than ever imagined and that once their self-assembly abilities are understood and co-opted, much interesting and natural technology will become available.

The Protein Array Project is dedicated to finding techniques by which proteins can be coerced to self-assemble into a wide variety of 2D assemblies with unique features (Nagayama, 1994). To achieve this goal, not only do the protein-protein interactions and self-assembling processes need to be elucidated, but suitable substrates, or fields of space, must be perfected. Rather than using biomembranes, like those found in nature, an attempt is being made to utilize mobile fluid layers (thin films) made on very clean substrate surfaces.

Six years ago we developed a technique to spread proteins on mercury (Yoshimura et al., 1990), which provides a very clean and flat surface. The present project is now

Molecular Manufacturing, edited by C. Nicolini
Plenum Press, New York, 1996

generalizing this concept. It is also developing techniques by which assembled protein crystalline films can be transferred to the surface of a silicon or other substrate in such a way that the ordering of protein crystals can be preserved.

Further, an attempt is being made to maintain the original function of proteins while controlling their manner of integration on the substrate surface. By doing this it might be possible to design protein array having critical catalytic sites at optimum positions and orientations on the surface. To prepare a protein array with such desired 2D patterns, the properties of protein-protein interactions are being changed by chemically modifying and genetically mutating amino acids on the protein surface and reshuffling domains using genetic engineering.

ASSEMBLING PRINCIPLES

The process of protein array (supramolecules to crystals) formation could be regulated by various physicochemical factors, ranging from the molecular assembling in the initial stage to the molecular orientation adjustment in the final stage. The governing factors for the assembling for the initial stage are usually long-range attractive forces as observed in nonspecific colloidal coagulation. When proteins are mutually separated by a distance longer than their own geometrical sizes, they behave like colloidal particles exhibiting a simple nonspecific interactions. On the other hand, short range forces become dominant, when proteins come to contact. In a shorter range they can recognize each other and find proper orientation to make specific complexes or crystals. In the actual protein array formation, both processes are equally important and may occur in a coupled manner.

Assembling Principle in Biological Supramolecules

As observed in life, component proteins in the supramolecules such as ribosomes and nucleosomes automatically gather to complete their assembly. This phenomenon, termed as self-assembly, is believed to be a direct consequence of molecular reordering harnessing a natural law, thermodynamics. Then the assembling can be divided to two processes: the diffusion-limited aggregation (random assembly) of individual protein molecules (Fig.1), and the subsequent step of molecular orientation adjustment by specific interactions (molecular recognition) in the assembled proteins. This combination of the random assembly and the molecular recognition among protein molecules is the reality of the self-assembly notion found in the living cells. Due to the limited cell volume (~1 μm in diameter) the random assembly step can avoid to be a bottleneck for the formation of supramolecules. For the reason, the whole process appears as if it can automatically run dominated by the

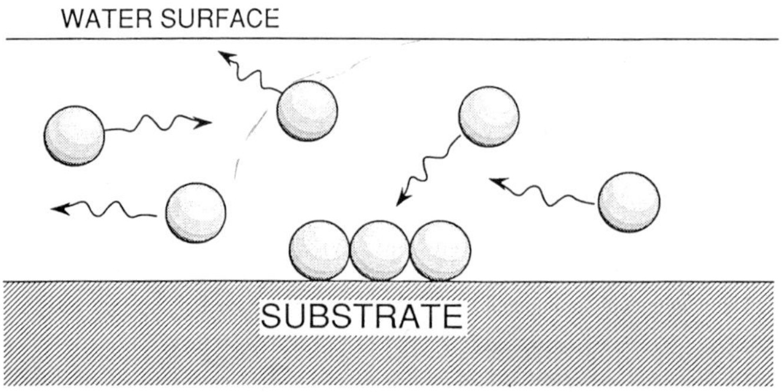

Figure 1. A schematic illustrating a slow growth of supramolecules driven by random assembling and molecular recognition in the dispersion phase. We call this process diffusion-limited self-assembly, a basic technology used in the living cells for the construction of large supramolecular architectures. (Taken from ref. 2)

assembly but not by the mutual recognition. This biological framework of the self-assembly is, however, inadequate for the array engineering which intends to explore an industrial scale of materials definitely larger than 1 mm, since the thermally driven diffusion and molecular recognition mechanism can not properly work on the large scale.

Assembling Principle in Two-Dimensional Protein Arrays

In contrast with the biological way of fabrication of supramolecules in the solution state, the 2D assembling of particles proceeds on planes of adequate substrate surfaces. The advantage by lowering the dimension of the working space from 3D to 2D is two-fold; 1): the reduction of possible patterns, say space groups of crystals from 254 to 17; 2): the use of thin layers of liquids or solids directly accessible and controllable from the remaining dimensions, the substrate under or the free space on the layer. The choice of an appropriate substrate surface then becomes very crucial for the fabrication of highly ordered protein monolayers. In biosystems protein 2D crystals are often embedded in biological membranes, such as cell membranes and liposomes (Michel, 1991). The assembling of protein molecules inside the membranes again seems to be controlled by the diffusion. To replace the bilayer membranes by artificial substrates, several different interfaces such as air-water, lipid-water, and water-solid have been employed (Jap et al., 1992). But, the same problem as described above (Fig.1) appears again in these surfaces, the diffusion-limited assembly. Our recent innovation for the protein 2D crystallization using the clean mercury surface (Yoshimura et al., 1990) or the air-water interface (Yoshimura et al., in press) has solved this problem. The technological breakthrough has come from two directions, the rapid spreading wetting of protein solutions and uniform stable thin films on the substrate surface. Instead of the high energy of the mercury surface, water cannot completely wet the surface. We found that the mercury surface changed to be completely wettable in the oxygen gas atmosphere and can spread the protein solution to a thin film (Yoshimura et al., 1990). We also found that the protein solutions rapidly spread on the subphase surface of higher density aqueous solution, when they were injected to the subphase (Yoshimura et al., in press). In the thin protein solution layer made on the surface, protein particles tend to assemble to make 2D crystalline arrays.

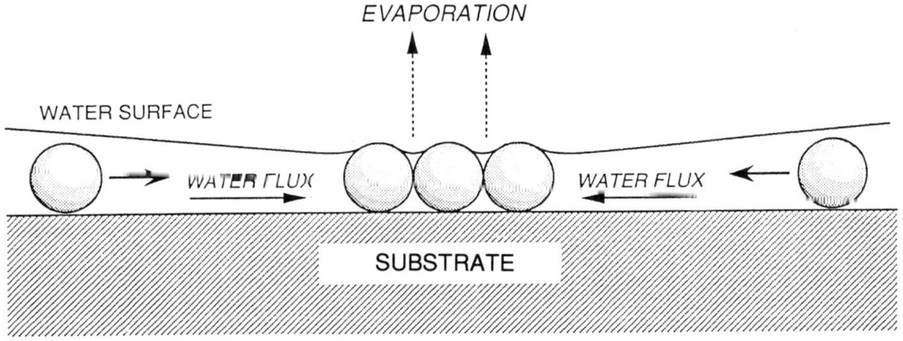

Figure 2. A schematic illustrating a rapid growth of particle arrays in a film. The water influx flowing into the array to wet it carries the particles to the array periphery where continuously evaporated water is compensated by the flow. The lateral capillary force acting between the particles and the array wall, then, presses and fixes them to the array boundary (see Kralchevsky et al., 1992a, 1992b; Paunov et al., 1992, 1993; Kralchevsky and Nagayama, 1994; Kralchevsky et al., 1994). A linear growth of the 2D array is expected in this mechanism (Dushkin et al., 1993). We call this process convective self-assembly, a basic technology used in the protein array fabrication. (Taken from Nagayama, 1994)

Model experiments using polystyrene latex particles well visible by light microscope and monitored by video-movie has suggested an intriguing mechanistic aspect of the array formation (Denkov et al., 1992; Nagayama, 1993). Namely we have found a directional particle motion carried by the water influx that flows into the existing crystal area to compensate the evaporated water there. This convective process extremely accelerates the array growth and facilitates making a large domain of 2D arrays in a short period. The flow speed of water and, hence, the speed of crystal growth is controlled by adjusting the evaporation rate as shown in Fig.2. When the evaporation rate was kept constant, the crystal grows linearly in proportion to time (Dushkin et al., 1993a). This directional assembly, termed as convective self-assembly, seems to be that occurring in the array formation of protein molecules, which makes a big contrast with the diffusion self-assembly harnessed in making biological supramolecules in the living cells. Using various sizes of polystyrene latex particles ranging form 50 nm to 2 μm, the new surface phenomena have intensively been studied in our project (Denkov et al., 1992; Nagayama, 1993; Dushkin et al., 1993a; Dushkin et al., 1993b; Dimitrov et al., 1994).

Another important factor which was found in our project is the sudden appearance of a strong attractive force between particles when particles protrude their heads from the solvent surface. We call this new force the lateral capillary force, which originates from the deformation of the liquid surface, supposed to be flat in the absence of particles (Kralchevsky et al., 1992a; Kralchevsky et al., 1992b; Paunov et al., 1992; Paunov et al., 1993; Kralchevsky and Nagayama, 1994; Kralchevsky et al., 1994). The attractive capillary force was found to have a very long-ranged as well as sizeable even for the nanometer sized particles (Dimitrov et al., 1994). This force becomes prominent at the last stage of the film drying and is, therefore, responsible to the closed packing of assembled particles favoring the hexagonal form of 2D crystals after drying.

STABLE FILMS ON THE FUNCTIONALIZED SURFACES

The essence of 2D particle crystallization lies in the stable wetting film made on the substrate surface as shown in Fig.2. The liquid film plays two important roles in making 2D

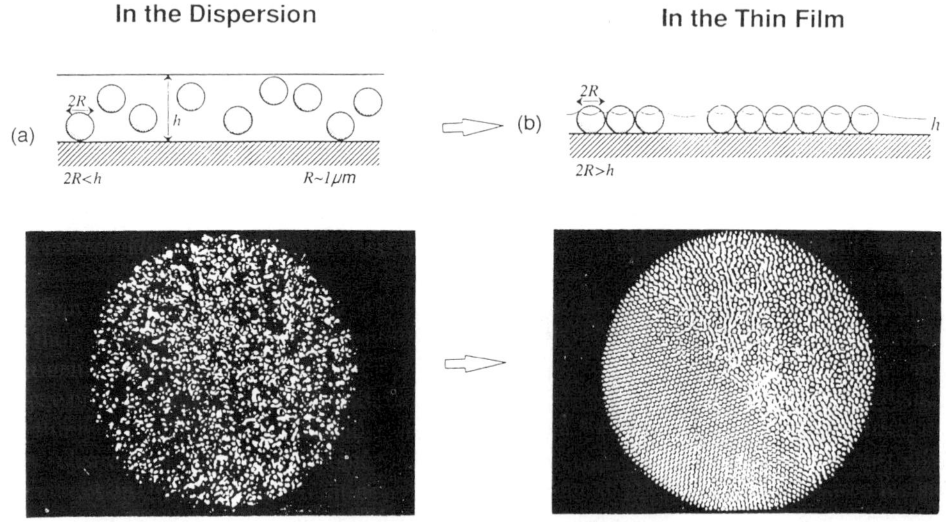

Figure 3. Two-dimensional array formation in the liquid film.(Adapted from Denkov et al., 1992).

particle arrays: 1): liquid flow carrying particles to the crystal boundary for growth; and 2): closed packing of particles by lateral capillary force induced at the film surface. In parallel to the usual 3D crystallization, these processes also proceed automatically by using the free energy difference before and after the reaction in the non-equilibrium state. This is schematically shown in Fig.3. First, particles undergo the Brownian motion in the thick film (Fig.3a). When the liquid layer becomes thin when compared with the particle size, an ordered 2D domain starts to form on the surface (Fig.3b). It is, however, a difficult task to create stable wetting films appropriate for the submicrometer or nanometer particles since the wetting films on the solid substrate are not as stable and can easily be ruptured when they become thin (Blake, 1984; Derjaguin, 1989). Film rupture is a complicated phenomenon depending on the roughness, wettability and chemical stability of the substrate surface. Cleaved mica or acid-rinsed glass provides a very wettable surface suitable for the array formation of colloidal particles bigger than 30 nm in diameter. Owing to the surface roughness, protein molecules or particles smaller than 30 nm in diameter have difficulty in making ordered 2D arrays on the solid surfaces. This is the major reason for using clean mercury surface (Yoshimura et al., 1990) or the subphase surface of glucose solution (Yoshimura et al., 1994).

(a) Particles undergo Brownian motion in the liquid layer whose thickness is much larger than the particle size.
(b) Particles start to assemble in the thin film whose thickness is comparable or slightly smaller than the particle size.(Adapted from Denkov et al., 1992).

We adopted three different thinning methods to sustain the liquid thin layer on substrates; 1): thinning by evaporation (Denkov et al., 1992; Dushkin et al., 1993a; Dushkin et al., 1993b; Dimitrov et al., 1994); 2): thinning by suction (Dimitrov et al., 1994); and 3): thinning by spreading wetting (Yoshimura et al., 1990; Yoshimura et al., 1994). For colloidal particles (polystyrene latex) larger than 30 nm, the three methods work well on the various substrates, *i.e.*, mica, glass, metal-coated mica, carbon-coated mica, metal-coated glass, and mercury. For the smaller particles, especially protein molecules, the combination of spreading-wetting and evaporation on the mercury or aqueous subphase surface can give satisfactory results for making crystalline films.

EXPERIMENTAL

Materials

Recombinant ferritin samples, wild type and a mutant of horse spleen L-apoferritin, were all prepared in this project. The mutant (Asp84Ser/Gln86Ser) is deficient of one negative charge at the cadmium ion binding site which is responsible for the connection of neighboring molecules by its ion bridge in the 3D crystal (Lawson et al., 1991). Special care had been given to remove protein aggregates and denatured protein molecules for the production of qualified 2D crystals or crystal domains of a large size. The protein solutions were, therefore, applied to gel permeation chromatography for fractionation before use.

Mercury method

The detailed description of the 2D crystallization apparatus using the mercury trough has been given in the original report (Yoshimura et al., 1990). Here only the experimental procedure is briefly described for the reader to understand the fabrication.

(1): After evacuating the chamber by a turbo vacuum pump, fill the trough (25 cm in diameter, 450 cm^2 surface area) with distilled mercury, (2): fill the chamber with pure oxygen

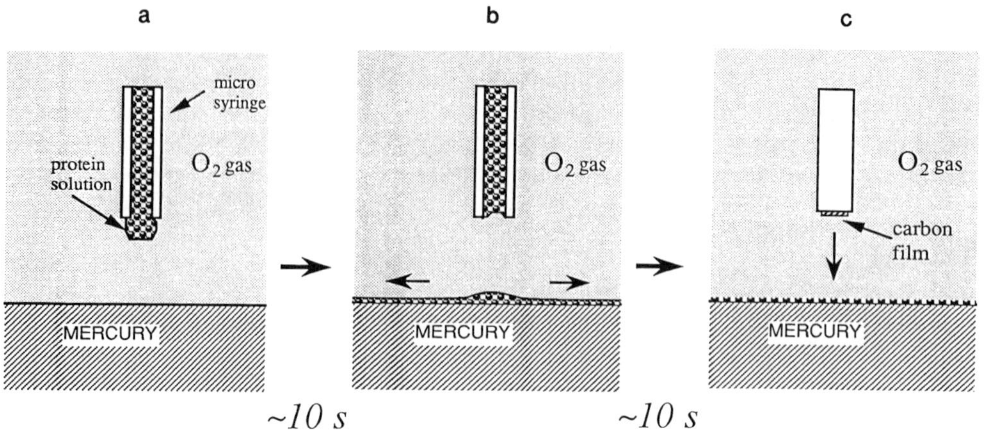

Figure 4. Schematics of the 2D crystallization of protein molecules on the mercury surface (mercury method) (a) Gently dripping of a small amount (1~5 μl) of protein solution (b) Rapid spreading of the solution to yield a stable thin film on mercury (c) Array formation of protein molecules during the drying and the transfer of the array to specimen grid for the electron microscopy. (Adapted from Nagayama et al., 1995).

gas having a pressure slightly higher than 1 atm (Fig.4a), (3): clean the mercury surface by sweeping the surface by two barriers and by sucking the mercury by a suction pump, (4): check the cleanness of the surface by ellipsometry, (5): set the spreading area by opening two barriers (usually 430 cm^2), (6): open the gate of the microsyringe and allow the solution (1 to 10 μl) to spread onto the mercury surface (Fig.4b). After a short maturing time (~1 min) posterior to spreading of protein solutions, the protein monolayer thus formed on the mercury surface is transferred to a carbon supporting film on the electron microscope grid (Fig.4c). The carbon supporting film have to be previously hydrophilized by ion bombardment to enhance the adhesion of protein arrays to the carbon film.

Aqueous Subphase Method

2D protein crystallization at the air-water interface on the aqueous subphase was carried out according to the procedure proposed by Yoshimura et al. (Yoshimura et al., 1994). (1): fill the subphase solution (0.5 ml, 10 mM MES (pH 5.7), 2% glucose with or without 10 mM $CdSO_4$ and 150 mM NaCl) into a Teflon trough (15 mm in diameter, 177 mm^2 surface area); (2): inject a protein solution (1 μl, 1 mg/ml) into the subphase solution (Fig.5a); (3): the protein solution which have a density lesser than the glucose solution of subphase, floats and rapidly spreads on the surface of the subphase (Fig.5b); (4): a part of protein molecules reaching to the water-air interface immediately unfolds to yield a thin monolayer film on the surface (Fig.5c); (5): the growth of apoferritin 2D crystals takes place under the unfolded protein film (Yoshimura et al., 1994) (Fig.5d); (6): after the 10 min incubation at the room temperature, place holey carbon grids on the surface of the subphase with the carbon side facing to the subphase (Fig.5d); (7): pick up the grids after the 5 min incubation and remove excess solution by blotting the grid surface with a filter paper; (8): To stabilize the protein array on the grids, deposit an additional carbon thin layer about a thickness 50 nm on the unfolded protein film.

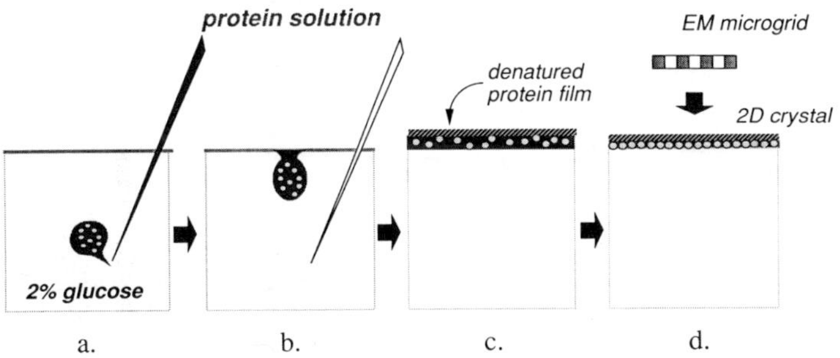

Figure 5. Schematics of the 2D crystallization of protein molecules on the surface of aqueous subphase of higher density (aqueous subphase method) (a) Gentle injection of protein solution (1 μl) into the subphase of 2% glucose (b) Floatation of the solution to the air-water interface (c) Spreading of the solution on the subphase surface and a film made of unfolded protein molecules (d) Array formation and 2D crystallization of the molecules in the thin film sandwiched by the unfolded protein film and the subphase surface. (Taken from Yoshimura et al., 1994).

Electron Microscopy and Image Processing

The protein specimen was negatively stained with 2% uranyl acetate and observed by transmission electron microscope (JEM-1200EXII, JEOL, Japan). Images of electron micrographs were digitized by using microdensitometer (PERKIN-ELMER model 1010M, USA). Computational image processing was performed with EMIDO software systems which were developed by this ERATO project according to the standard procedure employed for the analysis of 2D protein crystal (Amos et al., 1982).

Simulated projection images of apoferritin molecule, used for the comparison with the corresponding electron micrographs, were obtained with use of a program, GENSFC (M. R. C., UK) originally developed for X-ray crystallography. Structure factors for the recombinant horse L-apoferritin were calculated from the atomic coordinates of X-ray structure (Harrison, personal communication) weighted with Gaussian form factors (Maslen et al., 1992). From these structure factors, cross-sections perpendicular to the 2-, the 3-, and the 4-fold axes of the apoferritin molecule were constructed as 2D structure factors, respectively. The projection images along these axes were obtained from thus obtained 2D structure factors with filtering at 2.0 nm resolution and by performing the inverse Fourier transform.

RESULTS

2D Crystals of Wild Type Ferritin

Mercury Method. Two μl of ferritin solution (140~200 mg/ml in a solution of 0.15M NaCl, 7 wt% glucose, and 0.025mM HCl at 20 ûC) was spread on the mercury surface over an area of 430 cm^2. The molecular density was taken to get a closely packed ferritin state of 120 nm^2/molecule, calculated from the molecular dimension obtained by X-ray crystallography (Lawson et al., 1991). The thickness of ferritin film obtained by the ellipsometer for a series of experiments were 9±3 nm. The thickness of the film was found to be almost the same throughout the spread area (Yamaki et al., 1993).

In electron micrographs of protein films transferred to the carbon film, ferritin was found to form a large domain of hexagonally packed monolayer (~10 μm^2). In the domain, highly ordered 2D crystalline areas with a size of 1 μm^2 (unit cell parameters are a=b=12nm and γ=60°) were observed. From the intensity and phase of calculated diffraction spots, ferritin 2D crystal showed three-fold symmetry and mirror image (P3 ml). The same constant for lattice spacing and the same symmetry of peak intensity was observed in electron diffraction from one experiment to another.

The projection image of a ferritin crystal reconstructured without any symmetrization is shown in Fig.6a. The diffraction spots extending to the fifth order (corresponding to resolution was 2.0 nm) were taken for calculation. Since this 2D crystal was negatively stained with uranyl acetate, the resolution was limited to almost 2 nm. The protein part of horse spleen ferritin is composed of 24 subunits (MW=24×18,500) which are arranged in F432 symmetry to form a cubic molecule (Lawson et al., 1991). Since the projection along the (1, 1, 1) axis of a F432 space group is P3 ml, ferritin in the 2D crystal would be directed along this axis perpendicular to the mercury surface. The projection profile along the 3-fold symmetry axis, which was simulated based on the atomic coordinates of X-ray structure (Fig.6b), was similar to our projection image averaged over the 3-fold symmetry (Fig.6c).

Aqueous Subphase Method. One μl recombinant ferritin solution (1 mg/ml) was injected into an aqueous subphase of 2% glucose solution that had a higher density and surface tension than protein solution. The buoyancy of the protein solution made it rise to the surface where it spread quickly and smoothly. The subphase solution includes cadmium ions essential to yield qualified 2D (Yoshimura et al., 1994) and 3D crystals (Lawson et al., 1991) of wild type apoferritin. Without the ion, the wild type ferritin only yielded monolayers of amorphous aggregates.

To investigate the time course of formation of the unfolded protein film prior to the 2D crystallization process, recombinant apoferritin (1 mg/ml, 1μl) was spread on the same subphase and samples were taken after different time periods. The films were transferred to holey carbon films and negatively stained with 2 % uranyl acetate after carbon reinforcement. One minute after spreading, the holes of the carbon film were already observed to be covered with a smooth unfolded protein film. A few intact apoferritin molecules that were adsorbed onto this film were also observed. Three minutes after spreading, the number of adsorbed apoferritin molecules increased and small hexagonal arrays of ferritin were observed. Apoferritin formed large domains of hexagonal arrays ten minutes after spreading (Fig.7) (Yoshimura et al., 1994). These observations suggest that the unfolding of protein molecules took place first and subsequently intact protein molecules adsorbed on the film, leading finally to the formation of protein arrays.

To examine the crystallinity of 2D arrays of apoferritin, correlation averaging of the TEM images (Saxton and Baumeister, 1982) followed by displacement field analysis (DFA) (Dürr, 1991; Saxton et al., 1992) were accomplished. More than 4,500 correlation peaks with little displacement of unit cells were calculated. As like the case of 2D crystallization on the mercury surface (Yoshimura et al., 1990), the averaged image showed 3-fold symmetry and the molecules are considered to be oriented with their 3-fold axes perpendicular to the crystal plane. Because the lattice displacements also exhibit local distortions of unit cells, the distortion, deformation and rotation, of the unit cells in the array was analyzed by DFA. The histograms of the deformation and rotation that were obtained showed Gaussian distribution and the calculated standard deviations (σ) of magnification, elongation, and rotation were 0.60%, 0.42% and 0.30°, respectively. These small standard deviations reveal that the array is a crystal with only slight distortions. Multilayered 3D crystals were occasionally observed in the crystalline films prepared in the same experiment. The 3D crystals seemed to epitaxially grow starting from the 2D crystal monolayer layer by layer. This explains the cause that the crystal had the same projection images of hexagonal lattice as those of the 2D crystals.

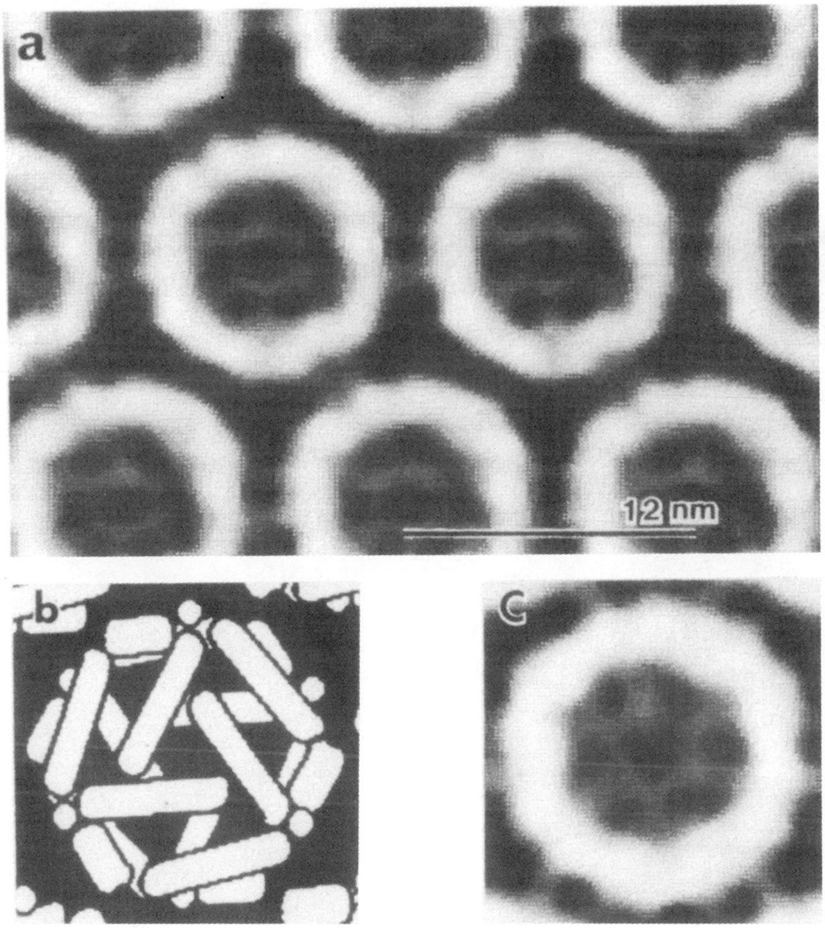

Figure 6. Two-dimensional array of ferrifin. (Taken from Yoshimura et al., 1990) (a) Image reconstruction of a ferritin crystal without symmetrization (b) schematic presentation of subunit arrangement predicted by X-ray crystallography (Lawson et al., 1991) (c) and the projected image with three-fold symmetrizadon (taken from Yoshimura et al., 1990).

2D Crystals of Mutant Ferritin (Takeda et al., 1995)

The mutant apoferritin lacking of the cadmium binding site, Asp84Ser/Gln86Ser, was found to rapidly crystallize in the bulk solution concentrated up to about 0.5 mg/ml. This crystallization was even observed during the purification. The crystal, once produced, became highly insoluble to the aqueous solution. Among a dozen of reagents tested, only 1M guanidine hydrochrolide solution successfully redissolved the crystal without protein denaturation. The mutant solution thus dissolved was recrystallized after removing guanidine by dialysis. The 3D crystals were small and thin (typically about $1\times1\times0.1\mu m$), and had a projection image of square lattice (a=b=18nm, $\gamma=90°$) as observed with a electron microscope. Most of the view fields were found to be covered by amorphous protein monolayer when the suspension of the mutant 3D micro crystals were spread on the subphase surface of glucose solution. But occasionally we had small 2D crystals of the mutant apoferritin reorganized from the 3D crystals. The 2D crystals showed two types of oblique

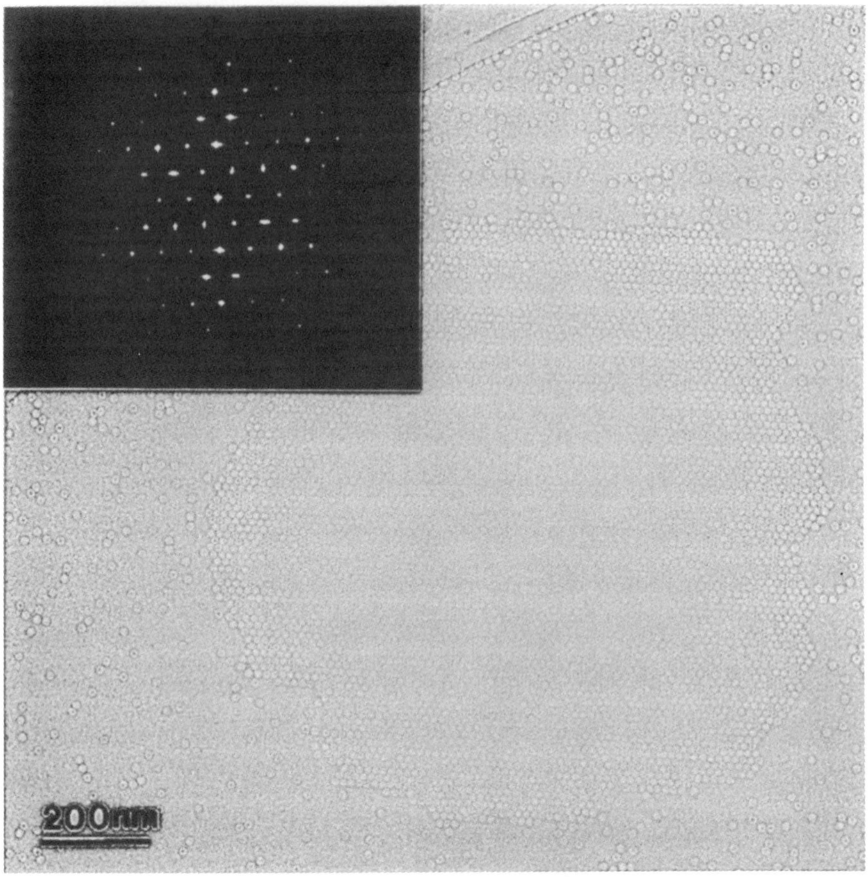

Figure 7. Two-dimensional array of apoferrifin. A monolayer film of unfolded ferritin molecules is first formed on the glucose solution. The intact apoferritins adsorbed to the unfolded ferritin film subsequently form a large 2D array (see the text). The inset is the power spectrum of the Fourier transform for the center area (560 x 560 nm^2). The diffraction spots extend to the sixth order (the signal to noise ratio (S/N) is 3.6 for the spot (6, 0). (taken from ref. 6).

lattice (a=13nm, b=15nm, γ=120° and a=b=13nm, γ=100°). The size of 2D crystal domain of the mutant was a few hundred times smaller than that of the wild type.

The projection images of two types of crystals of the mutant were reconstructed from the electron micrographs at the resolution of 2.0 nm. In the reconstructed images of the mutant, we observed mass distribution different from that of the wild type. Two walls with dense mass were observed in the surrounding of the molecule. A diametrical distribution of mass through the center of molecule was also observed. These images had the 2-fold symmetry.

Next, we compared the reconstructed images with projection images of apoferritin viewed along the 2-, the 3-, and the 4-fold axes which were calculated by computer based on the crystallographic structure. The reconstructed images of the mutant were only compatible to the projection viewed along the 2-fold axis. We concluded that the mutant molecule oriented with the 2-fold symmetry axis perpendicular the crystal plane.

DISCUSSIONS

Molecular Symmetry and Protein Size

More than a dozen different protein and protein complex samples were tested obeying our fabrication of the crystallization in thin film to date. One-third of the trials have resulted in satisfactory ordering in crystalline arrays. Those include water-soluble proteins, ferritin (Yoshimura et al., 1990, 1991), the water-soluble part of the H^+-ATPase (TF_1) of thermophilic bacterium (Yoshimura et al., 1990, 1989; Ishii et al., 1993), a membrane protein, LP-ring (bushing protein) of flagellar motor (Akiba et al., 1991), a chaperonin from thermophilic bacterium (Ishii et al., 1991) and a DNA-protein complex, bacteriophage (Yoshimura et al., 1991). TF_1 is a soluble component of the plasma membrane H^+-ATPase of thermophilic bacterium (PS3) with a molecular size of 10 nm and was crystallized in a form of the hexagonal lattice with lattice constants of a=b=10.0nm and γ=120° (Yoshimura et al., 1989). The membrane protein, a bushing of bacterial motor (LP-ring), with a sylindical shape of a diameter of 20 nm can grow to a 2D crystal with a hexagonal lattice on the mercury surface. The huge complex of proteins and nucleic acids, bacteriophage, was also crystallized in two dimensions. It is not surprising that such a large molecule (60 nm in diameter, about six times larger than ferritin or TF_1) can be crystallized, since much larger colloidal particles which have sizes from 50 nm to 2 μm have been proven to be two-dimensionally crystallized with the crystal form of hexagonal lattice (Denkov et al., 1992; Nagayama, 1993; Dushkin et al., 1993a; Dushkin et al., 1993b; Dimitrov et al., 1994).

Since the hexagonal lattice is a natural form for the closest packing of particles, the proteins with symmetrical shape such as p3, p6 and round can easily be transformed to that lattice type. This is really the case observed for the various protein samples hitherto tested. In the Table 1, we summarized the obtained results by relating the crystal quality reflected in the highest diffraction order to the molecular properties, size and symmetry. Roughly speaking, the better quality in crystal order is obtained for the larger proteins with higher symmetry. From the view point of the assembling mechanism as discussed next, the tendency revealing in the comparative study can be reasoned and justified.

Mechanism of 2D Crystallization

Separate experiments with model particles, polystyrene latex particles, have shed light on the underlying mechanism of protein 2D crystallization, particularly the initial stage of particle assembling (Denkov et al., 1992; Nagayama, 1993; Dushkin et al., 1993a; Dushkin et al., 1993b; Dimitrov et al., 1994). The use of model particles larger than protein molecules has allowed us to make *in situ* observation of the particle assembling process. In a series of analyses of the process recorded in the video-movie taken for the latex arrays formation, two important features have been extracted, which characterize the particle assembling within the liquid thin layer in general as shown in Fig.2 (Denkov et al., 1992; Nagayama, 1993; Dushkin et al., 1993a; Dushkin et al., 1993b; Dimitrov et al., 1994); 1): thin layers of particle suspensions of which thickness is comparable to the particle diameter; 2): particles transported by the water convective flow directing toward the array area. Though studies to prove the proposed mechanism for the protein array are still progressing, the effects and factors influential to array formation of protein molecules know to date can be well interpreted within the theoretical framework. For example, the wettability of substrate surface is important to make a thin and stable film of molecules or particles suspension. The humidity in the atmosphere should be low to stimulate the strong evaporation from the liquid thin layer, which drives the water flow carrying particles as explained in the convective

Table 1. Size, Symmetry, and Ordering.

		Size (kDa)	Molecular Symmetry	Diffraction Order	References
Water soluble proteins					
ferritin	(horse spleen)	400	P3 ml	6	3, 27
H^+–ATPase F1	(*Thermus thermophilus*)	340	P3	5	3, 28, 29
chaperonin	(*Thermus thermophilus*)	420	P1	4	31
myosin S1	(rabbit skeletal)	100		1	a
metallothionein	(rabbit liver)	8		b	a
Membrane proteins					
L–P ring	(*Salmonella*)	1,300	P6	9	30
Na^+, K^+–ATPase	(dog kidney)	120	P1	0	a
Protein-nucleic acid complex					
bacteriophage		>104	P6	11	27

a: not published
b: not transferred to the specimen grid of electron microscope

self-assembly mechanism (Fig.2). The requirement of glucose or glycerol in solution is partly interpreted as the enhanced film stability to avoid film rupturing. Recent study of the competitive binding between the glucose and proteins has also clarified the important role of the reagent to prevent the protein denaturation upon strong its adhesion to the mercury surface. The significance of high concentration of protein solution and small loaded volume of solution can be interpreted to obtain a thin film on mercury without being disturbed by the trough wall where a meniscus can easily be made and withdraw the solution from the surface area only leaving a fraction of the loaded protein sample. Finally the dependence of the quality of crystalline ordering on the protein size and symmetry is considered to be the direct consequence of the proposed array formation mechanism since liquid thin layers with larger thickness are easily prepared on the substrate surface and the round shape of particles can easily be packed to the hexagonal lattice.

Control of Crystal Forms and Molecular Orientations by Mutagenesis (Takeda et al., 1995)

Change distributions on protein surfaces must be one of the determining factors for the inter-protein interactions responsible for the protein crystallization. Now we have a few different 2D crystals made of wild type and mutant apoferritin molecules. In the mutant (Asp84Ser/Gln86Ser) one negative charge at the cadmium ion binding site, which connects the neighboring molecules by its ion bridge in the 3D crystal, was replaced to the neutral. Numerical calculation of electrostatic potential suggested that the mutation resulted in a significant change of the protein surface potential. In the presence of cadmium ion the wild type apoferritin yielded a type of 2D crystal with hexagonal lattice. On the other hand, the mutant apoferritin was crystallized to two different oblique forms in the absence of cadmium ion. Image reconstruction of the mutant crystals clarified that the mutant apoferritin molecules oriented with 2-fold axis perpendicular to the crystal plane in both crystal forms, while the wild type apoferritins oriented with the 3-fold axis perpendicular to the crystal plane. These results clearly indicate that the wild type and mutant apoferritins differ in their inter-protein interactions responsible for the crystalline order. Numerically obtained electrostatic potential compatibly explained the two possible crystal forms. The mutagenesis to a particular site on the protein surface can control the crystal forms and hence the molecular orientations in the protein 2D crystals.

ACKNOWLEDGEMENT

Project members involved in the reported study, Dr. H. Yoshimura, Dr. S. Endo, Dr. S. Takeda, Dr. N. D. Denkov, Dr. C. D. Dushkin, Dr. A. S. Dimitrov, K. Matsubara, and M. Yamaki are highly acknowledged. The author is grateful to Dr. I. B. Ivanov, Dr. P. A. Kralchevsky, and Dr. S. Ebina for their theoretical works and illuminative discussions. He wants to express his thanks to collaborators who kindly presented their relevant protein samples to him, Dr. N. Ishii, Dr. K. Yoshida, Dr. Y. Kagawa, Dr. T. Akiba, and Dr. K. Namba.

REFERENCES

Akiba, T., Yoshimura, H., and Namba, N., 1991, *Science* 252:1554-1546.

Alberts, B., et al., 1994, Molecular Biology of the Cell, 3rd Ed., Garland, N.Y.

Amos, L. A., Henderson, R. H., and Unwin, P. N. T., 1982, *Prog. Biophys. Mol. Biol.* 39:183-231.

Blake, T. D., 1984, in Surfactants (Ed. T. F. Tadros), Academic Press, London., 221.

Denkov, N. D., Velev, O. D., Kralchevsky, P. A., Ivanov, I. B., Yoshimura, H., and Nagayama, K., 1992, *Langmuir* 8:3183-3190.

Derjaguin, B. V., 1989, Theory of Stability of Colloids and Thin Films (translated by R. K. Johnson), Chapter 4, Consultants Bureau, New York.

Dimitrov, A. S., Dushkin, C. D., Ivanov, I. B., Yoshimura, H., and Nagayama, K., 1994, *Langmuir* 10:432-440.

Dürr, R., 1991, *Ultramicroscopy* 38:135.

Dushkin, C. D., Nagayama, K., Miwa, T., and Kralchevsky, P. A., 1993, *Langmuir* 9:3695-3701.

Dushkin, C. D., Yoshimura, H., and Nagayama, K., 1993, *Chem. Phys. Lett.* 204:455-460.

Ishii, N., Taguchi, H., Yoshida, M., Yoshimura, H., and Nagayama, K., 1991, *J. Biochem.* 110:905-908.

Ishii, N., Yoshimura, H., Nagayama, K., Kagawa, Y., and Yoshida, M., 1993, *J. Biochem.* 113:245-250.

Jap, B. K., Zulauf, M., Scheybani, T., Hefti, A., Baumeister, W., Aebi, U., and Engel, A., 1992, *Ultramicroscopy* 46:45-84.

Kralchevsky, P. A., and Nagayama, K., 1994, *Langmuir* 10:23-36.

Kralchevsky, P. A., et al., 1994, *J. Phys., Condens. Matter* 6:A395-A402.

Kralchevsky, P. A., Paunov, V. N., Ivanov, I. B., and Nagayama, K., 1992a, *J. Coll. Interface Sci.* 151:79-94.

Kralchevsky, P. A., Paunov, V. N., Denkov, N. D., Ivanov, I. B., and Nagayama, K., 1992b, *J. Colloid Interface Sci.* 155:420-437.

Lawson, D. M., et al., 1991, *Nature* 349:541-544.

Maslen, W. M., Fox, A. G., and O'Keefe, M. A., 1992, Internatonal Tables for Crystallography, Volume C (Wilson ed.) 476-511, Kluwer Academic, Netherlands.

Michel, H., 1991,Crystallization of Membrane Proteins, CRC Press, Boston.

Nagayama, K., 1993, *Phase Transitions* 45:185-203.

Nagayama, K., 1994, *Mateer. Sci. Engineer.* C1:87-94.

Nagayama, K., Takeda, S., Endo, S., and Yoshimura, H., 1995, *Jpn. J. Appl. Phys.*, in press.

Paunov, V. N., Kralchevsky, P. A., Denkov, N. D., and Nagayama, K., 1993, *J. Colloid Interface Sci.* 157:100-112.

Paunov, V. N., Kralchevsky, P. A., Denkov, N. D., Ivanov, I. B., and Nagayama, K., 1992, *Colloids Interface* 67:119-138.

Saxton, O. W., and Baumeister, W., 1982, *J. Microscopy* 127:127-138.

Saxton, O. W., Dürr, R., and Baumeister, W., 1992, *Ultramicroscopy* 46:287.

Takeda, S., Yoshimura, H., Endo, S., Takahashi, T., and Nagayama, K., *J. Mol. Biol.* submitted.

Yamaki, M., Matsubara, K., and Nagayama, K., 1993, *Langmuir* 9:3154-3158.

Yoshimura, H., Endo, S., Nagayama, K., and Matsumoto, M., 1991, *JEOL News* 29E:2-5.

Yoshimura, H., Matsumoto, M., Endo, S., and Nagayama, K., 1990,*Ultramicroscopy* 32:265-274.

Yoshimura, H., Matsumoto, M., Endo, S., Nagayama, K., and Kagawa, Y., 1989, *J. Biochem.* 106:958-960.

Yoshimura, H., Scheybani, T., Baumeister, W., and Nagayama, K., 1994 *Langmuir*,10: 3290-3295.

3

SCANNING PROBE MICROSCOPY AND NANOTECHNOLOGY

Victor A. Bykov

Zelenograd Research Institute for Physical Problems
Corporation MDT
Moscow, Russia

INTRODUCTION

It is presently a well-known fact (Adam, 1941; Blinov, 1988) that LB-films can be utilized as materials for biosensors (Petty, 1991), gate dielectrics for MOS technology (Matveeva and Bokov, 1992), high resolution plasma etching e-beam- and photo-resists for micro- and submicro-electronics (Matveeva and Bokov, 1992), pyroelectric sensors. LB-film materials with the properties of conductors (Troitsky, 1991), superconductors, semiconductors, dielectrics of various quality parameters are already available. LB-films are an extremely interesting object for the scanning probe microscopy (SPM) as well as for scanning tunneling (STM) and atomic force microscopy (AFM).

Taking into account: the ultra high stability of the LB-films based molecular structures at air interface and the possibility of production of nanometer scale elements at room temperature with topological connection of these elements to the chip, it is possible to understand the interest in such structures and in the technology of production of such structures - the so-called "LB-technology".

LB-TECHNOLOGY

LB-technology is the technology of deposition of molecular layers onto solid substrates and is by now well described in special literature (e.g. Roberts, 1990).

One point, however, requires special mention. If at the water-air interface a monolayer with surface pressure π_s is present and a hydrophobic lamina is being dipped through this interface into the water then surface will be deflected up to a contact angle Θ_{down}. After this the monolayer of the surfactant starts to absorb on hydrophobic substrate, thus minimizing the interface energy in the system. With the start of the taking out procedure the water surface will deflect to the opposite side until contact angle Θ_c reaches the value Θ_{up}. Only after this the second monomolecular layer of the surfactant starts to absorb on the hydrophilic

Molecular Manufacturing, edited by C. Nicolini
Plenum Press, New York, 1996

part of the lamina. However, the extremely important fact is that deposition process does not occur within the range of the contact angles $\Theta_{up} \leq \Theta_c \leq \Theta_{down}$.

The h_{down} and h_{up} are the equilibrium values of the capillary lowering/rising of the meniscus line during the LB-film deposition process. The h_{down} and h_{up} can be explained after the integration of well known Young-Laplace equation (Adamson, 1976):

$$h^2_{down/up} = a^2 * (1 - \sin\Theta_{down/up}) \quad (1),$$

where $a = 2*\gamma/\rho*g$ is the capillary constant; γ, ρ -the surface tension and density of the water and g - the acceleration of the gravity.

For LB-film deposition process $\Delta h = h_{up} + h_{down}$ is the value of the hysteresis for the deposition. In accordance with above equation Δh at the deposition process may be within $0.9 \div 1.8$ mm.

The fact that the amplitude of the hysteresis has a macroscopic value and exceeds by large the thickness of monolayers is extremely important for the technology of molecular layer deposition from the air-water interface. The fact that Δh is much larger in magnitude than both monolayers thickness and amplitude of water surface normal oscillations caused by capillary waves and accidental waves induced by external (e.g. floor and building) vibrations is precisely the fundamental reason why LB-method can be used as the basis for the technological process not conditioned by external factors. Thus, without any special precautions it is possible to produce complicated structures with the required sequence of the alternation of the monomolecular layers in the film.

SCANNING PROBE MICROSCOPY (SPM) OF LB-FILMS

With the development of the SPM (Binnig and Rohrer, 1982; Binnig et al., 1986), including scanning tunneling microscopy (STM) and scanning atomic force microscopy (AFM), we have the real possibility for the surface structure investigations with the extremely high resolution. However, usually the real surfaces at the air or vacuum interface have uncontrollable adsorption of layers and it is very difficult to interpret the results of the surface structure measurements.

The absorption ability of the surface depends upon the value of surface free energy of this surface and adsorption is the self-induced process of surface free energy minimization. The surfaces with the minimal surface free energy (like monocrystalline graphite surface or LB-films coating surfaces) do not exhibit uncontrollable adsorption of layers and are thus very convenient for SPM investigations.

Major part of known LB-films are dielectric materials and STM can be used for studying thin LB-films. Usually, the number of monolayers for STM investigation should not exceed three.

Only special substrates can be used for SPM structural studies of LB-films, with general requirement being high homogeneity at the molecular level as well as geometry and free surface energy homogeneity at the air and at the water interface. Moreover, for STM investigation the substrate must be conductive.

In our experiments we used the surface of the quick-shearing high oriented pyrolytic graphite (HOPG) with the angle of carbon plates disorientation less then 1°. It should be noted, that sometimes it is possible to observe superstructures on HOPG. The appearance of such structures (Fig.1) on an STM-image is due to the disorientation of the surface carbon plate which brings forth moiré patterns. Fig.1a presents an STM-image of HOPG and Fig.2 - an STM moiré pattern with the angle of the surface plate disorientation of approximately 16°.

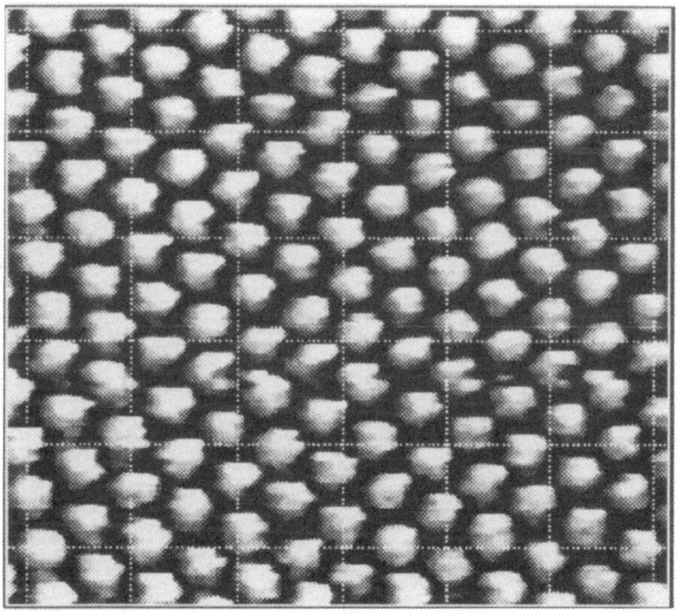

Figure 1. STM image of HOPG.

For STM investigations of the LB-films we used the devices S1_STM and S4_STM (joint production of companies MDT (Moscow, Russia) and ASSE-Z (Padova, Italy). The characteristics of these devices are indicated in Table 1.

Using these devices it is possible not only to investigate the surface structures but also to act on the surface mechanically and by electrical voltage pulses at any point of an STM-image, along a straight line and along a discontinuous trajectory.

For SPM study of 'soft' LB-films it is very important to have a value of drift less than 0.01 ÷ 0.02nm/min. The importance of this characteristic is due to the value of the orientation relaxation time τ_{or} of these films. For LB-film-like smectic liquid crystals $\tau_{or} \approx 50 \mu sec$. For reproducible data the observed τ_{or} must be much less than the time of data measurements τ_m. For the model S1_STM the single data byte measurement time is τ_{min}=20 μsec (period of measurement 80 μsec). According to the requirement $\tau_m >> \tau_{or}$, $\tau_m = N * \tau_{min}$, the number of measurements N in one point has to be not less than 20 ÷ 25. Then, for the measurement of {256*256 points, 0.015nm step} image the scan period will be 3 ÷ 3.5 min. Under this condition the scan size is 3.8*3.8 nm^2 and for the correct data to be obtained, it is necessary to have the resolution of approximately 0.05 nm. In order to satisfy these requirements the drift has to be less than 0.02nm/min. The drift requirements for the investigations of 'hard' films (like fullerens or aromatic molecules) are not so binding because it is possible to have a smaller integration time.

It is not so difficult to determine the integration time and to have the approximation of τ_{or}. The measurement conditions are good if the scan image has no 'scratches' along the scanning axis. If the STM-image contains a lot of scratches along the scanning axis (X or Y), the first step to do is to carry out measurements in the spectroscopy mode of a series of I-V-characteristics with removed tip (usually the voltage range of ±1V is adopted). After obtaining a stable I-V response, the value of the working voltage should be selected at which

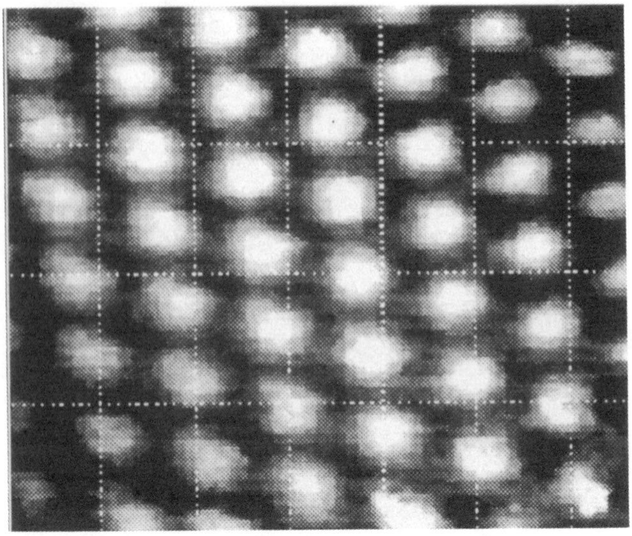

Figure 2. Super structure on HOPG.

the I-V response has the maximum stability. If in this case the scratches along the scanning axis are still present, the tunneling current should be decreased. Only after these simple procedures are carried out, the number of measurements in one specific point might be increased to obtaining correct and reproducible data and to calculate the approximation of τ_{or}.

Table 1. Comparative characteristics of S1_STM and S4_STM

CHARACTERISTICS	S1_STM	S4_STM
minimum X,Y step	0.015nm	0.02nm
minimum Z step	0.001nm	0.001nm
maximum scan range	4*4*1.2 μm^3	2*2*1.2 μm^3
registration current range	$10^{-13} \div 5*10^{-8}$ Å	$10^{-13} \div 5*10^{-8}$ Å
feedback supporting STM current range	$10^{-11} \div 10^{-8}$ Å	$10^{-11} \div 10^{-8}$ Å
potential definition range	±5V with step 2mV	±5V with step 2mV
types of spectroscopy	I(U,r,t),I(Z,r,t), dI/dZ(X,Y),dI/dU(X,Y)	I(U,r,t),I(Z,r,t), dI/dZ(X,Y)
Lithography: - litho acting voltage range - voltage pulse-duration	±200V,step 1V15 μsec	±200V,step 1V15 μsec

NANOTECHNOLOGY

The problems of stability and reproducibility of characteristics - the so-called "mesoscopic problems" - become a general problem for nanometer scale elements (Bykov et al., 1994).

In my opinion, nanometer scale nanoelectronics structures have to be constructed from materials that should posses the following properties for the sake of reproducibility of characteristics :

- The chemical bond system has to be completely saturated;
- The compositions of nanometer elements with the same functions must be absolutely identical to each over;
- The technology of nanometer element production has to be a self - organization technology.

To satisfy these requirements, nanometer elements must be molecules. I think that if nanoelectonics is to become a reality, it must happen with molecular nanoelectronics. But it is a question of the future.

However, already today it is possible to simulate processes and elements of the nanometer scale and to investigate the characteristics of these elements on the basis of existing materials.

The development of the process of the directed local action on nanometer scale structures is a general problem in this field of investigations. The solution of this problem has to cross to the investigations of characteristics of the topology connected systems of these nanometer scale elements.

Conductive LB-films are an extremely interesting object for experimental simulations of the nanoelectronics functional elements.

RESULTS AND DISCUSSIONS

Langmuir layers were formed and deposited by using a computer controlled LB-trough "LB-MDT-5000"(produced by Corporation MDT, Russia), which has two independent barriers and aluminum trough coated by teflon. Surface pressure measurements were made with accuracy of 0.02 mN/m.

The investigation of structures and modification of LB-films were done by STM S1_STM (produced by Corporation MDT, Russia). For modification mode we used $Pt_{0.8}Ir_{0.2}$ and $Pt_{0.8}Ro_{0.2}$ tips with and without isolating coating. The quality of tips was controlled by measurement of $I_t(z)$ dependence. We used tips if $Z(I_t)-Z(I_t/2)<0.5$nm.

Langmuir-Blodgett films of $C_{18}H_{37}$-BEDT-TTF were formed at the water-air interface. The concentration of $FeCl_3$ on the water subfase was 10^{-4} M. The deposition was done on HOPG (reorientation angle <30′) and monocrystalline sapphire by the horizontal lift method. After deposition of each layer the substrate was dried by centrifugation.

The STM structure studies were made on HOPG and sapphire substrates after deposition of 20 LB-layers. Fig.3 presents STM images of these films. Fig.3a presents the STM image of 20 LB layers of $FeCl_3$ doped $C_{18}H_{37}$-BEDT-TTF on HOPG. The conditions of STM measurements were: I_t=const.=0.2 nA, U=0.5 V. The periodical structure observed in this this STM-image has the period of about 0.4nm. Fig.3b presnts the STM image of 20 LB layers of $FeCl_3$ doped $C_{18}H_{37}$-BEDT-TTF on monocrystalline sapphire with Cr spray-coated electrode in form of contra-direction lines with the inter-line distance of 200 μm (Bykov et al., 1993). The conductivity of LB-films was controlled with a conventional tester.

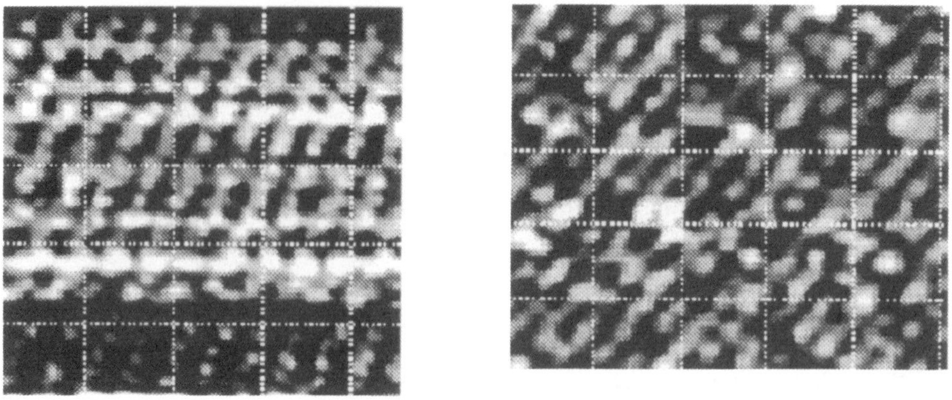

Figure 3. STM images of conductive films of 20 LB layers of $FeCl_3$-doped $C_{18}H_{37}$-BEDT-TTF on HOPG (a) and sapphire (b).

The conductivity was absent before LB-film deposition. The resistance of structures were 300 Ω ÷ 1.5 kΩ after deposition of 20 LB-layers. The STM tip was installed between Cr lines using optical microscope control. The conditions of STM measurements were: I_t=const.=0.15 nA, U=0.5V, scanning rate 30 nm/sec. The periodical structure observed had the same period of 0.4nm, but the quality of film deposited on HOPG was much better than on sapphire and the sizes of loose defect regions on HOPG were larger.

The local modifications study were made on conductive LB-films of $FeCl_3$-doped $C_{18}H_{37}$-BEDT-TTF (three LB-layers on HOPG) and non-conductive LB-layers of oligomer molecules of butyl-cyanacrylate (BCA) and chlorbutadiene (CBD) in combination $\{(BCA)_5(CBA)\}_n$ on HOPG substrates.

The modification of $FeCl_3$-doped $C_{18}H_{37}$-BEDT-TTF takes place if the amplitude of the voltage pulse (VP) is more than 3.5 V (the VP time is about 15 μsec). We observed two types of the VP-induced modifications: defects like a deep flop (Fig.4b) and defects like a crater (Fig.4a, Fig.5). Deep flop defects had the triangle form with the side of 4 nm - 5 nm in the plane. The depth of deep flop defects was 6 nm - 8 nm with practically vertical walls. The second type defects -"crater" - had the diameter of 40 nm - 50 nm and the depth of about 2 nm. Around the crater there was a swell with the height of about 2 nm.

The STS studies of crater defect regions (see Fig.5a,b) have the boundary fluctuations (WF) of work functions and the same WF at the center of crater and at non acting surfaces.

We assume that deep flop defects and crater defects have a quite different origin. Deep flop defects are mainly VP-induced dielectric points on a conductive layer. The depth of it is due to mechanical deformation of substrate which occured in order to have a current during the measurement. The crater defects are due to Coulomb charge destruction of conductive layer. Only crater defects take place on non-conductive materials.

Fig.6 presents STM images of LB film of three monomolecular layers of $\{(BCA)_5(CBA)\}_n$ on HOPG substrate. The conditions of modification were U_{mod}=10V, τ_{mod} ≈ 10 μsec. The conditions of STM measurements were I_t=const=0.2 nA, U=50 mV. On Fig.6a is the structure image immediately after action. A hill and and a hole with the same size can be observed. The vertical hole/hill sizes are approximately 1.1 nm and it corresponds to the

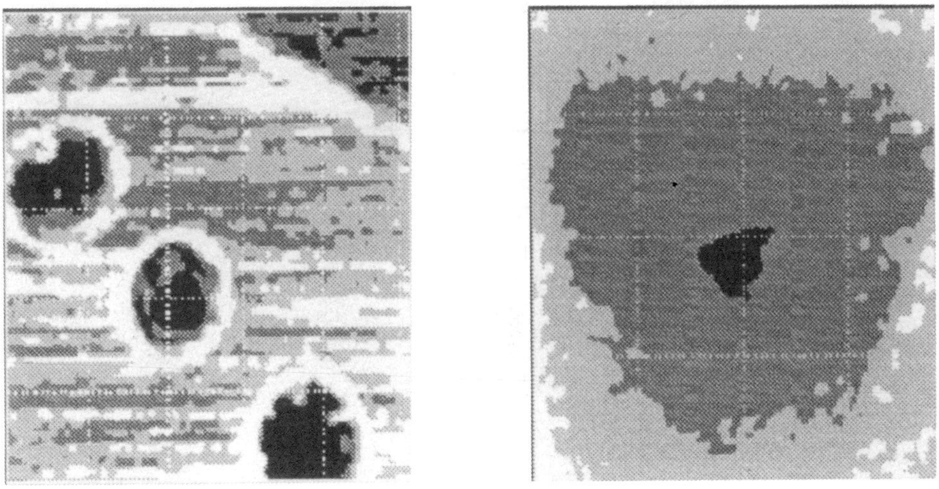

Figure 4. a) VP-induced crater defects; b) VP-induced deep flop defect.

thickness of one Langmuir layer of this material. On Fig.6b is the image of the same place after five scans. An additional hole in place of a hill is observed. This is a result of mechanical action between the tip and substrate. It should be noted that the hill (Fig.6a) and holes have approximately the same and anisotropic sizes. On Fig.6b it is possible to see the regularity with the form of structurized inclined rows. The directions of this row correspond to the hill/hole anisotropy. It is possible to suggest that on Fig.6b we have the molecular resolution of the oligomer LB-film and the hill and holes are the separated oligomer molecular clusters.

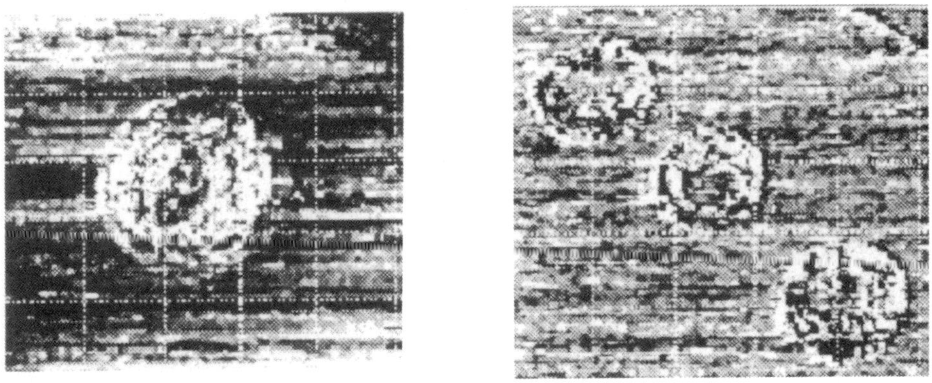

Figure 5. Scanning tunneling spectroscopy. $(\delta I_t/\delta Z)_{I=const}(X,Y)$ images of VP-induced crater defects. a)- after first litho act and b) the same film after three litho actions. The actions were done by the voltage amplitude 5 V, τ = 15 μsec.

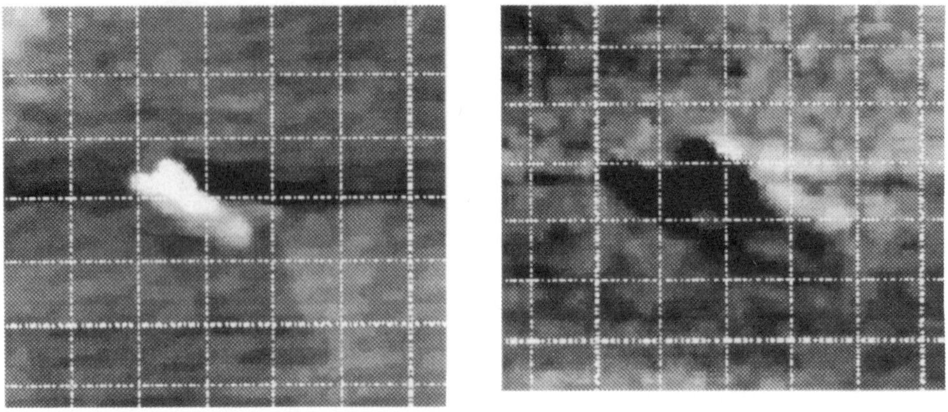

Figure 6. STM image of three layers of $\{(BCA)_5(CBA)\}_n$ after litho action. a) - immediately after action, b) - the same place after five scans.

C_{60} or fullerene is an extremely interesting organic material. An essential step in the utilization of this material is availability of technology of its molecular layer deposition and applicability of methods of LB-technology for this purpose. The C_{60} itself is not a surfactant and it is thus impossible to have LB-films from C_{60} in a straightforward way. However, since C_{60} is a water-insoluble material, it is possible to use the mixture of C_{60} with a special surfactant for preparation of uniform LB-films. Various types of surfactants could be used for this purpose but the idea of a matrix surfactant introduced by T.Berzina and V.Troitsky (Berzina et al., 1995) seems to be one of the best. The proposal was to use a double-tail

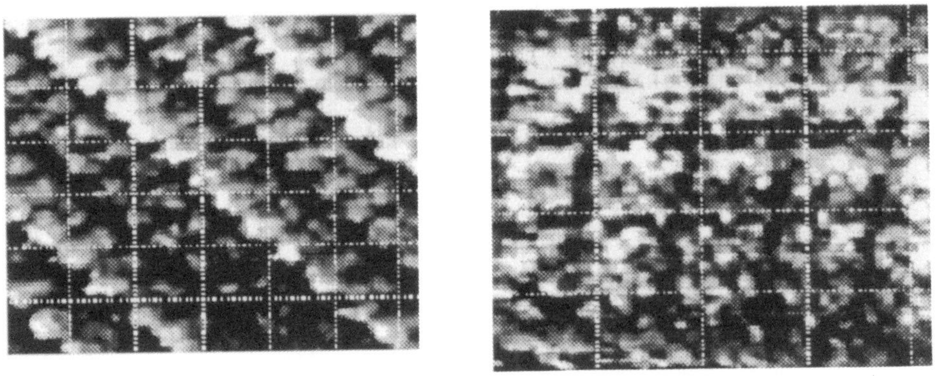

Figure 7. STM image of C_{60} films on HOPG produced by using LB-technology with $C_{17}H_{34}$-TTF-$(CH)_4$-TTF-$C_{17}H_{34}$.dissolved in chloroform. a) - 1 layer deposition, b) - three layers.

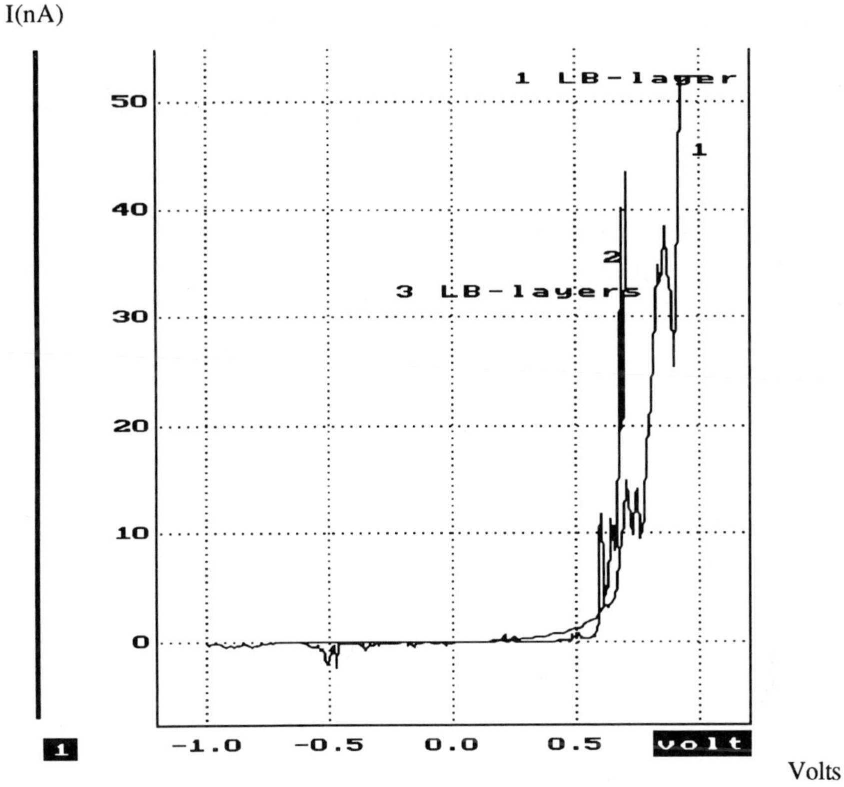

Figure 8. Voltage-Current characteristics of the system Tip($Pt_{0.8}Ir_{0.2}$)-LB-film C_{60}-HOPG.

surfactant with strucutral cavities large enough for the C_{60} to fit in. A suitable surfactant was synthesized by Y.Neilands laboratory at the Institute of Organic Chemistry in Riga (Latvia). The chemical formula of this surfactant is $C_{17}H_{34}$-TTF-$(CH)_4$-TTF-$C_{17}H_{34}$ and high quality LB-films can be obtained with application of equimolar mixtures of C_{60} and the surfactant.

Fig.7. presents STM images of molecular layers of C_{60} on HOPG (Petty, 1991). Highly ordered structure can be observed on the STM images after deposition of both one (Fig.7a) and three (Fig.7b) layers. The periods of structures corresponded to the C_{60} size (ª1nm). These results demonstrate the possibility of molecular layers deposition of C_{60} by means of LB-technology.

A very interesting effect on the Voltage-Current characteristics of the system Tip($Pt_{0.8}Ir_{0.2}$)-LB-film C_{60}-HOPG was observed (Bykov et al., 1993) and is presented in Fig.8. If U is more then 0.6 V (sign at the substrate) a strong increase of current with stepping instabilities is observed for any substrate with C_{60}. This phenomenon seems to be resonance tunneling.

CONCLUSION

Presently we are rather close to the solution of nanoelectronics problems. Interesting materials, technologies and instruments are already available to begin work on modeling of processes and elements. It is a remarkable fact that often efforts of a small (1-3 persons) research group are enough for obtaining complete results in the field of molecular nanoelectronics. But a lot of such groups are necessary for the development of this direction since the peculiarity of these studies is a very wide set of possible materials, the synthesis of which is rather expensive.

REFERENCES

Adam, N.K., 1941, The Physics and Chemistry of Surfaces.

Adamson, A.W., 1976, Physical Chemistry of surfaces Third Edition. A.Wiley-Interscience publication. N.Y., London, Sydney, Toronto.

Alexeev, A.S., Valiansky, S.I., Savransky, V.V., 1992, *ICPAS materials* 8:209-214.

Berzina T.S., Troitsky V.I., Neilands O. Ya., Sudmale I. V., Nicolini C., 1995, *Thin Solid Films* 256:186-191

Binnig, G., Rohrer, H., 1982, *Helv.Phys.Acta* 55:726.

Binnig, G., Quate, C.F., Gerber, Ch., 1986, *Phys.Rev.Lett.* 56:930.

Blinov, L.M., 1988, *Usp. Phis. Nauk (Russ.)* 155:443-480.

Bykov, V.A., Belyev, A.V., Eremchenko, M.D., Jgijgimontov, V.V., Ikonnikov, A.V., Katsour, S.F., Redchenko, V.V., Saounin, S.A., 1995, *Industrial microelectronics (Russ.)* in press.

Bykov, V.A., Emelynov, A.V., Poltoratsky, W.A., Samsonov, N.S., 1994, *Industrial microelectronics* (Russ.) 1

Bykov, V.A., Redchenko, V.V., Eremchenko, M.D., Ivanov, V.A., Mjagkov, I.V., 1993, Abstracts of Int. Conf. of scanning tunneling microscopy STM-93.Beijing, China, p.196.

Bykov, V.A., Redchenko, V.V., Shikin, S.A., Eremchenko, M.D., Novak, V.R., Mjagkov, I.V., 1993, Abstracts of Int. Conf. of scanning tunneling microscopy STM-93. Beijing, China, p.197.

Matveeva, N.K., and Bokov, Y.S., 1992, *Thin Solid Films* 210/211:477-479.

Matveeva, N.K., 1995, *Thin Solid Films*, in press.

Myagkov, I.V., 1989, Letters to the JTP (Russ.) 15:15-18

Myagkov, I.V., Novak, V.R., Lvov, Y.M., Teternik, G.A., 1989, Second Int.Conf. “Molecular Electronics and Biocomputers”, abstracts, Moscow, USSR, p.89-90.

Petty, M.C., 1991, *J.Biomed.Eng.* 13:209-214.

Roberts, G.G., 1989, *Ferroelectrics* 91:21-38.

Roberts, G.G., 1990, “Langmuir-Blodgett Films”, Plenum Press, New York.

Troitsky, V.I., 1991, *Macromol. Chem., Macromol. Symp.*46:223-227.

4

APPLICATION OF LANGMUIR-BLODGETT TECHNIQUE FOR DEPOSITING THIN FILMS OF LIPIDS FROM ARCHAEBACTERIUM

Tatiana S. Berzina,[1*] Vladimir I. Troitsky,[1*] Sergey Vakula,[1,5] Antonio Riccio,[1] Agata Gambacorta,[2] Mario De Rosa,[2] Silvia Dante,[3] Elisabetta Maccioni,[4] Franco Rustichelli,[3] Paola Accossato,[4] and Claudio Nicolini[4]

[1] Technobiochip
via Roma 28, Marciana (LI), Italy
[2] Istituto di Biochimica delle Macromolecole
II Università di Napoli, via Costantinopoli 16, Napoli ICMIB
Via Toiano 6, Arco Felice, Napoli, Italy
[3] Istituto di Scienze Fisiche
Università di Ancona, via Ranieri 65, Ancona, Italy
[4] Istituto di Biofisica
Università di Genova, via Giotto 2, Genova, Italy
[5] EL.B.A. Foundation
Via A. Moro 17, Marciana Marina, Italy

Extremophilic micro-organisms, Archaea in particular, have been the objects of intense studies (Kandler, 1992) because they grow in unusual habitats. Archaea, a phylogenetically coherent separate group different from Eubacteria and Eukarya, comprises a variety of extremophilic bacteria which live segregated into a few peculiar ecological niches characterized by extreme environmental parameters such as high temperature, acidic or alkaline pH values, and saturated salt solutions.

The archaeal membrane lipids possess unusual structural features (Gambacorta et al., 1994). They are based on the isoprenoid chains of different lengths with ether linkage to glycerol carbons of *sn*-2.3 configuration. That differs from the case of usual membrane lipids with fatty acyl chains, which are often unsaturated and are esterified to *sn*-1.2 carbon atoms of glycerol group. Study of the membrane lipids of thermophilic archaea such as *Sulfolobus solfataricus*, i.e. a micro-organism growing optimally at 87 °C and pH 3, is of great interest both from the theoretical and technological points of view . These lipids possess a bipolar structure which is characterized by the presence of two polar heads and a hydrophobic moiety

* Permanent address: Zelenograd Research Institute for Physical Problems, 103460 Moscow, Russia

Molecular Manufacturing, edited by C. Nicolini
Plenum Press, New York, 1996

of approximately double average length with respect to that of the aliphatic components of classic ester lipids. Due to such unique molecular structure lipids from thermophilic archaea may be promising for creation of artificial stable at extreme conditions media just destined by nature for incorporation of some functional molecules, proteins, for example.

The structure and polymorphism of a variety of lipids extracted from *S. solfataricus* have been investigated by X-Ray scattering technique (Gulik et al., 1985; Gulik et al., 1988) to understand the reasons of thermal stability. However, technical applications of such lipids are possible if the technology of producing highly organized structures on their basis is developed.

Such technology can be worked out using Langmuir-Blodgett (LB) method (Blodgett and Langmuir, 1937; Roberts, 1990) owing to its unique possibilities. This method enables one to produce thin uniform layers of surfactants on solid substrates of practically any thickness which is a multiple of the thickness of monomolecular layer. It is possible to create systems of complex structure where monolayers of different types of molecules possessing various properties are arranged in the required order across the thickness of the film, i.e. to accomplish "construction" at the molecular level (Kuhn, 1983). One can embed also different biological molecules into the monolayers or adsorb them on the latter to produce biologically active systems (Kozarac et al., 1988).

Monolayers of bipolar lipids of methanogenic archaebacterium *Methanospirillum hungatei* have been studied at the air-water interface by Tomoaiacotisel et al.(1992). U-shaped hydrocarbon chain configuration appears to be preferred to an extended chain form for these lipids. Dote et al.(1990) studied properties of the monolayers of two bipolar tetraethers produced from the lipids of thermoacidophilic archaebacterium. They did not find essential evidence for the U-shaped molecules and proposed an extended form to be realized.

In this work we have tried to find the way of preparation of highly uniform LB films using lipids extracted from *S. solfataricus* and their derivatives as well. Nine compounds have been studied. Strategy of chemical modifications of the native lipids to achieve high quality of deposition and stability of the films in water solutions is based on the results of study of the monolayers at the air-water interface and of LB films on solid substrates. Elaboration of the LB technology appears to be possible for six compounds. LB films of high uniformity comparable with that of usual films of fatty acid salts were deposited for two of them. Structure and morphology of the films have been studied by optical microscopy, electron microscopy, X-Ray and electron diffraction techniques. Mixed films of one archaea lipid with arachidic acid have been prepared and investigated as well. X-Ray diffraction technique was applied for investigations of thermal stability of two types of films.

Another goal of the work was to find the compounds, which enable us to form the structures with uniform distribution of considerable amount of some functional molecules to ensure high efficiency in carrying out the desired functions. Valinomycin, which facilitates transport of potassium ions through membranes (Crisp, 1946; Ries and Walker, 1961), was used in the work as the example of functional molecule. First of all, mixed films of two lipids with valinomycin have been prepared, but electron microscopy images showed always bulk precipitates of valinomycin in the multilayers. Finally, the problem has been solved by the alternate deposition in various sequences of lipid and valinomycin monolayers. Uniform stable in aqueous solutions films have been deposited by using such method.

EXPERIMENTAL

The compounds used in this work are different bipolar derivatives of dialkyl tetraethers (Fig. 1). Every formula shown in the Figure really describes the mixture of the compounds with isopranyl chains $C_{40}H_{72\text{-}80}$ of various structures (Gulik et al., 1985) which differ by the number of cyclopentane rings in the chains. Native compounds Archae 2 (monosubstituted glycerol-dialkyl-nonitol tetraether with β-D-glucopyranosyl group), Archae 5 (monosubstituted glycerol-dialkyl-glycerol tetraether with phosphomyoinositol group), and Archae 6 (bisubstituted glycerol-dialkyl-nonitol tetraether with β-D-glucopyranosyl and phosphomyoinositol groups) have been separated from total lipid extract (TLE), purified, and characterized as it was reported previously (De Rosa et al., 1986). Two lipid species, i.e. symmetric glycerol-dialkyl-glycerol tetraether (GDGT) and asymmetric glycerol-dialkyl-nonitol tetraether (Archae 1, GDNT in the Refs. by Gulik et al. (1985) and Dote et al. (1990)), have been produced from TLE according to the procedure described by De Rosa et al. (1983). The GDGT was used as an intermediate compound for the synthesis of the Archae 3, 4, 7-9 ones.

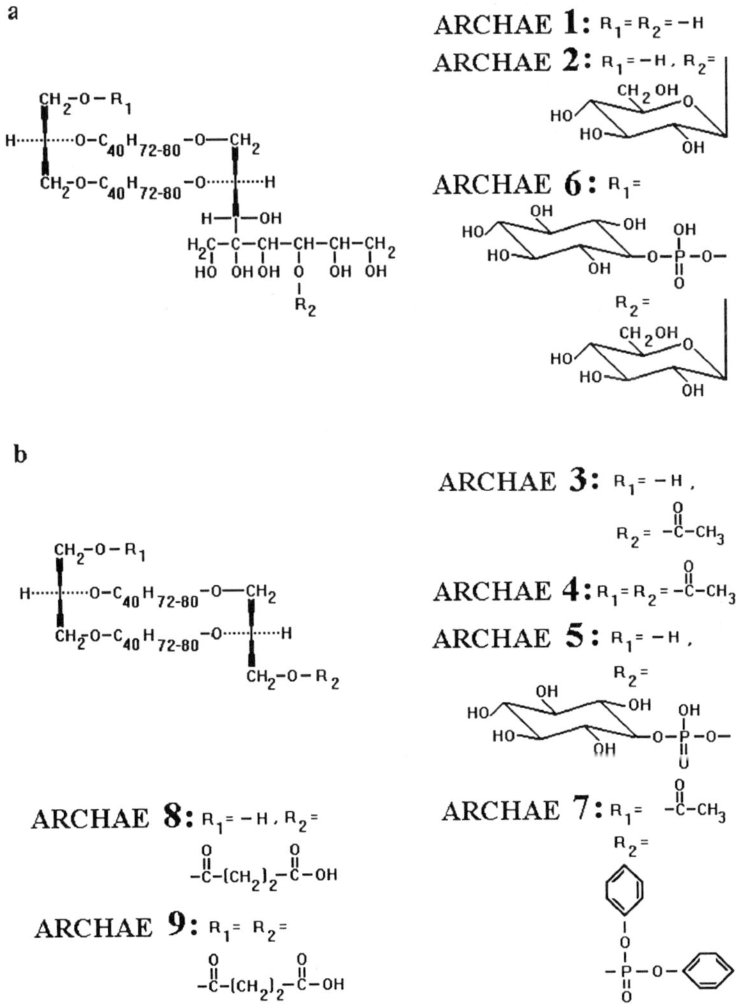

Figure 1. Formulae of the native compounds and their derivatives: (a) group of the lipids based on the glycerol-dialkyl-nonitol tetraether (GDNT); (b) group of the lipids based on the glycerol-dialkyl-glucerol tetraether (GDGT).

To obtain the Archae 3 lipid (bisubstituted GDGT with acetyle groups) mixture of the GDGT with the excess of acetic anhydride in pyridine was stirred for 12 hours at room temperature. The reaction was monitored by thin-layer chromatography (TLC) on silica gel (R_F 0.7 for the GDGT and 0.86 for the Archae 3 in chloroform). The residue of acetic anhydride was destroyed by methanol. The procedure yields the Archae 3 as a main component. The product was mildly hydrolyzed in 0.05N methanolic hydrochloride at room temperature for 4 hours to result in the Archae 4 lipid (monosubstituted GDGT with acetyle group). The TLC on silica gel shows the spots with R_F 0.86 and 0.55 for the Archae 3 and 4 respectively. The products were chromatographed on a silica gel column with n-hexane-ethyl acetate mixture as eluent (9:1 and 8:2 v/v for the Archae 3 and 4 respectively). The Archae 4 shows two sets of signals of methine and C-1 *sn* methylene carbons at 78.36 and 63.03 ppm (glycerol) as well as at 76.56 and 64.11 ppm (acetylated glycerol) in the ^{13}C-NMR spectrum. The diacetylated product, i.e. the Archae 3, does not show the set of signals at 76.56 and 64.11 ppm.

The Archae 7 lipid (bisubstituted GDGT with acetyle and diphenylphosphate groups) was synthesized from the Archae 4 one. The mixture of the latter with diphenylphosphoryl chloride in anhydrous pyridine was stirred at room temperature for 24 hours and monitored by TLC on silica gel (R_F 0.77 for the Archae 7 and 0.55 for the Archae 4 in chloroform). Upon completion of the reaction the product was filtered with silica gel column to remove organic salts and then was concentrated under reduced pressure. The ^{13}C-NMR spectrum shows C-1 *sn* signal of glycerol with acetate group at 64.12 ppm and C-1 *sn* signal of phosphorylated glycerol at 68.81 ppm. In addition, there are the signals of the phenyl groups in the range of 120-130 ppm.

To produce the Archae 8 lipid (monosubstituted GDGT with succinyl group) the equimolar mixture of the GDGT with succinic anhydride in pyridine was stirred for 12 hours at room temperature and the reaction was examined by TLC on silica gel. Then the mixture was acidified with aqueous 0.1N HCl solution and extracted with ethyl ether. The compound was purified by chromatography on a silica gel column with chloroform-methanol (9:1 v/v) as eluent. The product shows one spot with R_F 0.63 on TLC in the solvent system chloroform-methanol (9:1 v/v). Two signals of methine carbons of glycerol and succinilated glycerol at 78.40 and 76.54 ppm as well as two signals of C-1 *sn* of primary hydroxyl group and esterified one of glycerol moieties respectively at 64.34 and 62.96 ppm appear in the ^{13}C-NMR spectrum.

Synthesis of the Archae 9 lipid (bisubstituted GDGT with succinyl groups) was based on the reaction of GDGT with the excess of succinic anhydride. Mixture of the compounds in pyridine was stirred for 12 hours at room temperature and the reaction was monitored by TLC on silica gel. The mixture was acidified with aqueous hydrochloride 0.05N solution and extracted with ethyl ether. The product was purified by chromatography on a silica gel RP-18 column with chloroform-methanol (1:1 v/v) as eluent. The compound gives one spot with R_F 0.3 on TLC in the solvent system chloroform-methanol (9:1 v/v). The ^{13}C-NMR spectrum shows one signal of methine carbons at 76.48 ppm and one signal of C-1 *sn* carbons at 64.44 ppm.

Technological work on the elaboration of LB film deposition included selection of the solvent composition, check of the possibility of monolayer transferring by horizontal and vertical lift techniques, change of substrate treatment and composition of water sub-phase, variation of deposition parameters such as surface pressure, dipping speed, and pause duration for film drying in upper position. Silicon and sapphire substrates were used at this stage of the work. To obtain hydrophilic surfaces the substrates were treated by sulphuric acid or by air plasma discharge. To make silicon substrates hydrophobic the treatment by dimethyldichlorosilane or by hydrofluoric acid was used. For many compounds attempts of deposition were done at several pH values in the range of 3-9. Water subphases containing barium or cadmium ions in the concentration of 10^{-4} M were used as well. Study of monolayers and deposition of the films were carried out with LB-MDT System. Here we present the optimum conditions, which have been finally used for forming the monolayers at the air-water interface and transferring them onto the substrates after numerous variations of the parameters to achieve the best possible spreading of the compounds, stability of the monolayers, and especially the best possible quality of deposition.

RESULTS AND DISCUSSION

First we have studied the films of native lipids Archae 6 and 2 (Fig. 1a) based on GDNT (Archae 1) and Archae 5 (Fig. 1b) based on GDGT. The films of satisfactory uniformity of the native lipid Archae 6 can be deposited only with horizontal lift technique. Electron micrograph in Fig. 2a demonstrates this fact although the morphology is worse than that of usual LB films of fatty acid salts. Deposition of the Archae 6 with vertical lift technique is quite difficult in spite of variations of the conditions of film formation. The film is transferred by chance from some places of the substrate onto water subphase during the dipping acts resulting in the appearance of non-uniform areas. On the other hand, in the places where the film remains its morphology is not worse then that in Fig. 2a. Thus, really process of deposition takes place in a normal way, layer by layer, but the energies of interaction of the monolayers with water and with each other seem to be of approximately the same values, which result in a transferring the peaces of the film back onto water surface.

The Archae 6 films possess well organized structure in the direction normal to the substrate plane. X-ray small-angle diffraction curve (Fig. 3a) shows five Bragg reflections what is rather surprising because in reality the compound is the mixture of several components with different number of cyclopentane rings in the $C_{40}H_{72\text{-}80}$ isoprenoid chains (Gulik et al., 1985). Moreover, electron diffraction pattern shows practically amorphous film in the layer plane, i.e. although molecules do not form crystallites the thickness of the monolayer is fixed with high precision. The structure does not depend on technique of deposition. Period in the normal direction is equal to 5.02 nm and deposition of Y-type takes place. Thus, thickness of the monolayer equals 2.51 nm. Surface pressure-area isotherm shows the area per molecule of 1.00 ± 0.03 nm^2. Morphology of the film in Fig. 2a proves satisfactory spreading of the monolayer at the air-water interface. Under such conditions U-shape of the molecule with approximately vertical arrangement of the hydrophobic tails is preferred because minimum possible area per one chain (S_{min}) is equal to approximately 0.27 nm^2 (Gulik et al., 1985), and total cross-section of four closely packed chains equals 1.04 nm^2. Evaluations show that thickness of the monolayer in this case is in the agreement with the value of period. Indeed, the length of isopranyl chain (Gulik et al., 1985) is of about 4 nm and minimum dimension of two hydrophilic groups in the direction normal to the layer plane can not be less than 1.0-1.2 nm. The last value is roughly evaluated as a ratio of calculated volume of the groups (Gulik et al., 1985; Kitaigorodsky, 1971) and the area per molecule. The experimental data show also that the bend of isopranyl chains somewhere in the middle position happens in such a manner that the chains appear to be closely packed in the monolayer. Strictly speaking, such situation can not be distinguished from that with strongly tilted molecules in the extended form. But the extended structure of molecule is hardly probable. First of all, both hydrophilic groups interact strongly with water and there are no sufficient reasons to one of them to be pushed out from aqueous subphase. Then, the tilt angles of the chain axes must be of approximately 57°, which is unusual for the monolayer with sharp surface pressure-area isotherm and high value of collapse pressure. At last, the area of the cross-section of the molecule in the extended form parallel to the water surface is of about 5.00 nm^2. Approximately at this value of area per molecule the surface pressure

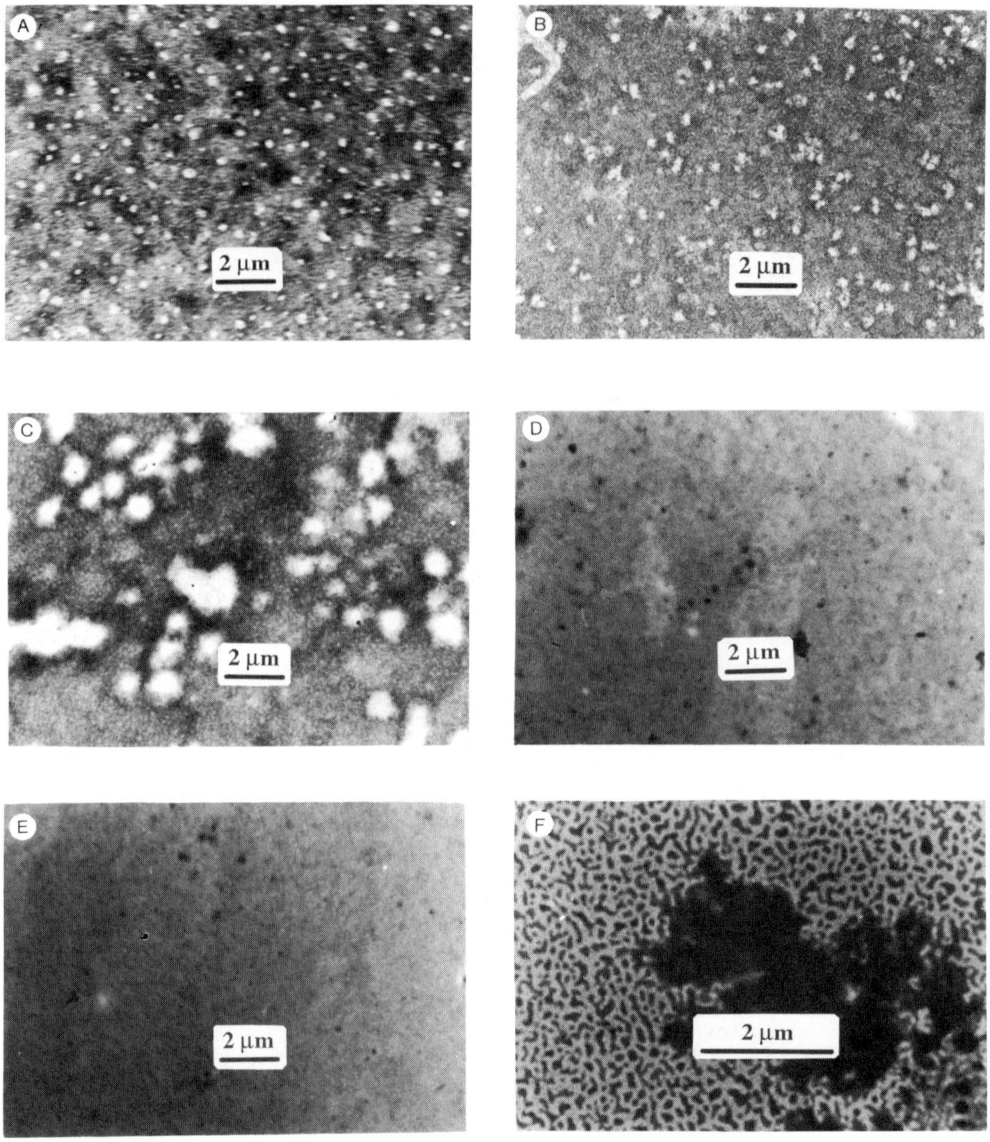

Figure 2. Transmission electron micrographs of multilayers of the archaeal lipids and their derivatives deposited onto collodion films by horizontal lift (hl) or vertical lift (vl) techniques : (a) 15 monolayers of the Archae 6 transferred from the surface of pure water at 27 mN/m (hl); (b) 29 monolayers of the Archae 2 deposited at surface pressure of 27 mN/m (vl) and pH 5.0; (c) 15 monolayers of the Archae 5 deposited from the surface of pure water at 23 mN/m (hl); (d) 25 monolayers of the Archae 2 mixture with arachidic acid (1:10) before treatment in hexane deposited at 27 mN/m (vl) and pH 7.0; (e) 15 monolayers of the Archae 9 deposited from the surface of 10^{-4}M Ba acetate solution at surface pressure of 26 mN/m (vl) and pH 7.0; (f) 20 monolayers of the Archae 5 mixture with valinomycin (1:1) deposited at 23 mN/m (vl) from the surface of pure water onto the substrate covered by 3 monolayers of Ba behenate.

should increase because molecules begin to interact, but really surface pressure is equal to zero up to 1.2 nm^2. As it is shown below the monolayers and the films of the compounds studied possess approximately the same structural properties. In particular, similar parame-

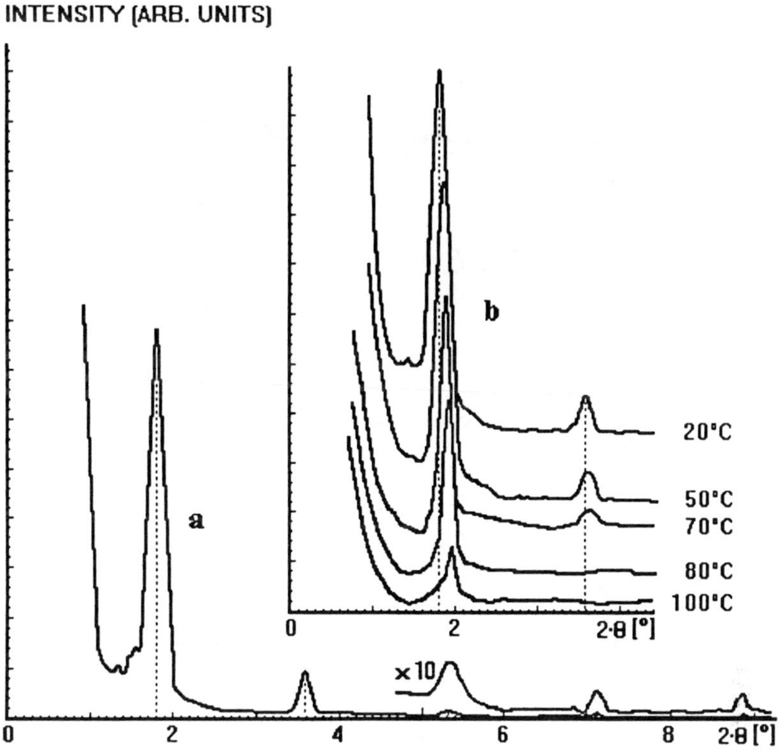

Figure 3. Small-angle X-Ray diffraction curves of 50 monolayers of the Archae 6 deposited by horizontal lift technique from the surface of pure water onto hydrophobic silicon substrate at room temperature (D = 5.02 nm) (a) and after heating (b).

ters are obtained for the compounds with two equal hydrophilic groups. That gives additional arguments in favour of the U-shaped molecules. The Archae 6 films appear to be very thermostable. Bragg reflections in the small-angle X-Ray diffraction curves do not disappear up to the temperature of more than 100°C (Fig. 3b). At the same time annealing of the sample takes place during the heating, so that the period of the structure changes first from 5.02 to 4.50 nm and does not return to the initial value after cooling.

The Archae 2 lipid can be produced from the Archae 6 one by hydrolysis of phosphodiester bond, but in this work the native extracted and purified compound was used. The Archae 2 multilayer of approximately the same morphology as that of the Archae 6 film can be deposited with horizontal lift technique. The character of deposition by vertical lift technique is in general also the same. But contrary to the Archae 6 lipid the change of pH value slightly influences the quality of the Archae 2 film although this influence can not be explained from the chemical point of view. Example of rather satisfactory deposition by vertical lift technique is presented in the electron micrograph (Fig. 2b). However, the reproducibility of deposition quality is bad and the results are unpredictable. Surface pressure-area isotherm shows the area per molecule of about 1.20 nm^2. The deposition of Y-type takes place and period in the normal direction calculated from X-ray diffraction pattern equals 4.70 nm what can be explained again by the model of U-shaped molecule. Electron diffraction pattern shows practically amorphous film in the layer plane. According to X-Ray data the Archae 2 films are thermostable up to the temperature of 65°C. Annealing of the Archae 2 film takes place when heating the sample for the first time as in the case of the Archae 6 one. The value of period changes from 4.70 to 4.40 nm.

Surface pressure-area isotherm of the Archae 5 native lipid shows the area per molecule of about 1.00 nm^2 and a selection of the solvents gives the possibility to increase it up to 1.10 nm^2. Although this value is practically equal to S_{min} the further studies show that problems of spreading were not solved completely for this compound because non-uniform areas and some bulk crystallites are detected in the deposited films as it is demonstrated by electron micrograph and electron diffraction pattern in Fig. 2c. The film is disordered in the layer plane, but precipitated crystallites give point like reflections in the electron diffraction pattern. The best average transfer ratio of 0.8 was achieved when the deposition was fulfilled onto hydrophobic substrates. Practically no transfer is observed when depositing onto hydrophilic substrates. However, non-uniform areas in the form of strips are observed always for the best films also. Such behaviour is typical when adhesion of the film to the substrate is insufficient. To increase adhesion the substrates were covered before the deposition of Archae 5 films by four monolayers of barium behenate. In this case the average transfer ratio increases up to 0.95.

Unfortunately, the Archae 2, 5, and 6 films are unstable in water. In a short time pieces of the film begin to break away from the substrate. That is not surprising because the molecules contain hydrophilic groups, which can form strong hydrogen bonds with water molecules, and, in fact, under native conditions these molecules form membranes surrounded by aqueous solution.

To improve reproducibility of the Archae 2 film deposition as well as morphology of the multilayer another technological solution was proposed. Since the Archae 2 is insoluble in hexane we have deposited first the Archae 2 mixed film with a surfactant compound soluble in hexane. Then this component was removed from the film by the treatment in hexane. In our case arachidic acid was used as a soluble component of the mixture, in fact, the LB film of pure arachidic acid is removed completely by hexane from the substrate in few seconds. To achieve comparable values of areas occupied by molecules of both types in the monolayer the mixtures with molar ratios of the Archae 2 and arachidic acid of 1:5 and 1:10 were chosen for film deposition because area per arachidic acid molecule (Roberts, 1990) is equal to 0.20 nm^2. Transferring such mixed monolayers onto the substrates appears to be reproducible and uniformity of the film is better than that of the Archae 2 (Fig. 2d). X-ray diffraction curve shows Bragg reflections from arachidic acid only. The Archae 2 molecules seem to be dissolved in the matrix and does not form separate phase.

After immersing the sample in hexane the thickness of the film decreases approximately proportional to the content of arachidic acid according to the change of interference colour of the film on silicon substrate. It is remarkable that uniformity of the multilayer does not deteriorate under such treatment. Electron microscopy image remains approximately the same as that in the Fig. 2d. However, X-ray diffraction patterns show always the presence of separate arachidic acid phase in the sample after treatment. Probably, due to interaction with the Archae 2 molecules some small aggregates of arachidic acid can not be removed from multilayer. The formation of some insoluble salt with uncontrolled impurity dissolved in water can be the other reason for this result. Indeed, mixed films are deposited at pH 7.0 when the definite part of arachidic acid molecules is dissociated while successful formation of the multilayer of pure arachidic acid is possible at pH 4.5-5.0. In principle, such a method is suitable for depositing the uniform LB films of lipids from archaebacterium because small rest of arachidic acid can not influence strongly the properties of the films.

Modified Lipids

The Archae 1 lipid (GDNT) produced by hydrolysis of TLE in methanolic HCl can be considered as the core lipid derivative of the Archae 2 and 6 ones. Its deposition occurs approximately in the same manner as the deposition of the Archae 2. Area per molecule is equal to 1.20 nm^2 and a period diminishes with respect to that of Archae 2 films up to 4.00 nm. Electron diffraction patterns show only week broad rings. Comparison of the presented data with the appropriate values for the Archae 2 and 6 films shows that the change of the main features of molecular packing is not expected. Morphology of the deposited film can be considered as satisfactory, which is confirmed by optical and electron microscopy studies, but reproducibility of the results is bad. The modified lipid Archae1 appears to be unstable in water as well.

To overcome the problems of deposition and instability of the LB films studied before several semisynthetic derivatives of the archaeal lipids with different types of polar head groups have been produced. Semisynthetic derivatives Archae 3, 4, and 7 obtained from GDGT lipid are unsuitable for formation of the LB films. The monolayers of the Archae 3 and 4 possess very low collapse pressure of 18 and 10 mN/m respectively and are quite unstable. The films deposited by horizontal lift technique are non-uniform and disordered, the small-angle X-Ray diffraction curves do not show any Bragg reflections. Monolayers of the Archae 7 possess a low collapse pressure of 19 mN/m although they are stable at some conditions. The deposition of the Archae 7 can be carried out by horizontal lift technique only, but uniform films have not been obtained, and change of the deposition pressure does

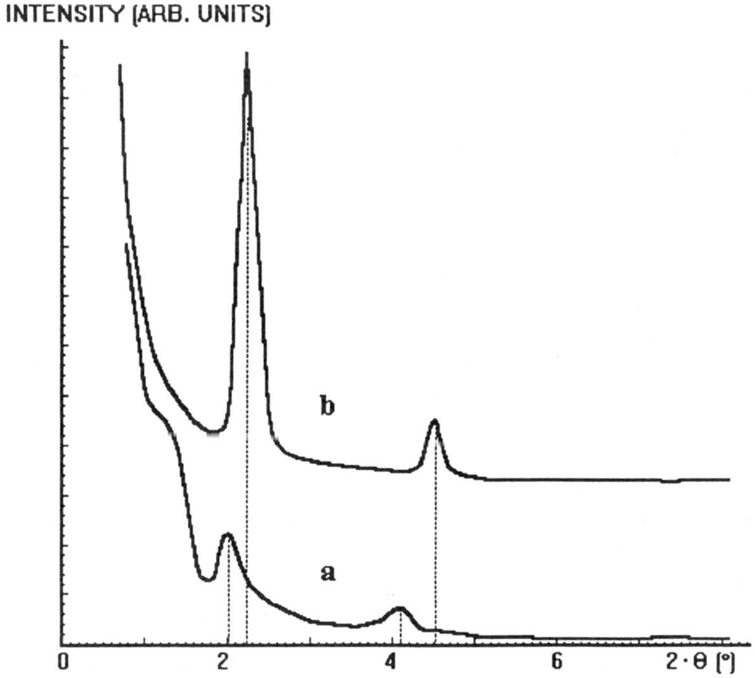

Figure 4. Curves of X-Ray small-angle scattering for 90 monolayers of the Archae 8 (a) and for 60 monolayers of the Archae 9 (b) deposited by vertical lift technique onto hydrophobic silicon substrates from the surface of 10^{-4}M Ba acetate solution at pH 7.0. Values of period D in the normal direction are equal to 4.40 and 3.97 nm for the Archae 8 and 9 multilayers recpectively.

not improve film morphology. Small-angle X-Ray diffraction curve of the Archae 7 film consists of numerous Bragg peaks and electron diffraction pattern shows also point like reflections indicating that bulk crystallites are transferred onto the substrate from the surface of water. Because of impossibility to obtain ideal monolayers at the air-water interface and to deposit the uniform LB films the experimental data do not give sufficient confirmation on the U-shape of the Archae 3, 4, and 7 molecules. However, rough evaluations from the surface pressure-area isotherms taking into account instability of the monolayers show that the U-shape is probable again.

The Archae 8 and 9 lipids are mono- and bisubstituted derivatives of GDGT respectively with succinyl groups (Fig. 1b). The introduction of such groups into the molecule enables one to influence strongly the properties of monolayers by the variation of water subphase composition. Actually it appears that the highly uniform LB films of barium salts of these compounds can be deposited easily by vertical lift technique. Areas per molecule of 1.45 and 1.47 nm^2 and periods in the normal direction of 4.40 and 3.97 nm (Fig. 4) for the Archae 8 and 9 respectively give the preference to the U-shaped molecules with tilted to the layer plane isopranyl chains. Reproducibility of the results is considerably better for the Archae 9. In this case transfer ratios are equal usually to 1.0 ± 0.2 while during the deposition of Archae 8 film they may become considerable lower than 1 for some monolayers. Morphology of the Archae 9 films is shown in the Fig. 2e. The attractive property of these films is stability in water solutions caused by salt formation.

Mixtures of Lipids with Valinomycin

The possibility to deposit rather uniform films of the Archae 5 lipid with vertical lift technique ensured us the use of such monolayers as a matrix for dissolution of valinomycin. Valinomycin films themselves are quite unstable and recrystallize during a few days as a rule. For this reason, the search for stable matrix for dissolution of valinomycin in considerable concentrations is important. Although the possibility for valinomycin to function in

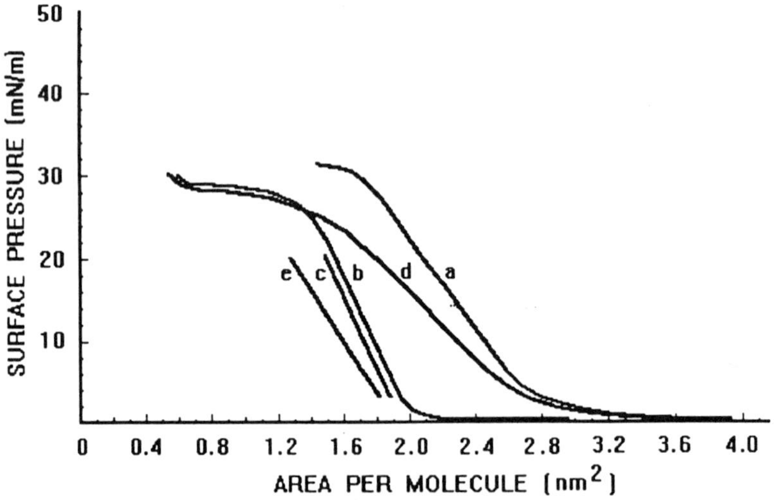

Figure 5. Surface pressure-area isotherms of the monolayers of valinomycin (a), mixture of the Archae 5 with valinomycin in molar ratio of 1:1 (b), and mixture of the Archae 9 with valinomycin in molar ratio of 1:0.5 (d) spread at the surface of pure water. Solid lines c and d show the calculated positions of the isotherms b and d respectively. Areas per molecule are equal to 2.84, 2.00, and 2.87 nm^2 for the curves a, b, and d.

such films, i.e. to adsorb the ions selectively, was not studied in the present work it is clear that the higher concentrations will provide the higher sensitivity values. The system on the basis of the Archae 5 can be considered only as a model for checking such possibility because the Archae 5 films are unstable in water. However, we would like to carry out such experiments for both native compound and its derivative.

First studies of properties of the mixed monolayers at the air-water interface made us hope that complete dissolution of valinomycin in the Archae 5 matrix up to very high concentration is possible because location of corresponding surface pressure-area isotherm was very close to ideal position (Fig. 5). However, electron micrographs (Fig. 2f) showed always bulk precipitates of valinomycin. The same situation takes place also at considerably low concentrations of valinomycin. This contradiction between the experimental data can be caused, for example, by different orientations of valinomycin crystallites in the matrix and in monocomponent valinomycin films. Area per valinomycin molecule dissolved in the lipid matrix can be considerably more. In any case, the conclusion can be done that uniform distribution of valinomycin molecules with rather high concentrations of the latters is hardly probable for the mixed films of the Archae 5 and valinomycin.

The Archae 9 has been chosen also as a matrix for dissolution of valinomycin because of stability of the Archae 9 barium salt LB films in aqueous solutions. Surface pressure-area isotherm in Fig. 5 shows that distribution of valinomycin molecules is more uniform with respect to the case of the Archae 5 matrix since the curve for the mixture is shifted considerably to the region of higher area per molecule values. However, precipitates similar to those in Fig. 2f are observed always in the electron micrographs of the multilayers of the Archae 9 barium salt also.

Finally, to avoid precipitation of valinomycin in the film of the Archae 9 barium salt the alternating monolayers of two compounds were deposited according to two programs, i.e. ABABAB... and AABBAABB..., where A and B are valinomycin and lipid monolayers, respectively. Transfer ratios for both sequences of alternation were practically equal to unity and no precipitation of bulk crystallites was found with optical microscopy. Observation of the film of the ABABAB... type deposited onto sapphire substrate shows that recrystallization happens in several days probably because of weak adhesion of the film to hydrophilic surface. Indeed, the films of ABABAB...and AABBAABB... types deposited onto hydrophobic silicon substrates are stable during several months at least. The films of alternating monolayers deposited onto hydrophobic surfaces are stable in water solutions as well. Structural studies of such "superlattices" are in progress. However, it is obvious that the method used enables us the possibility to distribute uniformly considerable amount of the functional molecules in the lipid matrix.

CONCLUSION

Application of the LB technique appears to be successful for depositing uniform films of natural and semisynthetic archaea lipids although the molecules used possess an unusual bipolar architecture. This possibility seems to be provided by U-shape of the molecules at the air-water interface with close packing of isopranyl chains at the same time, so that behaviour of the molecules at the surface and their interaction with water are similar to those for usual surfactant molecules.

Instability of the films of native compounds as well as of some their derivatives in aqueous solutions is caused by the presence of hydrophilic groups, which interact with water stronger than with each other. This interaction can not be changed practically when varying the composition of water subphase. Introduction of carboxyl groups into hydrophilic parts of the molecules changes the situation in cardinal manner. Interaction of the monolayers with each other becomes very strong due to salt formation with metal ions dissolved in water subphase. The films become stable in aqueous solutions and the quality of deposition is improved considerably.

The LB technique enables us also to achieve uniform distribution of some functional molecules, such as valinomycin, with high concentration in the lipid matrix. This is provided by depositing the films of alternating monolayers of different types.

ACKNOWLEDGEMENT

This work has been supported by a grant within the National Research Progam "Technologies for Bioelectronics" of the Italian Ministry of University of Scientific and Technological Research given to Technobiochip, Marciana (LI), Italy.

REFERENCES

Blodgett, K. and Langmuir, I., 1937, *Phys.Rev.,* 51:964.
Crisp, D.J., 1946, *J.Colloid Sci.,* 1:49.
De Rosa, M., Gambacorta, A., Nicolaus, B., Chappe, B., and Albrecht, P., 1983, *Biochim.Biophys.Acta,* 753:249.
De Rosa, M., Gambacorta, A.,and Gliozzi, A., 1986, *Microbiol.Rev.,* 50:70.
Dote, J.L., Barger, W.R., Behroozi, F., Chang, E.L., Lo, S.-L., Montague, C.E., and Nagumo, M., 1990, *Langmuir,* 6:1017.
Gambacorta, A., Trincone, A., Nicolaus, B., Lama, L., and De Rosa, M., 1994, *System. Appl. Microbiol.,* 16:518.
Gulik, A., Luzzati, V., De Rosa, M., and Gambacorta, A., 1985, *J.Mol.Biol.,* 182:131.
Gulik, A., Luzzati, V., De Rosa, M., and Gambacorta, A., 1988, *J.Mol.Biol.,* 201:429.
Howarth, V.A., Petty, M.C., Davies, G.H., and Yarwood, J., 1988, *Thin Solid Films,* 160:483.
Kandler, O., 1992, Where next with Archaebacteria?, in: "Archaebacteria: Biochemistry and Biotechnology", M.J. Danson, D.W. Hough, and G.C. Lunt, eds., Portland Press, London.
Kitaigorodsky, A.I., 1971, "Molecular Crystals and Molecules", Academic, New York.
Kozarac, Z., Dhathathreyan, A., and Möbius, D., 1988, *FEBS Lett.,* 229:372.
Kuhn, H., 1983, *Thin Solid Films,* 99:1.
Ries, H.E.,Jr. and Walker, D.C., 1961, *J.Colloid Sci.,* 16:361.
Roberts, G.G., 1990, "Langmuir-Blodgett Films", Plenum Press, New York.
Tomaiacotisel, M., Chifu, E., Zsako, J., Mocanu, A., Quinn, P.J., and Kates, M., 1992, *Chemistry and Physics of Lipids,* 63:31.

5

INDUSTRIAL-SCALE PRODUCTION OF THERMOSTABLE ENZYMES: THE MODEL-SYSTEM OF THE β-GLYCOSIDASE FROM *SULFOLOBUS SOLFATARICUS*

Marco Moracci, Rosa La Montagna, Luisa Capalbo,[1] Alessandra Morana,[1] Antonio Ottombrino,[1] Roberto Nucci, Mario De Rosa,[1] Enzo Martegani,[2] Maria Ciaramella, and Mosè Rossi

Institute of Protein Biochemistry and Enzymology
Via Marconi 10, 80125 Naples, Italy
[1] Istituto di Biochimica delle Macromolecole
Università di Napoli, Italy
[2] Dipartimento di Fisiologia e Biochimica Generali
sez. di Biochimica Comparata, Università di Milano, Italy

The current relevance of enzymes in biotechnology is well known and many examples exist of applications of these catalysts in different fields (i.e. food and chemicals industries, agriculture, medicine). Since industrial use often revealed to be denaturing for these molecules, several techniques have been used to improve the stability of enzymes in the operative conditions (protein engineering, protein modification and immobilization). However, since this approach resulted some times unsuccessful for the intrinsic fragility of the molecules, particular attention has been focused on enzymes from extreme thermophilic bacteria which grow between 70°C and 100°C and over. All the micro-organisms living at temperatures over 70°C belong to the new taxonomic unit of Archaea, recognised as a super-kingdom (Woese *et al.*, 1990). Whereas conventional enzymes are irreversibly inactivated by heat, the enzymes from extremophiles not only show great stability but also enhanced activity in the presence of common protein denaturants such as heat, detergents, organic solvents and proteolytic enzymes. As a consequence, these molecules have considerable industrial potentialities giving better results at "extreme" operational conditions. However, despite the enthusiasm induced by the use of thermophilic enzymes in biotechnology in middle-eighties, in recent years it has been demonstrated that their properties resulted to be sometimes too "extreme" to allow an effective application in industrial processes. In fact, some practical adversities (modification of pre-existing plants, high temperature-working reactors, incompatibility of reagents and products with high temperatures) can represent a serious limitation for their application. For these reasons these enzymes can be an ideal and unique answer to specific problems, but cannot be considered a general solution in

Molecular Manufacturing, edited by C. Nicolini
Plenum Press, New York, 1996

biotechnology (Cowan, 1992). A further restriction to their direct use in biotechnology is the moderated amount available of these proteins, since their purification from thermophilic sources is expensive and time consuming. Hence, the gene cloning and expression in heterologous mesophilic hosts such as *Escherichia coli*, *Bacillus subtilis* and yeast is advisable.

Among the enzymes with the widest potentialities in biotechnology there are the β-glycosidases (EC 3.2.1.x). This group of enzymes, which catalyse the hydrolysis of glycosidic bonds between a large variety of saccharides, includes β-gluco, β-galacto and phospho-β–galactosidases. This wide substrate specificity allows their utilisation in several fields and, in particular, β-glycosidases from thermophilic micro-organisms can represent an ideal tool for some applications in chemistry, biochemistry and molecular biology. The saccharification of cellulosic materials, the hydrolysis of lactose at high temperature in dairy products, the use as gene-reporter in molecular biology and the synthesis of drugs and chemicals in pharmaceutical industry by reversed-catalysed reactions represent some applicative examples under evaluation.

One of the enzymes under detailed study in our laboratory is the β-glycosidase from the thermoacidophilic archaeon *Sulfolobus solfataricus* (Sβ-gly). This enzyme was purified, characterized in detail and crystallised and its gene cloned and expressed in eukaryotic cells and in *E. coli* (Moracci *et al.*, 1994). From these studies, new perspectives for the utilization of the β-glycosidase from *Sulfolobus solfataricus* can be gained.

We present here a review of these findings particularly concerned with the increased biotechnological applications of the enzyme by heterologous expression. Since some of the Sβ-gly features, such as wide substrate specificity and enhanced stability and activity at high temperatures, are maintained in its recombinant form, this enzyme represents an excellent example of how the biotechnological potentialities of a thermophilic enzyme can be improved by its cloning and expression in mesophilic hosts.

STRUCTURAL AND KINETIC FEATURES OF THE β-GLYCOSIDASE FROM *SULFOLOBUS SOLFATARICUS*

A β-galactosidase activity was purified for the first time in our laboratory from cellular extracts of the extreme thermoacidophilic archaeon *Sulfolobus solfataricus*, strain MT-4 (Pisani *et al.*, 1990). The enzyme is a tetramer of four identical subunits with a total native molecular weight of 240,000 Da (60,000 Da per subunit). Only one cysteine per subunit not involved in disulphuric bridges could be detected, hence, its relevant resistance to heat and other protein denaturing agents, such as proteases and SDS, was presumably due only to weak interactions (Tab. I).

More recently, a new purification procedure and an extensive substrate characterisation of the enzyme was performed (Nucci *et al.*, 1993). This study allowed a new definition of the enzyme and opened new perspectives for its utilisation. The enzyme is able to catalyse the hydrolysis of a large number of disaccharides derived from the β-linking of the glucose with aryl, alkyl or 6-carbon-atoms sugars. Furthermore, in the presence of β-linked glucose oligomers from two to five glucose units, the enzyme produces at high temperature glucose from the non-reducing end of the oligosaccharide. Consequently, it was classified as a β-glycosidase with remarkable exo-glucosidase activity and it was designed Sβ-gly (Tab. II). From our results, the most effective substrates were, in this order, aryl-β-gluco, -fuco and -galactopyranosides and the active site involved in the catalysis appears to be the same (for details see Nucci *et al.,* 1993). This indication is also confirmed by the similar thermal stability and thermal activity of both the β-gluco and β-galactosidase

Table 1. Structural and kinetic features of *S.solfataricus* β–glycosidase

Optimal pH(β-gal activity)	6.5
Optimal T (°C)	> 80
Isoelectric point	4.5
Half life at 75°C	> 24 hrs
Number of subunits	4
Molecular Weight native	240 kDa
Molecular Weightmonomer	60 kDa
Number of cysteines per subunit	1

activities. As a consequence, the latter is normally used in routine assays following the Sβ-gly expression and purification.

The wide substrate specificity of Sβ-gly includes the ability of the enzyme to synthesise different glycosides by trans-glycosidase activity, using several carbohydrates and alcohols as donors and acceptors respectively. Primary, secondary, but not tertiary alcohols are good acceptors expecially if used in molar excess (Moracci *et al.*, 1994). Furthermore, noticeable enantioselectivity was found using phenyl-β-galactosides as donors and various 1,2-diols as acceptors (Trincone *et al.*, 1991a, b). These results confirmed the great versatility of Sβ-gly and the possibility of using it not only in the synthesis of different glycosides but also in the study of the stereochemistry of the enzyme.

In order to understand the structural basis of the extreme thermostability and thermophilicity of Sβ–gly, we are attempting to determine its three-dimensional structure by X-ray crystallography. Using protein purified from *S.solfataricus,* crystals with low solvent content (27%), producing useful diffraction to 2.5 Å resolution were obtained (Pearl *et al.*, 1993). Furthermore, we have evidence that the enzyme maintains its activity in the crystallised form (Moracci *et al.*, 1994).

Table 2. Kinetic constants of *S.solfataricus* β-glycosidase

Substrate	K_M (mM)	k_{cat} (s^{-1})	k_{cat}/K_M (s^{-1} mM^{-1})
p-nph-β-D-Fuc*p*	1.0	772	772
p-nph-β-D-Glc*p*	0.5	542	1084
o-nph-β-D-Glc*p*	3.5	1096	313
p-nph-β-D-Gal*p*	4.7	1020	217
o-nph-β-D-Gal*p*	4.0	1239	309
p-nph-β-D-Xyl*p*	4.0	284	71
laminaribiose[a,b]	1.0	908	908
β-gentiobiose[a,b]	100	1360	14
cellobiose[a,b]	30	1333	44
cellotriose[b]	3.0	197	66
cellotetraose[b]	1.7	584	343
cellopentaose[b]	2.4	270	112
p-nph-β-D-cellobioside[b]	0.3	503	1676

[a]Kinetic values for these substrates were calculated taking into account that two glucose equivalents were produced for every glucosidic bond hydrolyzed.
[b]Hydrolysis products were detected as glucose and K_M values are shown as glucose equivalents.
p/o-nph-β-D-Glc*p* (etc), *para/orto*-nitrophenyl-β-D-Glucopyranoside (etc); Fuc*p*, Fucopyranose; Gal*p*, Galactopyranose; Xyl*p*, Xylopyranose.(This table was from Nucci et al. 1993)

CLONING AND EXPRESSION OF THE β-GLYCOSIDASE FROM *SULFOLOBUS SOLFATARICUS* IN *ESCHERICHIA COLI*

Because of the special properties described in the previous section, the Sβ-gly shows considerable biotechnological potentialities and represents a good model to search for general rules improving the enzyme stability. However, as for any other extremophile, the extreme growth conditions of *S. solfataricus* and its low biomass yields, represent serious adversities to obtain quantitative amounts of enzyme for basic-research studies and for industrial applications. Cloning and over-expression of thermophilic genes in mesophilic hosts is now a common tool to overcome (at least in part) these problems. However, the choice of a particular expression system depends on a variety of different parameters and on the objectives of the expression itself. We have expressed the Sβ-gly in *Escherichia coli* and in the yeast *Saccharomyces cerevisiae* and we have found both systems convenient for different applications. Expression in *E. coli* allows production of proteins essentially useful for basic-research by simple growing in laboratory scale. Expression in yeast shows noticeable advantages for the industrial production of the enzyme.

The gene coding for the Sβ-gly was named *lacS* (Cubellis *et al.*, 1990) for historical reasons, since the enzyme was first described for its β-galactosidase activity. However, the *lacS* gene fails to show significant sequence homology with the *Escherichia coli lacZ* gene, while it shows striking homology with the glycosyl hydrolases family I (Henrissat, 1991) including several enzymes both from eubacteria and eukaryotes. Member of his family is the human enzyme lactase/phlorizin hydrolase whose deficiency is responsible for the genetic disturbance hypolactasia.

The over-expression of the *lacS* gene in *E.coli* was obtained cloning it in a plasmid under the control of the β10 gene promoter of the T7 bacteriophage. When this recombinant vector is introduced in the appropriate *E. coli* strain the gene is transcribed, after induction with IPTG, by the T7 RNA polymerase present in the host genome. Consequently the enzyme was over-expressed in its functional form and the purification procedure resulted much simpler and faster. In fact, crude extracts were simply heated at 80°C for 20 min, the denatured proteins were eliminated by centrifugation and all the Sβ-gly was recovered in the supernatant. This treatment allowed also the complete elimination of the thermally unstable endogenous β-galactosidase activity. Subsequent conventional chromatography gave a purified protein with the same specific activity than the enzyme from *S. solfataricus* but with better final purification fold and yield. A typical purification gave more than 1.5 mg/lt culture of homogeneous enzyme which represent 100-150 β-gal units/gr of cell in the thermoprecipitated fraction. As a comparison, in *S. solfataricus* crude extracts normally 15-20 β-gal units/gr were found. Hence, *lac S* gene is expressed in *E. coli* almost ten times more than in *S. solfataricus*.

This thermoprecipitation procedure allows the use of the Sβ-gly for all the applications where it is not required the protein in homogeneous form. In fact, after only one simple and fast purification step the enzyme shows already high specific activity (34 units/mg at 75°C) and all the other contaminating activities have been heat-inactivated.

The Sβ-gly expressed in *E. coli* has been extensively characterised structurally and kinetically, in order to evaluate whether it maintains the features of its native form (Moracci *et al.*, manuscript in preparation).

EXPRESSION OF THE β-GLYCOSIDASE FROM *SULFOLOBUS SOLFATARICUS* IN THE YEAST *SACCHAROMYCES CEREVISIAE*

Recently, expression in yeast became popular expecially for the large scale production of proteins from eukaryotic source which require post-translational modifications and the secretion in culture medium. Nevertheless, yeast shows noticeable advantages also for the production of protein with applicative potentialities, in fact it allows high biomass production and the use of fed-batch systems. Furthermore, since yeast is a GRAS (Generally Recognised As Safe) organism, it represents an ideal pilot system for the production of enzymes for food technology industry. Hence, the expression in the yeast *Saccharomyces cerevisiae* of the β-glycosidase from *Sulfolobus solfataricus* was performed in order to obtain large amount of stable biocatalyst befitting for food industry processes.

The functional expression of the enzyme was obtained using a plasmid (pYE87) in which the *lacS* gene was under the control of a yeast galactose-inducible promoter (UAS_{GAL}) (Moracci *et al.*, 1992). Furthermore, to evaluate the use of heat-inducible promoters for the expression of biotechnological enzymes, we test if a sequence, called poly-HSE (Heat Shock Element) could be an appropriate expression tool for our thermostable β-glycosidase. This sequence is capable of mimicking a heat-inducible Upstream Activating Sequence (UAS) in *S. cerevisiae*. From a plasmid in which UAS_{GAL} is substituted by the poly-HSE, we obtained the expression of the gene. Consequently, Sβ-gly was expressed in yeast from both these promoters, the classical galactose inducible promoter and a new heterologous heat-inducible promoter.

The expressed protein was purified and partially characterised: it was the first protein from an extreme thermophile expressed in functional form in yeast and showing the same thermostability and thermophilicity.

The amount of protein produced using both promoters is comparable and comprises about 1% of total yeast proteins. However, this level of expression is similar to that found in *S. solfataricus* itself (15-20 units/gr. cells) and we have evidence that some post-transcriptional event prevents higher expression of the enzyme (Moracci *et al.*, 1992).

The poly-HSE is convenient because the induction is obtained just by temperature shift and does not require a particular growth medium, while induction of the GAL promoter occurs in expensive media containing galactose as the sole carbon source, on which cells grow slowly. Furthermore the poly-HSE driven induction is very rapid and efficient if compared with galactose (45 min. versus 24 hours). However, poly-HSE is not as tightly controlled as the GAL promoter, since the basal level of protein produced at room temperature is relatively high. This aspect could represent a problem for expression of proteins toxic for the cell, but this is not the case for thermophilic proteins inactive at low temperatures.

In the following sections are described the strategies for the scaling up of the Sβ-gly production and for the optimisation of its expression in yeast.

FERMENTATION AND AUTOLYSIS OF RECOMBINANT YEAST

To improve the yields of the expression in yeast of Sβ-gly we optimised the growth conditions of the micro-organism. The scaling up experiment for Sβ-gly expression was performed growing a yeast strain containing the plasmid pYE87, in a 100 lt fermentor containing rich medium with galactose as inductor. During the fermentation we followed the β-galactosidase activity to arrest the growth at the highest level of enzyme expression. The growth was followed by monitoring the optical density of the suspension at 540 nm and the enzyme expression assaying the β-galactosidase activity on whole cells. As shown (Fig 1), there is an increase of the enzyme production during the growth, and the maximum is reached in the late stationary phase.

The recovery of the thermophilic expressed enzyme from the biomass was easily achieved by autolysis. When the yeast culture was incubated at 37°C, the enzyme was slowly released into the medium by autolysis of the cells reaching a maximal yield after 72 hours with a high specific activity (Fig. 2A). Interestingly, the protein concentration first increased

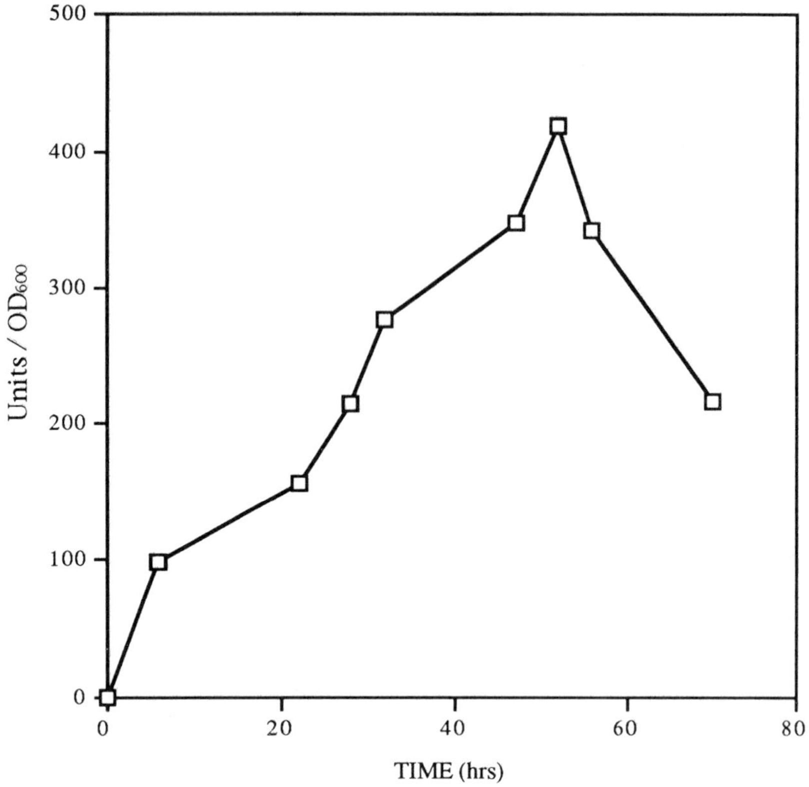

Figure 1. Sβ-gly expression during the growth of *Saccharomyces cerevisiae.* β-Galactosidase assays were performed on whole cells.

with the time and after 72 hours decreased up to 1/5 of maximum reached value. This phenomenon could be explained by the selective action of endogenous proteases on the host proteins and by the Sβ-gly resistance to the hydrolysis.

Since yeast autolysis can be improved by treatment with detergents which accelerates cells permeabilization, we tested the release of the Sβ-gly at 37°C in presence of Triton X-100 that was reported to be one of the most efficient. As shown in Fig. 2B, though the enzyme was released faster, after 72 hours its specific activity was 6-10 times lower if compared to autolysis without the detergent. Probably, yeast proteases are less active in the presence of the detergent. Furthermore, the optimal autolysis temperature was confirmed to be 37°C since at higher and lower temperatures the release of β-glycosidase resulted much slower.

Tab. III summarises the comparison of β-glycosidase production from *S. solfataricus* and recombinant *S. cerevisiae*. Although the activity of the enzyme for gr of cells is not very different, the enzymatic volumetric yield, in yeast, is about 35-fold higher than that obtained from the thermophilic microorganism.

We believe that the strategy described here could be adopted for the recovery of other thermophilic proteins expressed in yeast as it could be employed on a large scale plan, it is easily performed and it gives extracts at an higher degree of purity as compared to conventional crude extracts. Nevertheless, further improvement in yield could be achieved by increasing the number of cells which maintain the plasmid in the culture and by secreting the enzyme in the culture medium.

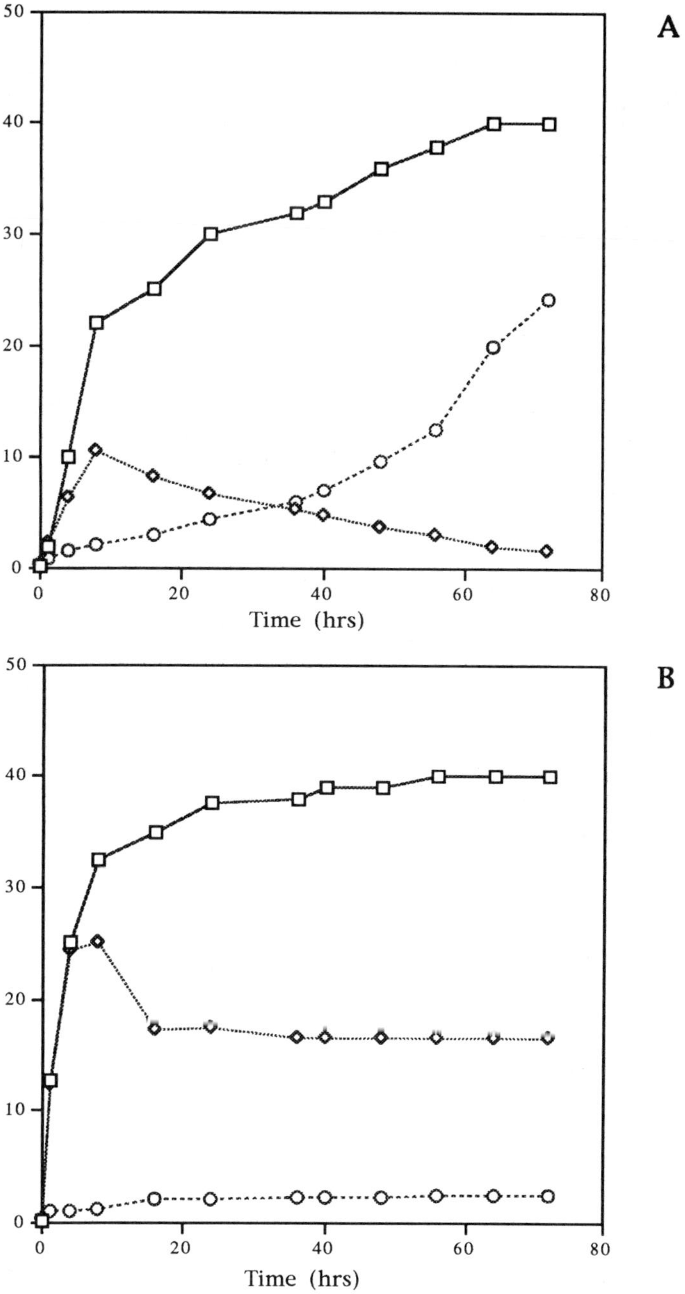

Figure 2. *S. cerevisiae* cells autolysis at 37°C in the absence of detergent (A) and in the presence of Triton X-100 (B). Sβ-gly units/g (—); total proteins (×—×); Sβ-gly specific activity (o—o).

Table 3. β-glycosidase production from *S. solfataricus* and *S. cerevisiae*

	S.solfataricus	*S. cerevisiae*
Fermentation conditions°		
Temperature	87	30
pH	3.5	4.8
Time (hrs)	96	50
Yield		
Biomass (g/l)	1.5	13
Enzyme (units/g)*	10	40
Enzyme (units/l)**	15	520

°In 100 l fermentor
*Enzymatic activity for g of wet cells
**Enzymatic activity in 1.0 l of homogenate or autolysate

EXPRESSION OF Sβ-GLY BY AUTOSELECTING PLASMID

Important factors affecting the expression of cloned heterologous genes, besides trascription and translation efficiency, are plasmid stability and copy number. The expression vector pYE87, contains the origin of replication and regulatory sequences of the 2μ circle, the most suitable and versatile expression vector in *S. cerevisiae* since it is relatively stable and maintained at high copy number (from 30 to 200 copies/cell). However, the stability of such plasmids is reasonably high only if cells are maintained under selective pressure (Futcher, 1988). During continous growth under non-selective conditions, plasmid stability has been found to drop significantly and the fraction of plasmidless cells increases exponentially with time. As a consequence, this expression system is difficult to scale-up as minimal media are not suitable for production volumes growth for the high costs and low biomass yield.

In order to improve the Sβ-gly production and to develop an expression system genetically stable on industrial scale, allowing the use of rich, economical media, eventually deriving from the recycling of industrial refuse, we focused our attention on autoselection systems. Here, the expression vector harbors a gene essential in any growth condition that complements a lethal mutation in the host; the vector is therefore maintained independently of the medium. Few autoselection systems have been already developed in yeast, for our aims we used a novel autoselection system based on the FBA1 gene, coding for the glycolitic enzyme fructose-1, 6-bisphosphate aldolase, that is essential in any growth condition (Compagno *et al.*, 1993). In this system, a yeast strain lacking the chromosomal copy of the FBA1 gene is transformed with the expression vector which contains a functional copy of the same gene. Hence, only cells retaining the vector are viable and the plasmid is completely stable even after 50 generations in rich medium. We have constructed an autoselection-expression vector named pYG5 containing a functional copy of the FBA1 gene and the *S. solfataricus lacS* gene under the control of the galactose-inducible UAS_{GAL}. pYG5 was transformed into the yeast strain WA6, that contains the chromosomal copy of the FBA1 gene disrupted, consequently, the only functional copy of the FBA1 gene was carried by the vector pYG5. This new yeast strain was selected and used for stability tests and expression experiments.

In order to measure the plasmid stability, cultures of the strains WA6/pYG5 and as control, KY117/pYE87, were grown in media, either rich or minimal, containing either glucose or galactose as carbon source. After 65 hours growth, cells were plated on glucose rich and minimal media with the appropriate supplements. In Fig. 3 the stability of the strains

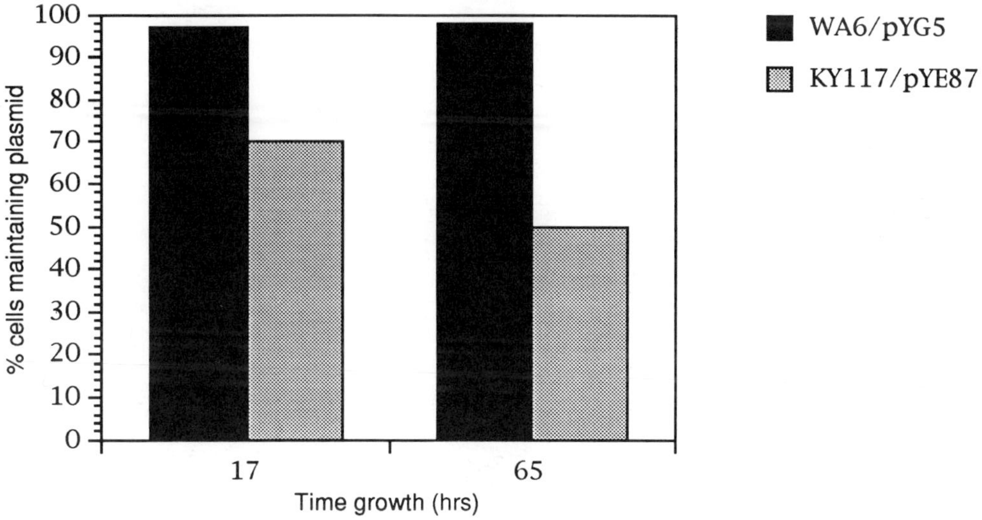

Figure 3. Comparison of the Sβ-gly expression-plasmids stability in *S. cerevisiae* using rich medium supplemented with galactose.

in galactose containing rich medium is reported. As expected, plasmid pYE87 showed an intrinsic instability even in minimal media (data not shown). The instability in minimal medium is the effect, at the cell division, of the accumulation in the daughter cells of enough product of several auxotrophic markers able to support the growth for several generations. Consequently, using this conventional vector this phenomenon is much more evident in rich

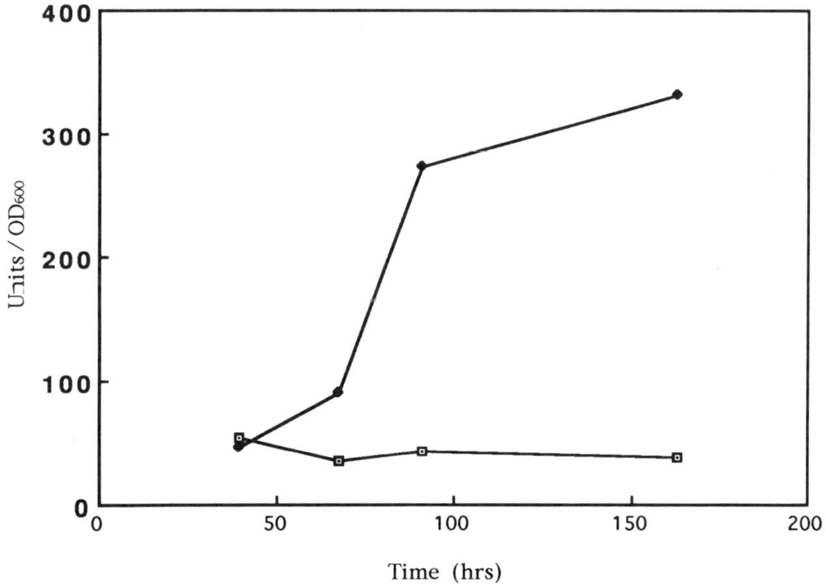

Figure 4. Sβ-gly expression by autoselecting strain pYG5/WA6 (_—_) and conventional strain pYE87/KY117 (—) in *S. cerevisiae* . β-Galactosidase assays were performed on whole cells.

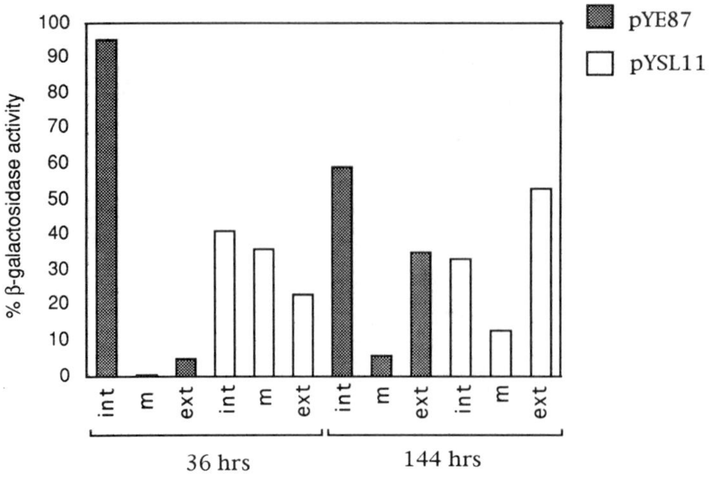

Figure 5. Sβ-gly expression in *S. cerevisiae* driven by secretion (pYSL11) and conventional (pYE87) expressing vectors. *Int*; *m*; and *ext* notations represent intracellular, membrane and extracellular fractions, respectively.

media (Fig. 3), where the amount of cells in the culture that have lost the plasmid increases with time and plasmid is lost at high frequency.

The stability of the autoselection-expression plasmid is significatively higher: almost 100% of the colonies contain the plasmid after 65 hours of growth in rich medium (Fig. 3).

Cultures grown in liquid minimal medium containing galactose were assayed for thermophilic β-galactosidase activity in order to follow the Sβ-gly expression. Assays were performed on whole permeabilized cells (Fig 4). Interestingly, the enzyme activity of the strain WA6/pYG5 is at least 5 fold higher than that obtained in the strain KY117/pYE87.

SECRETION OF THE Sβ-GLY IN THE CULTURE MEDIUM

Fundamental research efforts on the study of the mechanism of secretion in *S. cerevisiae* greatly improved in recent years especially for the noticeable advantages offered by protein secretion in applied science. In fact, protein secretion represents a unique tool for the expression of molecules unstable, toxic for the host or requiring for their biological activity to cross the eukaryotic post-translational pathway. Furthermore, secretion in yeast allows the recovery of the product directly from the culture medium, greatly improving the purification procedure. A further advantage offered by the secretion of the Sβ-gly in *S. cerevisiae* is the easy recovery of the enzyme by simple heating of the medium. In our experiments we used a secretion-expression vector called pYSL11 derived from YEpsec vector (Baldari *et al.*, 1987) where *lac S* gene is cloned in frame with the yeast *Kluvyromyces lactis* killer toxin and under the control of UAS_{GAL} promoter. This vector was used to transform *S. cerevisiae* strain KY117 and the presence of Sβ-gly in intracellular, membrane and extracellular fractions was monitored following its β-galactosidase activity. Thermophilic and thermostable enzymatic activity could be detected in all fractions after 36 hours growth. In contrast, as expected, cells harboring the expression vector pYE87 presented the large majority of the β-gal activity (95%) in the intracellular fraction (Fig 5). To test if the secretion was a time-dependent process, β-gal activity was monitored after 144 hours of growth. This experiment could be performed assuming that Sβ-gly is resistant to protease action. As expected, not only the enzyme resulted stable, but also its concentration resulted significantly decreased in the periplasmic space and correspondingly increased in the medium.

This evidence confirms that the presence in the medium of Sβ-gly, although not highly efficient, is due to an active, cell-driven process and it is not the effect of aspecific cell lysis phenomena. Furthermore, the secretion by yeast cell of oligomeric proteins is not common and other enzymes, like the β-galactosidase of *E. coli,* are retained in the periplasmic space.

CONCLUSIONS

This review describes a series of strategies used to optimise the expression of a thermophilic enzyme with relevant biotechnological potentialities in mesophilic hosts. We demonstrated that it is possible to express in functional form large quantities of the β-glycosidase from *S. solfataricus* either in *E. coli* or in *S. cerevisiae*. The use of yeast as expression host, allowed the easy scale-up of the cell culture and the optimisation of the expression. We are confident that the development of the fermentation techniques, the plasmid maintainance and the secretion will allow a decisive improvement of the applicability of Sβ-gly.

ACKNOWLEDGEMENTS

This work was supported by CNR Target Project Biotechnology and Bioinstrumentation, by CNR Progetto Finalizzato Ingegneria Genetica and by EC project: "Biotechnology of extremophiles" contract n. BIO2-CT93-0274. L.C. was supported by a CNR-MISM Fellowship of the Istituto di Scienze dell'Alimentazione CNR-Avellino.

REFERENCES

Baldari, C., Murray, J.A.H., Ghiara, P., Cesareni, G., and Galeotti, C.L., 1987, *EMBO J.* 6:229-234.

Compagno, C., Tura, A., Ranzi, B.M., and Martegani, E., 1993, *Biotechnol. Progr.* 9:594-599.

Cowan, D.A., 1992, In Molecular Biology and Biotechnology of Extremophiles, R.A. Herbert, R.J. Sharp ed., Chapman and Hall, New York, 1-43.

Cubellis, M.V., Rozzo, C., Montecucchi, P., and Rossi, M., 1990, *Gene* 94:89-94.

Futcher, A.B., 1988, *Yeast* 4:27-40.

Henrissat, B., 1991, *Biochem. J.* 280:309-316.

Moracci, M., Ciaramella, M., Nucci, R., Pearl, L.H., Sanderson, I., Trincone, A., and Rossi, M., 1994, *Biocatalysis* 11:89-103.

Moracci, M., La Volpe, A., Pulitzer, J.F., Rossi, M., and Ciaramella, M., 1992, *J. Bacteriol.* 174:873-82.

Nucci, R., Moracci, M., Vaccaro, C., Vespa, N., and Rossi, M., 1993, *Biotechnol. Appl. Biochem.* 17:239-50.

Pearl, L.H., Hemmings, A.M., Nucci, R., and Rossi, R., 1993, *J. Mol. Biol.* 229:558-560.

Pisani, F.M., Rella, R., Raia, C.A., Rozzo, C., Nucci, R., Gambacorta, A., De Rosa, M., and Rossi, M., 1990, *Eur. J. Biochem.* 187:321-328.

Trincone, A., Nicolaus, B., Lama, L., and Gambacorta, A., 1991a, *J. Chem. Soc. Perkin Trans.* 1:2841-2844.

Trincone, A., Nicolaus, B., Lama, L., Morzillo, P., De Rosa, M., and Gambacorta, A., 1991b, *Biotechnol. Lett.* 13:235-240.

Woese, C.R., Kandler, O., and Wheelis, M.L., 1990, *Proc. Natl. Acad. Sci. USA* 87:4576-4579.

6

PROTEIN ENGINEERING OF ALBEBETIN, A *DE NOVO* PROTEIN WITH A PREDESIGNED THREE-DIMENSIONAL STRUCTURE

D. A. Dolgikh,[1] A. N. Fedorov,[2] A. V. Finkelstein,[2] O. B. Ptitsyn,[2]
K. G. Skryabin,[3] and M. P. Kirpichnikov[1,3]

[1] Institute of Molecular Biology, Russian Academy of Sciences
117984 Moscow, Russia
[2] Institute of Protein Research, Russian Academy of Sciences
142292 Pushchino, Moscow Region, Russia
[3] Centre "Bioengineering", Russian Academy of Sciences
117984 Moscow, Russia

INTRODUCTION

The design of *de novo* proteins which ensures the desirable three-dimensional (3-D) structures is the best way to test our understanding of the principles of protein structure and protein folding. The first successful engineering of a *de novo* protein, namely four-helix bundle, has been reported seven years ago (Regan and DeGrado, 1988) and then a number of attempts have been made in this direction during the past years (for a recent review see Sander, 1994). Most of the designed structures mimicked the 3-D structures of some natural proteins. In 1988 four of us have reported a design of a *de novo* protein albebetin (Figure 1) with a predetermined 3-D fold consisting of a four-stranded antiparallel β-sheet with one side screened by two α-helices (Finkelstein et al., 1988). This fold does not contradict any structural rule but has been never observed up to now in natural proteins. It consists of two identical αββ units, therefore we have called our *de novo* protein albebetin. We have expressed albebetin in a cell-free translational system and in *E.coli* and investigated it by various physical and chemical techniques (Dolgikh et al., 1991, 1992; Fedorov et al., 1992; Chemeris et al., 1994). It has been shown that the *de novo* protein is relatively stable and nearly as compact as natural proteins, it unfolds cooperatively at high urea concentrations and possess structural features of a definite structure consistent with the predesigned one. An eight amino acids biologically active human interferon fragment was attached to albebetin in order to obtain a *de novo* protein with a grafted biological function (Dolgikh et al., 1993).

This paper summarizes the results of our work on design, expression and investigation of albebetin and its derivative incorporating a grafted biological activity.

Molecular Manufacturing, edited by C. Nicolini
Plenum Press, New York, 1996

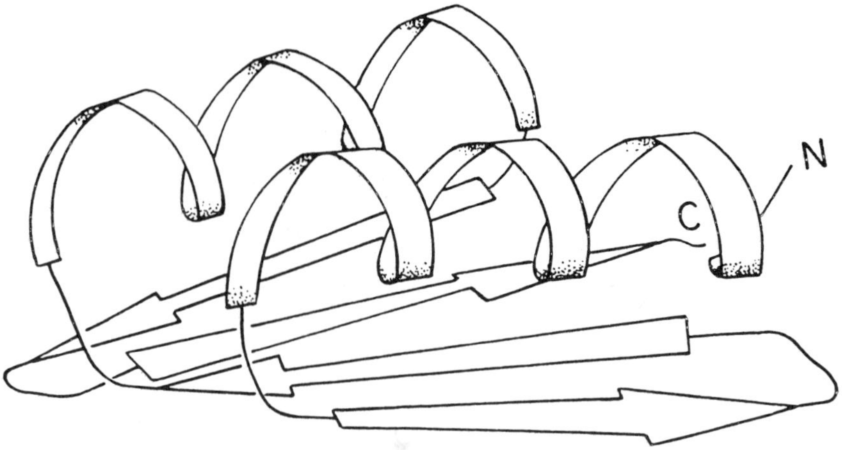

Figure 1. The predesigned three-dimensional structure of the *de novo* protein albebetin.

ALBEBETIN DESIGN

Design Principles

A small protein with a given secondary structure has a rather limited set of potentially stable folding patterns which do not contradict some general physical rules (Finkelstein & Ptitsyn, 1987). This set of folds is especially narrow when the loops are short. Then we can hope to provide a predominant stability of one of them, even though our estimates of stabilities are rather crude. Therefore the designed fold has a pronounced secondary structure and short loops (see Figure 1). To stabilize it, we have to design such an amino acid sequence that (i) the desired secondary structure is so stable intrinsically that any fold with another secondary structure cannot be stable; (ii) the secondary structure segments have the hydrophobic surfaces which can form a hydrophobic core; (iii) the desired tertiary fold is more stable than any other one with the same secondary structure. Moreover, the surface of the designed protein must be very hydrophilic to prevent its aggregation.

Design of Secondary Structure

α-Helices and β-regions have been designed on the basis of the physical theory of protein secondary structure (Ptitsyn and Finkelstein, 1980, 1983; Finkelstein, 1983; Finkelstein et al., 1991) This theory includes a set of stability parameters determined by the local interactions in protein chains (Finkelstein et al., 1991) or by the interactions of α- and β-regions with the rest of the protein (Ptitsyn and Finkelstein, 1983). All these data were incorporated into the computer program ALB (Finkelstein, 1983) which has been used for practical computations.

To ensure the formation of four β-strands and two α-helices with continuous non-polar surfaces in proper chain regions we inserted valines in i, i+2, i+4 positions in β-regions and leucine in i, i+3, i+4, i+7 in α-regions. In addition we put aspartic and glutamic acids, as well as prolines, near the N-termini of α-helices while arginines and glycines were put near their C-termini. α-Helices and β-strands were designed to have the same length of 1.8 nm and α- and β-regions were separated by α- and β-breaking glycines and prolines.

These connections had minimal lengths in order to minimize the number of possible structures.

Design of Tertiary Structure

Analysis of all possible topologies of a protein consisting of two α-helices and four β-strands has been described earlier (see Chemeris et al., 1994). It has been shown that the predesigned topology (Figure 1) of albebetin is more energetically favorable than the other ones if the protein has only short connections between the elements of regular secondary structure and the strands β_2 and β_4 are inside the antiparallel β-sheet. To ensure this topology the strands β_1 and β_3 were designed to be more polar than β_2 and β_4. To this end histidine residues of β_2 and β_4, which are almost uncharged at neutral pH, were replaced in β_1 and β_3 by negatively charged glutamic acids. The polar character of the protein surface were checked carefully in order to prevent protein from aggregation and the density of packing in albebetin globule was analyzed using CPK space-filling models and then by computer graphics, using the program "Insight II" by Biosym.

```
           EcoRI       BamHI                                     PvuII
GATCGAATTC ATG GAT CCG GGC GAC CCA GAA TGC CTG GAG CAG CTG CTG
                                                  10
           Met Asp Pro Gly Asp Pro Glu Cys Leu Glu Gln Leu Leu
                                                    α

                                                      KpnI
CGT CGC CTG GGC GGT TCC GTA GAA GTT GAA GTG ACT GGT GGT ACC GTA
                                20
Arg Arg Leu Gly Gly Ser Val Glu Val Glu Val Thr Gly Gly Thr Val
                                β

                    BspMII
CAC GTT GAA GTG TCT CCG GAA GAT CCG GGC GAC CCA GAA TGC CTG GAG
30                                      40
His Val Glu Val Ser Pro Glu Asp Pro Gly Asp Pro Glu Cys Leu Glu
    β                                                       α

PvuII
CAG CTG CTG CGT CGC CTG GGC GGT TCC GTA GAA GTT GAA GTG ACT GGT
                 50                                     60
Gln Leu Leu Arg Arg Leu Gly Gly Ser Val Glu Val Glu Val Thr Gly
                                            β

KpnI                          BspMII                    HindIII
GGT ACC GTA CAC GTT GAA GTG TCT CCG GAA GAT CGT TAG TAA AAGCTT
                                70
Gly Thr Val His Val Glu Val Ser Pro Glu Asp Arg Stop
                β
```

Figure 2. Amino acid sequence of albebetin and the nucleotide sequence of the corresponding gene. The regions with α-helices and β-structure and restriction sites are indicated.

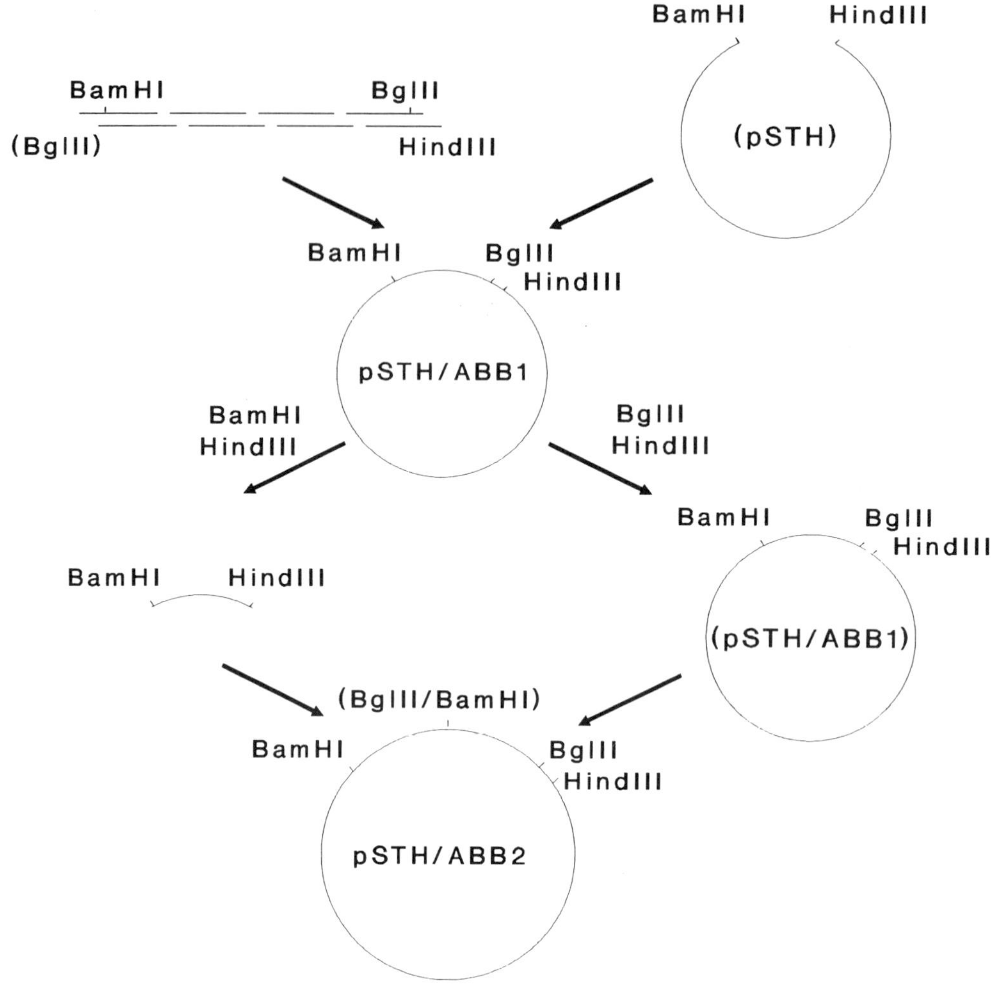

Figure 3. Cloning of the albebetin gene (see text).

Gene Design

A designed amino acid sequence of albebetin and nucleotide sequence of the corresponding gene is shown in Figure 2. A nucleotide sequence of the gene has been obtained from the amino acid one taking into account the codon usage frequencies in *E.coli*. Several restriction sites were introduced in the sequence in order to simplify the cloning procedure and further modifications of the gene. The albebetin structure consists of two repeated parts $\alpha\beta\beta$, therefore we decided to sinthesize only half of its gene and then to clone and duplicate it in a plasmid. The scheme used for obtaining the albebetin gene is shown in Figure 3. Eight oligonucleotides (31 to 37 bases long) that correspond to the direct and reverse chain of the half of the gene were synthesized and phosphorylated (excluding the 5′ end in the direct and reverse chain), mixed and annealed. The resultant four double-stranded fragments have overlapping sticky ends formed of 6 nucleotides. The first fragment begins with a protruding 5′ end GATC, which is complementary to the sticky end of BglII recognition site, whereas the last fragment ends with a sticky HindIII site. Plasmid pSTH that carries the growth hormone gene was used as vector, and the fragments were cloned into the mentioned restriction sites in the region corresponding to the C-terminal part of the growth hormone. Half of the albebetin gene was excised from the BamHI and HindIII sites of the obtained plasmid pSTH/ABB1 (Figure 3) and cloned into the sites BglII (with sticky end similar to that of BamHI site) and HindIII. As a result, plasmid pSTH/ABB2 bearing the complete albebetin gene was obtained. Further on the gene was modified in such a manner that the C-terminal leucine residue was replaced with Arg in order to increase the hydrophilicity of the protein and to prevent the possibility of its aggregation. For this purpose, a short double-stranded fragment consisting of two specially synthesized 15 bases long oligonucleotides with appropriate sticky ends and the proper substitution was inserted into BglII and HindIII sites of the gene. The modified albebetin gene does not have a BglII restriction site and this final variant was shown in Figure 2.

For albebetin expression in *E.coli* the albebetin gene has been subcloned into pET8c plasmid of a T7 promoter/expression system (Studier et al., 1990). We have used also the regulatory region of pSTH plasmid (tryptophan promoter) eliminating the growth hormone coding part from pSTH/ABB2 plasmid. In both cases we could not detect any reasonable expression of the protein. Therefore we decided to express albebetin in vitro in a cell-free translation system and elaborated a new approach to its structure testing. Later on we expressed albebetin also as a fusion with maltose-binding protein (see below). For expression of albebetin in a cell-free translational system the gene was cloned into EcoRI-HindIII sites of pSPT19 plasmid (Pharmacia) under control of the bacteriophage SP6 RNA polymerase promoter. The resulted plasmid was called pSPT/ABB3 and it was used for expression and for construction of the gene of modified albebetin variant carrying an interferon fragment.

ALBEBETIN EXPRESSION AND INVESTIGATION

Expression *In Vitro*

The cell-free wheat germ translation system has been optimized for the most efficient translation of the mRNA coding for albebetin. mRNA was transcripted from the DNA template by SP6 polymerase according to the general procedure (Melton et al., 1984) with modifications described in (Gurevich et al., 1991). Translation of the mRNA was done using the plasmid pSPT/ABB3 during 1 hour at 25 C in the cell-free wheat germ translation system in a total volume of 25 ml according to (Erickson and Blobel, 1983) with some modifications (S.Yu. Ovodov et al., unpublished data). $[^3H]$ Leucine was used as a specific radiolabel and the radiolabeled polypeptide was revealed by fluorography. The radioactivity incorporated into the polypeptide was counted after hot trichloracetic acid hydrolysis followed by precipitation. The obtained polypeptide has a molecular weight of 8 kDa according to SDS gel electrophoresis, which is close to the value 7.756 kDa calculated from the amino acid content.

A New Approach to Structure Testing

For structure testing of albebetin we elaborated a new approach that uses only nanogram amounts of a radiolabeled protein synthesized in a cell-free translation system without its previous purification. In this approach, radio labels are used not just to check the presence of the *de novo* protein, but also to probe protein structure by a number of separating techniques (like size-exclusion chromatography or urea-gradient electrophoresis) which do not demand isolation of *de novo* (radiolabeled) protein from the other (not labeled) ones. We suppose that such an approach provides a very effective method of preliminary study of

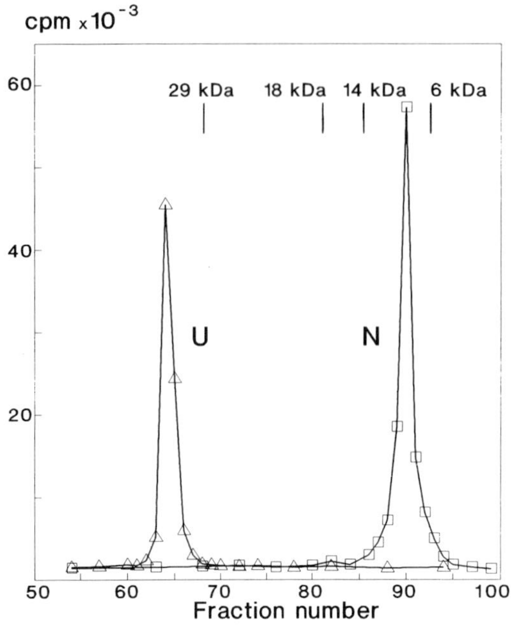

Figure 4. FPLC chromatography of albebetin in 20 mM HEPES, 200mM potassium acetate buffer, pH 7.6 ("native" condition, N) and in the same buffer containing 9 M urea (U). The positions of reference proteins and their molecular masses (kDa) are indicated.

artificial and mutant proteins which permits to find out whether a protein has a cooperative compact structure and decide whether it is worthwhile to synthesize it in preparative amounts.

Structural Properties of Albebetin

In order to characterize the compactness of the synthesized polypeptide, the total translation mixture was subjected to size-exclusion chromatography (Ackers, 1970) in "native" and "denaturing" conditions. The term "native" means that the translation system was dissolved by neutral (pH 7.6) buffer at room temperature without added denaturants. The data (Figure 4) show that the elution volume of albebetin under these conditions corresponds to the elution volume of a native globular protein with molecular weight 8.5 kDa. At the same time its elution volume in 9M urea is much smaller. This indicates that albebetin is monomeric under these conditions and that it is nearly as compact as native globular proteins. These changes in the elution volume can be interpreted in a terms of protein hydrodynamic Stokes radius (Ackers, 1970; Uversky, 1993). The increase of urea concentration from 0 to 9 M results in increase of Stokes radius of albebetin from 1.7 nm to 2.4 nm which is close to the value which can be expected for a complete unfolding of natural proteins with a similar molecular weights (Uversky, 1993, Uversky and Ptitsyn, 1994).

Another test for the compactness of a protein molecule is its stability against proteolysis (Lang and Schmid, 1986). In order to study albebetin with this technique we included two sites for trypsinolysis into the designed amino acid sequence, the arginine tandems 14-15 and 49-50 (see Figure 2). We have compared the kinetics of albebetin trypsinolysis at native conditions and in the presence of 4 M urea. The results (see Fedorov et al., 1992) show that the kinetics of albebetin digestion in native conditions is substantially (by about an order of a magnitude) lower than that in 4 M urea. The real difference in stability between "native" and unfolded albebetin should be at least two orders of magnitude, as albebetin in 4 M urea is at the very beginning of its unfolding and trypsin is twice less active in 4 M urea than in native conditions (Harris, 1956).

To check how stable is a compact structure of albebetin and how cooperatively is it destroyed, we have used electrophoresis in a horizontal urea gradient (Creighton, 1980) which is one of the most visual methods to show the cooperativity of urea-induced unfolding of proteins. The results for unfolding of albebetin have been described earlier (Fedorov et al., 1992). It has been shown that the transition curve has sigmoidal shape, typical for a cooperative processes and the protein is stable against urea-induced unfolding: the midpoint of transition is about 5.5 M, and the unfolding is not complete even at 8 M urea. These data show that the stability of albebetin against unfolding by urea is typical for the normal globular proteins (cf. Creighton, 1979) and the cooperativity of transition is also normal for such a molecular weight.

The experiments described above show that albebetin is compact at "native" conditions and can be cooperatively unfolded by urea. However, these experiments cannot distinguish between a compact but shapeless molecules and the molecules which have some fixed features of the 3-D structure. To obtain at least a preliminary answer to this question, we have studied the efficiency and the order of proteolytic attack on different target residues in albebetin at "native" conditions. As it was mentioned above, albebetin has two sites for trypsinolysis (see Figure 2) - the arginine tandems 14-15 and 49-50. Theoretically, this must lead to appearance of the following fragments: 1-13, 1-48, 16-48, 16-73 and 51-73 under a deep trypsin attack. The accumulation of all these fragments was really observed (see Fedorov et al., 1992). The analysis of degradation kinetics of large fragments (1-48 and 16-73) shows a significant difference in sensitivity of the arginine sites to proteolysis (Fedorov et al., 1992). As the sequences around the arginine tandems at these two fragments are identical, this difference in sensitivity can be attributed only to the difference in

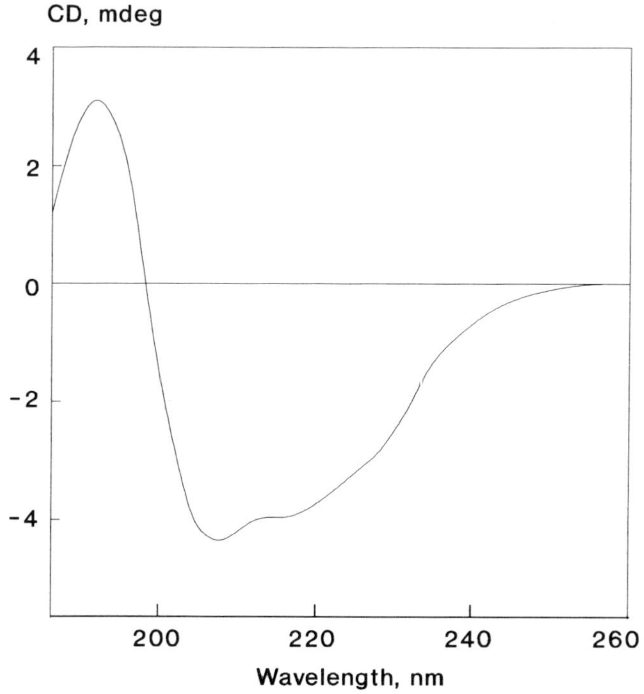

Figure 5. Circular dichroism spectrum of albebetin.

surrounding of these sites in space, i.e. to the existence of some 3-D structure in albebetin and even in its large fragments.

All the results described above were obtained using only nanogram amounts of a radiolabeled albebetin synthesized in a cell-free translation system without its previous purification. Recently we expressed the protein in *E.coli* as a fusion protein with maltose binding protein in order to test albebetin structure in more detail. The albebetin gene was cloned in plasmid pMALc (New England Biolabs) after the N-terminal part of the maltose binding protein gene. Expression, purification and cleavage of the fusion protein with factor Xa protease was performed according to the NEB manual as well as the final affine protein purification using the FPLC column chromatography on amylose binding resin. The protein expressed in *E.coli* have been tested by SDS/polyacrylamide gel and urea-gradient electrophoresis, size-exclusion chromatography and revealed the same properties as albebetin obtained in a cell-free translational system.

We studied secondary structure of albebetin using circular dichroism (CD) spectroscopy in far ultraviolet region. The CD spectrum of albebetin is shown in Figure 5. The spectrum reflects a high content of secondary structure and is similar to the spectra of natural proteins containing α- and β-structures. The quantitative analysis of the spectrum made by us using the approach of Provencher and Gloekner (1981) has shown that albebetin contains 28% of α- and 40% of β-structure. These values are close to the predesigned ones (30% and 36%, respectively). This indicates that our design of the secondary structure was successful and one can suppose that tertiary structure of albebetin also corresponds to the predesigned one.

Thus the experiments shows that albebetin is compact and structured and possess secondary structure similar to the predesigned one. These data are not enough to conclude whether this *de novo* protein has a rigid tertiary structure typical for native natural proteins or it is in the molten globule state like most of *de novo* proteins designed up to now. The more detailed study like NMR or X-ray analysis can clarify the question. (NMR study of albebetin is in progress in the laboratory of Dr. C.Dobson, Oxford University.) Nevertheless the data obtained allowed us to modify albebetin basing on its predesigned structure, namely to introduce a biological activity into the *de novo* protein.

A *DE NOVO* PROTEIN WITH A GRAFTED BIOLOGICAL FUNCTION

Principles of Grafting

To transfer a biological activity to a *de novo* protein one can try to imitate a natural protein active center in the protein. However this is a rather difficult, especially in the case with a spatially arranged active center consisting of elements separated along the sequence (like enzyme catalytic sites) - the exact 3-D structure of a *de novo* protein is not known, and, as a consequence, it is impossible to reproduce the original spatial structure of such a center. On the other hand, the contiguous functionally important site possessing biological activity *per se* as a peptide, may be simply inserted into a *de novo* protein. It is only necessary to check whether the overall structure of chimerical protein is maintained and whether the introduced sequence is able to form the conformation required for a biological function.

There are two obvious ways for the active sites to be inserted into a protein: replacement and attachment. According to the first approach, one could find a protein sequence, which is most homologous (or, maybe, analogous - i.e. similar in the arrangement of hydrophobic/hydrophilic amino acids) to the active site of interest, and to insert it, replacing the inherent protein sequence. The closer the homology - the more is, in principle, the probability of successful design. Nevertheless it was shown that even the same sequences could acquire various conformations in different proteins (Argos, 1987). The second approach is to insert the active sequence into an exposed part of the molecule. In this case we do not substitute any protein sequence, assuming that the local conformational changes in the exposed site would not seriously disturb the protein structure and the inserted sequence would have enough flexibility to acquire the active conformation. The loop regions and the termini of the protein molecule are suitable for this kind of a design.

Biological Activity to Be Grafted to Albebetin

In search for a biological activity which could be transferred to albebetin we have used the following requirements. (i) The existence of an easy test for the detection of this biological activity. (ii) Since albebetin is an acidic protein, a basic functional fragment would be preferable. (iii) The active fragment should be relatively short not to disturb the albebetin structure. We have compared the amino acid sequence of albebetin with the database (Gabrielian et al., 1992) of more than 150 functionally important fragments of proteins described in the literature, but no satisfactory homology with any known functional site was found. As a result, the only suitable variant to construct chimerical albebetin with a biological activity was the attachment of an active fragment to the protein exposed site. The octapeptide LKEKKYSP (Leu-Lys-Glu-Lys-Lys-Tyr-Ser-Pro) was chosen as a candidate for such incorporation. This octapeptide (Zav'yalov et al., 1991), corresponding to the human interferon α_2 sequence 131-138, binds at a high affinity to murine thymocyte receptors and activates the thymocyte blast transformation at a concentration as low as 10^{-11} M. It is known that the fragment 131-138 is located in a loop region of the interferon molecule and has a considerably reduced affinity to the receptor in comparison to that of the peptide itself (Zav'yalov et al., 1991). On the other hand it has been shown that the receptor binding properties of this peptide disappear after addition of amino acids to its N-terminus and reduce after addition of Cys residue to the C-terminus (Zav'yalov et al., 1991).Taking this into account, we have chosen attachment of this octapeptide to albebetin N-terminus as the way of its incorporation into the protein molecule. It is obvious that the octapeptide LKEKKYSP fulfills all the requirements listed above: its binding to receptors is simply tested due to the high affinity and sensitive assay, it has basic character and is relatively short and highly hydrophilic thus we could anticipate that it would not interfere with the predetermined 3-D fold of the *de novo* protein.

Synthesis of Albebetin with an Incorporated Interferon Fragment and Its Structural Testing

The gene of albebetin with interferon fragment was obtained by introducing of a synthesized double-strand oligonucleotide fragment (32 bases in a strand) corresponding to the amino acid sequence LKEKKYSP into EcoRI-BamHI sites of the albebetin gene (see Figure 2). The interferon fragment was included into albebetin after its first methionine residue. Synthesis of chimerical albebetin was carried out in the cell-free expression system at conditions, described above for albebetin. The obtained polypeptide has a relative molecular mass of 9 kDa according to SDS/polyacrylamide gel electrophoresis which is close to the value 8.729 kDa calculated from the amino acid content. To check whether the protein is monomeric and compact we have studied it by size exclusion FPLC in "native" conditions and in the presence of 9 M urea. Using the calibrated gel filtration column we have determined the hydrodinamic dimensions (Stokes radius values) of the molecule. The

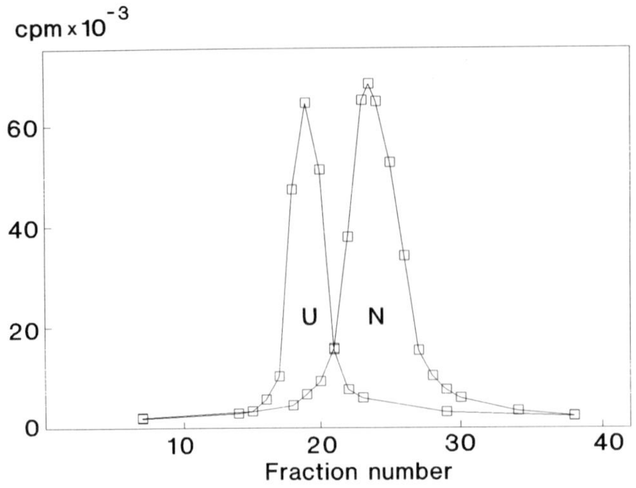

Figure 6. FPLC chromatography of albebetin with an incorporated interferon fragment in 20 mM HEPES, 200 mM potassium acetate buffer, pH 7.6 ("native" condition, N) and in the same buffer containing 9 M urea (U).

data (Figure 6) show that albebetin incorporating the interferon fragment is a monomeric protein and its Stokes radius in "native" conditions is considerably less than that in 9 M urea. The comparison of the Stokes radius obtained in "native" conditions with those for other globular proteins (Uversky, 1993) confirms that the chimerical protein is as compact as native globular proteins. It should be noted that its Stokes radius in "native" conditions (1.7 nm) does no exceed that for albebetin itself, i.e. according to this parameter the chimerical protein is at least as compact as albebetin. In 9 M urea the Stokes radius of albebetin incorporating the interferon fragment (2.5 nm) slightly exceeds that for albebetin in accordance with the molecular mass increase.

To check the stability of the chimerical protein we have studied its unfolding by urea using FPLC chromatography and urea-gradient electrophoresis. It has been shown (Corbett and Roche, 1984) that if the exchange rate between the native and the unfolded states of protein is slow in comparison with the characteristic time of the chromatography two well separated elution peaks (corresponding to compact and unfolded protein molecules) are observed within the transition region. This is the case for urea-induced unfolding of albebetin with an incorporated interferon fragment (data not shown). Figure 7 shows the dependence of the relative area of elution peak which corresponds to the unfolded molecules on urea concentration. The increase of the relative number of unfolded molecules (relative area of unfolded molecules peak) in the range of urea concentration from 3 to 7 M reflects cooperative unfolding of the chimerical protein and confirms that in the absence of urea this

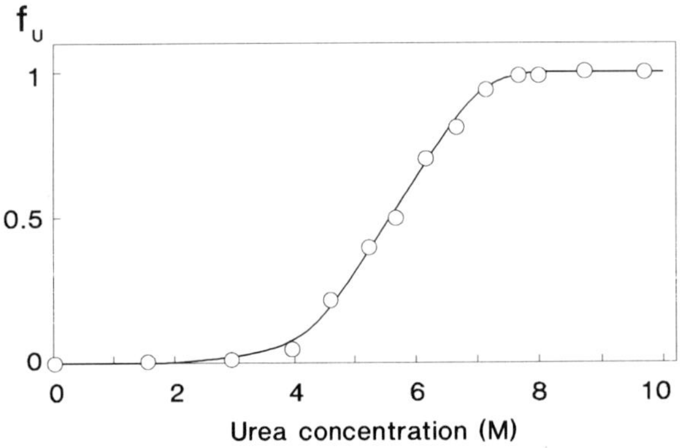

Figure 7. Urea-induced unfolding of albebetin with an incorporated interferon fragment. The unfolding curve was monitored by changes of the relative area fu of the FPLC elution peak which corresponds to the unfolded (non-compact) protein molecules.

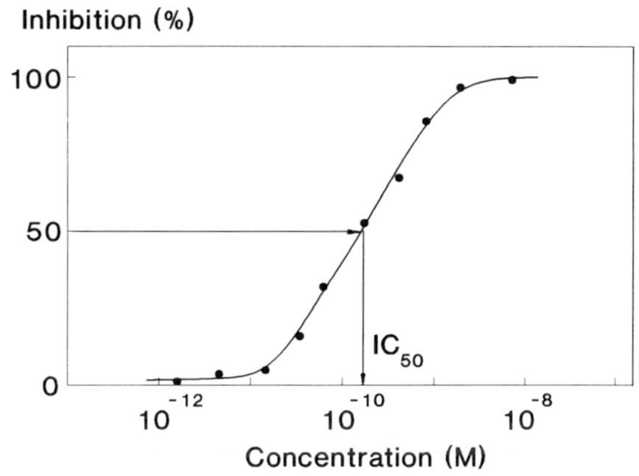

Figure 8. Inhibitory effect of albebetin with an incorporated interferon fragment on binding of ^{135}I-labeled interferon octapeptide (4.4 * 10^{-11}M) to the murine thymocyte. The arrow indicates the graphic mode of estimation of the IC_{50} value (see text).

protein has a compact structure. The similar result was also obtained by urea-gradient electrophoresis (data not shown). It should be noted that the chimerical protein stability against unfolding by urea is typical for the normal globular proteins and the cooperativity of the unfolding transition is also normal for such molecular mass (Uversky and Ptitsyn, 1994). According to these data we can suppose that the inserted interferon fragment does not disturb significantly predesigned three-dimensional structure of the *de novo* protein.

Biological Activity of Albebetin with an Incorporated Interferon Fragment

To testify the affinity of albebetin with an incorporated interferon fragment to murine thymocyte receptors we have studied the protein inhibitory effect on the binding of radiolabeled (^{135}I) octapeptide LKEKKYSP to the receptors. Earlier (Zav'yalov et al., 1991) Scatchard analysis of the binding revealed only one type of receptor for the octapeptide on the thymocyte cell (Scatchard plot represented a direct line). The data on the octapeptide binding inhibition are shown in Figure 8. One can see that the chimerical protein effectively competes with the octapeptide for thymocyte receptor. Using these date we have determined the value of IC_{50} (concentration of inhibitor which caused 50 per cent inhibition) and the inhibition constant according to Cheng and Prusoff, 1973. As it follows from Table 1, the chimerical protein has a high affinity to murine thymocyte receptor and binds it even more effectively than interferon itself.

The testing of the chimerical protein blast transforming activity confirms its high biological effect on the thymocyte blast transformation reaction. Figure 9 shows activation of this reaction by albebetin with an incorporated interferon fragment in comparison with

Table 1. Parameters of binding of albebetin with an incorporated interferon fragment, human interferon α_2 and the octapeptide interferon fragment LKEKKYSP to the thymocyte receptors

Polypeptide	K_d (M)	IC_{50} (M)	K_i (M)
Albebetin with interferon fragment	not determined	1.6 x 10^{-10}	1.4 x 10^{-11}
Human interferon $_2$	2.5 x 10^{-10}	9.9 x 10^{-9}	8.6 x 10^{-10}
Octapeptide LKEKKYSP	4.2 x 10^{-12}*	-	-

*Taken from (Zav'yalov et al.,1991)

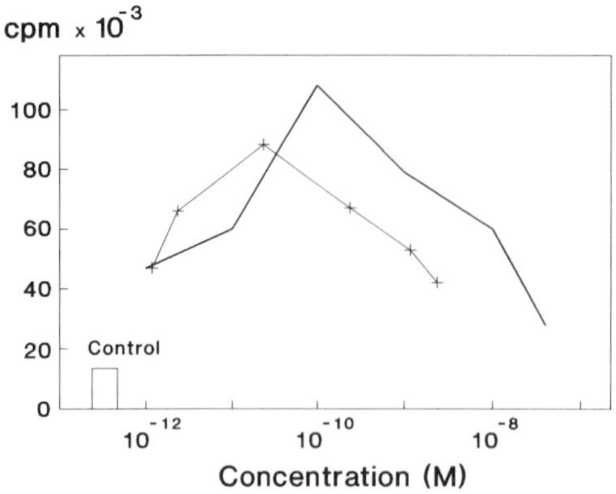

Figure 9. Activation of the murine thymocyte blast transformation by albebetin incorporating the interferon fragment (-+-) in comparison with that by human interferon α_2 (—). The radioactivity of [^{3}H] thymidine, absorbed by the thymocyte cells (cpm, counts per minute), is plotted as the ordinate and the protein molar concentration is plotted as the abscissa. The control value corresponds to the blast transformation reaction without adding of proteins.

that by human interferon α_2. One can see that albebetin with an incorporated interferon fragment induces the thymocyte blast transformation more efficiently than interferon at a concentration as low as 10^{-11}M.

Further experiments are required for studying in more details the structural properties of chimerical protein and its biological activity. However, the data presented allow us to assume that we have the *de novo* protein with the predetermined tertiary structure and biological activity. We think the *de novo* proteins may be used as the carriers of various biological activities. This work demonstrated a possibility to transfer the biological function mediated by short contiguous peptide to the *de novo* protein albebetin. It has been shown that contiguous amino acid sequences may imitate even those biological activities which are mediated by the spatially arranged functional sites and a technique was elaborated capable of searching for such sequences (see Wetzel, 1991). It includes incorporation of the random oligopeptides into the surface protein of filamentous phages and the functional sequences selection by their affinity to the specific receptors. This technique expands the sphere of possible biological activities which may be transferred from one protein to another including the *de novo* proteins.

ACKNOWLEDGMENTS

The design of a gene corresponding to the desired amino acid sequence has been performed by Dr. A.A. Shulga. Dr. B.K. Chernov have synthesized oligonucleotides for the gene. A substantial part of the beginning of this work has been done in the laboratory of Professor Yu.B.Alakhov. A part of the work on the cell-free expression and investigation of a radiolabeled albebetin has been performed by V.V.Chemeris and Dr. V.N.Uversky. Dr. G.Nyakatura has participated in albebetin expression in *E.coli*. Dr. A.E.Gabrielian has proposed a functionally active interferon fragment to introduce into albebetin and Drs. E.V.Navolotskaya and V.P.Zav'yalov have participated in analysis of the biological activity. Albebetin expression as a fusion protein and its structural testing has been performed in the laboratories of Dr.H.Bloecker and Professor D.Schomburg. We cordially thank all of them for their valuable contributions.

This work was partly supported by a grant from UNIDO/ICGEB and a fellowship from the Alexander von Humboldt Foundation for one of us (D.A.D.)

REFERENCES

Ackers, G.L., 1970, *Adv. Protein Chem.* 24: 343-443.

Argos, P., 1987, *J. Mol.Biol.* 197: 331-348.

Chemeris, V.V., Dolgikh,D.A., Fedorov, A.N., Finkelstein, A.V., Kirpichnikov, M.P., Uversky, V.N. and Ptitsyn, O.B., 1994, *Protein Engineering* 7: 1041-1052.

Cheng, Y.C. and Prusoff, W.H., 1973, *Biochem. Pharmacol.* 22: 3099-3108.

Corbett, R.J.T. and Roche, R.S., 1984, *Biochemistry* 23: 1888-1894.

Creighton, T.E., 1979, *J. Mol. Biol.* 129: 235-264.

Creighton, T.E., 1980, *J. Mol. Biol.* 137: 61-80.

Dolgikh, D.A., Fedorov, A.N., Chemeris, V.V., Chernov, B.K., Finkelstein, A.V., Schulga, A.A., Alakhov, Yu.B., Kirpichnikov, M.P. and Ptitsyn, O.B., 1991, *Doklady Akad. Nauk SSSR (Russian)* 320: 1266-1269.

Dolgikh, D.A., Gabrielian A.E., Navolotskaya, E.V., Chemeris, V.V., and Kirpichnikov, M.P., 1993, *Biofizika (Russian)* 38: 67-74.

Erickson, A.H. and Blobel, G., 1983, *Meth. Enzymol.* 96: 38-49.

Dolgikh,D.A., Kirpichnikov, M.P., Ptitsyn, O.B., Fedorov, A.N., Finkelstein, A.V. and Chemeris, V.V., 1992, *Molekulyarnaya Biologya (Russian)* 26: 1242-1250.

Fedorov, A.N., Dolgikh, D.A., Chemeris, V.V., Chernov, B.K., Finkelstein, A.V., Schulga, A.A., Alakhov, Yu.B., Kirpichnikov, M.P. and Ptitsyn, O.B., 1992, *J. Mol.Biol.* 225: 927-931.

Finkelstein, A.V., 1983, Program "ALB" for protein & polypeptide secondary structure calculation and prediction. Deposited at the Brookhaven Protein Data Bank, Upton, N.Y., U.S.A., at the EMBL Database, Heidelberg, FRG; distributed by request.

Finkelstein, A.V., Badretdinov, A.Yu. and Ptitsyn, O.B., 1991, *Proteins: Structure, Function and Genetics* 10: 287-299.

Finkelstein, A.V., Kirpichnikov, M.P., Ptitsyn, O.B., and Skryabin, K.G., 1988, *Vestnik AN SSSR (Russian)* 3: 102-111.

Finkelstein, A.V. and Ptitsyn, O.B., 1987, *Progr. Biophys. and Mol.Biol.* 50: 171-190.

Gabrielian, A.E., Dolgikh, D.A., Ivanova, N.B. and Kirpichnikov, M.P., 1992, In: *Proc. 2nd Russian-Israel Symposium on Peptides and Proteins, Moscow.*

Gurevich, V.V., Pokrovskaya, I.D., Obukhova, T.A. and Zozulya, S.A., 1991, *Analytical Biochemistry* 195: 207-213.

Harris, J.I., 1956, *Nature* 177: 471-473.

Lang, K. and Schmid, F.X., 1986, *Eur. J. Biochem.* 159: 275-281.

Melton, D.A., Krieg, P.A., Rebagliati, M.R., Maniatis, T, Zinn, K. and Green, M.R., 1984, *Nucl. Acids Res.* 2: 7035-7056.

Ptitsyn, O.B. and Finkelstein, A.V., 1980, *Quart. Rev. Biophys.* 13: 339-386.

Ptitsyn, O.B. and Finkelstein, A.V., 1983, *Biopolymers* 22: 15-25.

Provencher, S.W. and Gloeckner, J., 1981, *Biochemistry* 20: 33-37.

Regan, L. and DeGrado, W.F., 1988, *Science* 241: 976-978.

Sander, C., 1994, *Trends in Biotechnology* 12:163-167.

Studier, F.W., Rosenberg, A.H., Dunn, J.J. and Dubendorff, J.W., 1990, *Meth. Enzymol.* 185: 60-89.

Uversky, V.N., 1993, *Biochemistry* 32: 13288-13298.

Uversky, V.N. and Ptitsyn, O.B., 1994, *Biochemistry* 33: 2782-2791.

Wetzel, R., 1991, *Protein Engineering* 4: 371-374.

Zav'yalov, V.P., Navolotskaya, E.V., Abramov, V.M., Galaktionov, V.G., Isaev, I.S., Kaurov, O.A., Kozhich, A.T., Mayorov, V.A., Prusakov, A.N., Vasilenko, R.N. and Volodina E.Yu., 1991. *FEBS Lett.* 278: 187-189.

7

NMR STRUCTURE DETERMINATION AND RATIONAL DRUG DESIGN

Kurt Wüthrich

Institut für Molekularbiologie und Biophysik
Eidgenössische Technische Hochschule-Hönggerberg
CH-8093 Zürich, Switzerland

INTRODUCTION

In biological and biomedical research, the area of structural biology, in particular atomic resolution studies of the three-dimensional structure of biological macromolecules and their intermolecular interactions, has never before had as central a role and enjoyed as much popularity as today. This is mainly due to the facts that with the use of DNA recombination and chemical synthesis, a virtually unlimited array of different polypeptide sequences can be generated, and that the relations of the resulting primary structures with corresponding biological functions can only be rationalized through knowledge of the three-dimensional structure. In addition to its key role in academic institutions, structure determination of proteins, nucleic acids and other classes of biomolecules has therefore become a discipline that is actively pursued also in profit-oriented organizations, especially in the major pharmaceutical companies. Although X-ray diffraction with single crystals has for two decades, until 1984, been unique in its potential for efficient determination of three-dimensional molecular structures at atomic resolution, it presently shares this role with nuclear magnetic resonance (NMR) spectroscopy in solution. During the period 1990-1993, 607 new X-ray crystal structures, 179 new NMR solution structures, and 2 structures determined by other methods were published (Hendrickson and Wüthrich, 1991; 1992; 1993; 1994). This article presents a description of the NMR approach for three-dimensional structure determination of biological macromolecules and a discussion of the impact of molecular structural biology in biomedical research.

THE PRACTICE OF MACROMOLECULAR STRUCTURE DETERMINATION BY NMR

Fig. 1 presents an overview of the course of a NMR structure determination. Individual steps are the sample preparation, the NMR measurements, the assignment of the NMR lines to individual atoms in the polymer chain, the collection of conformational

Molecular Manufacturing, edited by C. Nicolini
Plenum Press, New York, 1996

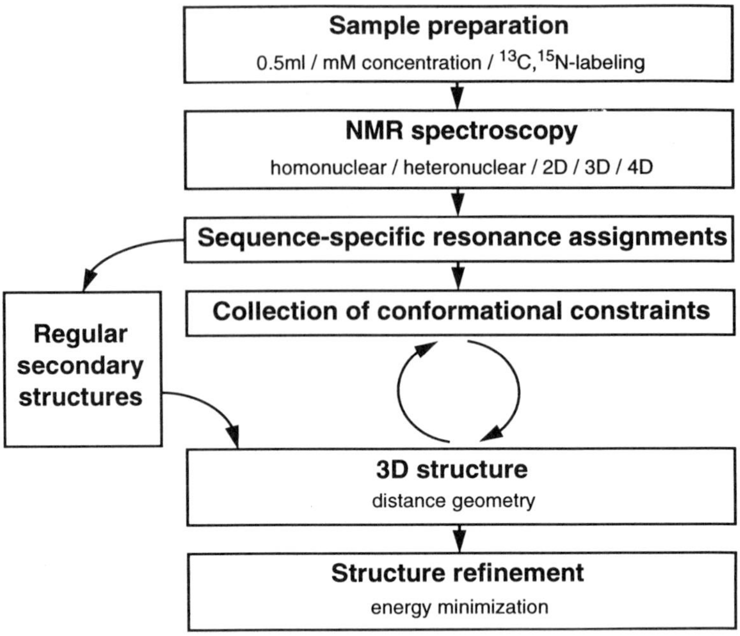

Figure 1. Diagram outlining the course of a macromolecular structure determination by NMR in solution (see text).

constraints, the structure calculation and structure refinements. Although the entire structure determination might conceivably at some future stage be handled as a single, computer-supported process, in present practice the sequence of steps does correspond largely to the flow diagram of Fig. 1. As is also indicated in Fig. 1, it is a special feature of protein structure determination by NMR that the secondary polypeptide structure, including the connections between individual segments of regular secondary structure, may be known early-on from the data used for obtaining the resonance assignments, actually before the structure calculation is even started.

Sample Preparation

As even minor macromolecular impurities may be detrimental to the outcome of a structure determination, the samples used must be highly homogeneous. The material is dissolved in 0.5 ml of water, and the ionic strength, pH and temperature may be adjusted so as to ensure near-physiological conditions (for proteins it is advantageous to work in the slightly acidic pH range from 3 to 5). The concentration should be about 1 mM, ideally 2-5 mM, so that, for example, 5-25 mg of a protein with molecular weight 10,000 should be available for a structure determination. Although this concentration is high relative to that of most proteins in their physiological milieu, it is not far from the total protein concentration in typical body fluids. For molecular sizes above approximately 10,000, quite often also for smaller sizes, the NMR study will have to include the preparation of compounds enriched with ^{15}N and/or ^{13}C. Uniformly isotope-labeled recombinant proteins are obtained with commonly available techniques. For RNA, labeling techniques have been introduced only recently, and they are still used only by a few groups. For DNA, labeling techniques are still in the process of being worked out.

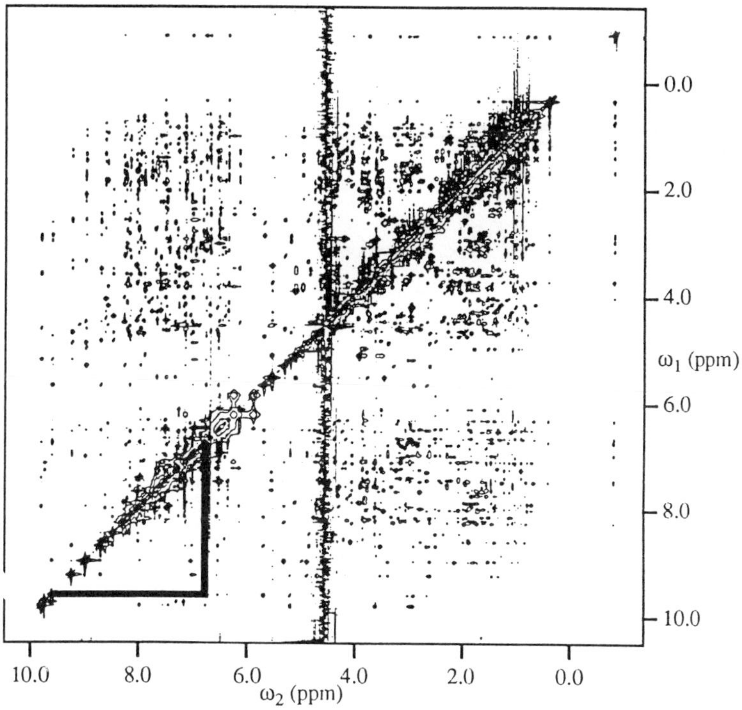

Figure 2. Two-dimensional [^{1}H,^{1}H]-NOESY (nuclear Overhauser enhancement spectroscopy) of the protein Toxin K (protein concentration = 10 mM, solvent = 90% H_2O/10% ^2H_2O, pH = 4.6, T = 36 °C, ^{1}H frequency = 600 MHz, mixing time = 40 ms). The horizontal line and the vertical line through a cross peak in the lower left indicate how the cross peaks correlate pairs of hydrogen atoms (represented by the chemical shift on the diagonal) which are related by a NOE.

Being able to work with solutions is a great asset of the NMR method. However, working with solutions also has its inherent potential difficulties. Most of all, many biological macromolecules tend to self-aggregate at the concentrations needed for NMR spectroscopy, which has slowed down many projects. Furthermore, in the course of an investigation it may be non-trivial to achieve identical solution conditions in different NMR samples of the same compound, which typically results in small chemical shift differences that slow down the combined analysis of different NMR spectra.

NMR Spectroscopy with Biological Macromolecules

The selection of NMR experiments for macromolecular structure determination is dictated mainly by two requirements: (i) Hundreds or even thousands of resonance lines in the ^{1}H NMR spectrum of a biological macromolecule must be observed as separate peaks, and selective correlations between pairs of spins (or groups of equivalent spins, for example, the three protons in a methyl group) need to be established. (ii) The maximum possible number of ^{1}H-^{1}H nuclear Overhauser effects (NOEs) must be identified, since the information needed for three-dimensional structure determination comes primarily from NOE distance constraints. In present practice these demands are met by multidimensional NMR (Ernst et al., 1987; Wüthrich, 1986).

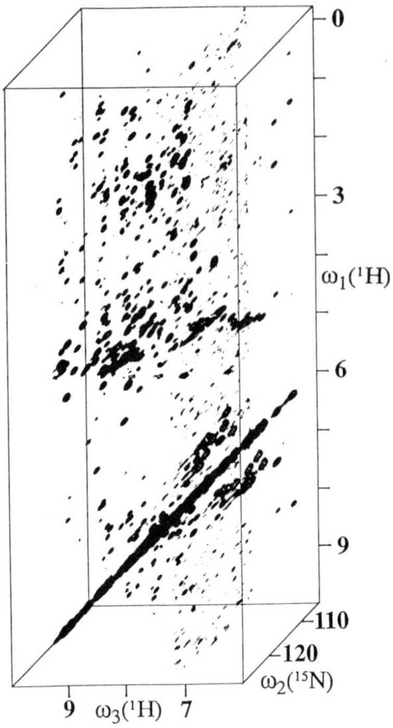

Figure 3. 3D ^{15}N-correlated [^1H,^1H]-NOESY spectrum of oxidized E. coli glutaredoxin (protein concentration = 1.5 mM, solvent = 90% H_2O/10% ^2H_2O, pH = 7.0, T = 28 °C, ^1H frequency = 500 mHz, mixing time = 100 ms). The protein was uniformly labeled with ^{15}N to the extent of 3 95%. (From Sodano et al., 1991).

The figure 2 may serve to illustrate the problems to be solved.In this two-dimensional (2D) ^1H NMR spectrum the peaks on the diagonal from the upper right to the lower left correspond to the one-dimensional (1D) ^1H NMR spectrum, and only a small percentage of the resonance lines are well separated so that there is extensive overlap of multiple resonances in most spectral regions. The additional "cross peaks" in the two-dimensional frequency plane outside of the diagonal, which manifest correlations between pairs of diagonal peaks, are, however, much better resolved than the resonance lines in the 1D NMR spectrum. The 2D ^1H NMR experiment can thus yield selective ^1H-^1H correlations also in crowded spectral regions. Depending on the type of experiment performed, different kinds of spin-spin correlations are manifested by the cross peaks (Ernst et al., 1987; Wüthrich, 1986).

With increasing size of the molecule studied and concomitant increase of the number of NMR peaks, it becomes more and more difficult to resolve and assign individual resonances in homonuclear 2D ^1H NMR spectra. In heteronuclear threee-dimensional (3D) or four-dimensional (4D) spectra recorded with recombinant proteins that are uniformly labeled with ^{15}N and/or ^{13}C, the peaks are spread out in a third and possibly fourth dimension along the ^{15}N and ^{13}C chemical shift axes (for a recent review, see Bax and Grzesiek, 1993). As an illustration, Fig. 3 shows a 3D ^{15}N-correlated [^1H,^1H]-NOESY spectrum, where the NMR peaks are spread out along a third axis which corresponds to the NMR frequencies of the ^{15}N spins in the ^{15}N-labeled protein. As a result, the same number of NMR peaks that would be observed in a single plane in the corresponding 2D [^1H,^1H]- NOESY spectrum (Fig. 2) are distributed among multiple ^1H-^1H planes, typically 64 or 128. The ensuing further improved separation of the peaks in the individual planes is indispensable for work with larger proteins, and the preparation of isotope-labeled proteins as well as the use of higher-dimensional, heteronuclear NMR experiments are now general practice, where the

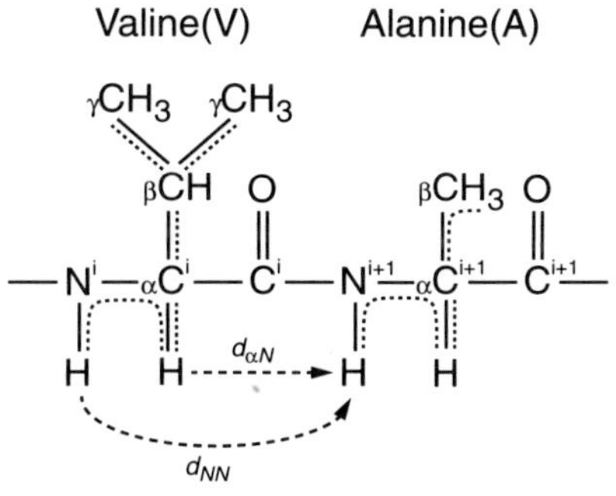

Figure 4. Sequential resonance assignment based on ^1H-^1H NOEs. In the dipeptide segment -Val-Ala-, the dotted lines indicate ^1H-^1H relations which can be established by scalar, through-bond spin-spin couplings. The broken arrows indicate relations between protons in sequentially neighboring residues, which can be established by NOESY cross peaks manifesting short sequential distances $d_{\alpha N}$ (between αCH and the amide proton of the following residue) and d_{NN} (between the amide protons of neighboring residues).

key purpose of the heteronuclear NMR experiments is to obtain the maximum possible number of individually assigned ^1H-^1H NOEs in the system studied.

Resonance Assignments

Obtaining resonance assignments in biopolymers is non-trivial for reasons that may best be rationalized by comparison with NMR applications in other fields. For example, in studies of organic-chemical compounds, resonance assignments are typically based on the identification of unique "spin systems", which are characteristic for the covalent molecular structure. In homonuclear ^1H NMR, each spin system includes all those hydrogen atoms that are related by scalar spin-spin couplings (e.g., Wüthrich, 1986). In macromolecules this encompasses those groups of hydrogen atoms that are separated in the chemical structure by steps of at most three covalent bonds. For most amino acid residues the ensemble of all hydrogen atoms thus forms one spin system. For example, as is indicated by the dotted lines in Fig. 4, the three equivalent methyl protons of alanine are separated by three bonds from the α-proton, which in turn relates to the amide proton via three bonds, so that a continuous network of scalar couplings can be established. Exeptions are the four aromatic amino acid side chains and some peripheral labile side chain protons, where the connection to the aliphatic hydrogen atoms of the same side chain must be established via ^1H-^1H NOEs (Wüthrich, 1986). Fig. 4 also shows that alanine and valine can readily be distinguished, since their 1H spin systems consist of 3 lines, and 5 lines, respectively. In a typical NMR application in organic chemistry, the assignment problem would be solved at this stage. In contrast, since proteins invariably contain more than one residue of most of the 20 common amino acids, the identification of the groups of scalar-coupled spins belonging to individual amino acid residues is in general not sufficient to define a unique sequence location.

The aforementionned situation is illustrated in Fig. 5 with the *lac* repressor DNA-binding domain (Zuiderweg et al., 1983). This polypeptide contains 8 valyl residues in the positions 4, 9, 15, 20, 23, 24, 30 and 38. Although the 8 valyl spin systems could be identified, there is no way to assign individual sequence positions from this information alone. This intrinsic problem was solved with the sequential assignment strategy (Dubs et al., 1979; Billeter et al., 1982; Wagner and Wüthrich, 1982): Unique segments of two or several sequentially adjoining amino acid residues are identified by NMR experiments, and then by comparison with the chemically determined amino acid sequence attributed to discrete positions in the polypeptide chain. The desired relations between protons in sequentially neighboring amino acid residues i and $i + 1$ can be established by NOEs manifesting close approach between αCH_i and NH_{i+1} ($d_{\alpha N}$), NH_i and NH_{i+1} (d_{NN}) (Fig. 4), and possibly βCH_i and NH_{i+1} ($d_{\beta N}$) (Wüthrich, 1986). For Xxx-Pro dipeptide segments, corresponding connectivities are observed to δCH_2 of Pro in the place of the amide proton. In Fig. 4, for example,

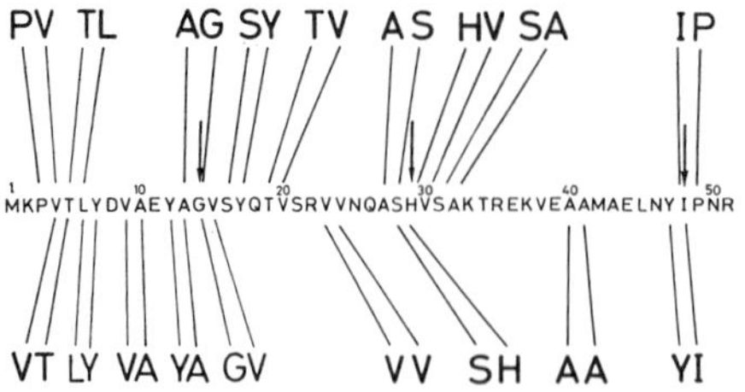

Figure 5. Amino acid sequence of the DNA-binding domain of residues 1-51 of the *E. coli lac* repressor with identification of unique residues (arrows), and indication of selected unique dipeptide segments that are useful for obtaining sequence-specific resonance assignments. (From Wüthrich, 1986).

observation of a NOESY cross peak corresponding to $d_{\alpha N}$ or d_{NN} tells us that the protein studied contains a dipeptide segment Val-Ala. This dipeptide is then matched against the independently known amino acid sequence. If the latter contains Val-Ala only once, the assignment problem is solved. In the *lac* repressor DNA-binding domain, the Val-Ala resonances could thus unambiguously be assigned to the residues in positions 9 and 10 (Fig. 5). Otherwise, to distinguish between multiple Val-Ala sites, tri- or tetrapeptide segments including Val-Ala must be identified by NMR and matched against the amino acid sequence. Since in globular proteins with up to 200 amino acid residues the probability is small that identical tri- or tetrapeptide segments occur more than once, the procedure is generally applicable (Wüthrich, 1986). For small proteins with up to about 100 amino acid residues, NOE-based sequential assignments can rely entirely on homonuclear ^{1}H NMR experiments, and for somewhat bigger proteins with up to about 200 residues they can be established based on the improved resolution of 3D NMR experiments with ^{15}N-labeled proteins (Fig. 3). No prior knowledge of the polypeptide conformation is needed, since at least one of the two distances $d_{\alpha N}$ or d_{NN} (Fig. 4) is always sufficiently short to be observed by NOEs (Billeter et al., 1982). A further attractive feature of this approach is that the identification of the sequential NOEs forms an integral part of the data collection for the protein structure determination (see below).

Sequential assignments can alternatively be obtained entirely via heteronuclear scalar couplings. This intuitively straightforward approach became a practical technique only with the availability of recombinant isotope-labeled proteins in the late 1980s. For a polypeptide chain that is uniformly labeled with ^{15}N and ^{13}C, the entire macromolecular structure can be represented as a single heteronuclear spin system, since there are observably large scalar couplings also between atoms in neighboring amino acid residues (Fig. 6). Using 3D and 4D heteronuclear triple-resonance experiments (e.g., Bax and Grzesiek, 1993; Szyperski et al., 1993), the resonance lines of sufficiently large fragments of the polypeptide structure are grouped together to enable sequence-specific resonance assignments (e.g., Clore and Gronenborn, 1991). The largest proteins for which the polypeptide backbone has been assigned using ^{13}C/^{15}N -double-labeling and triple-resonance experiments to correlate groups of atoms via the spin-spin couplings of Fig. 6 are in the range 250 to 300 residues.

For nucleic acids, assignment procedures were largely patterned after those used for proteins. In contrast to the situation with polypeptide chains, assignments using intra-nu-

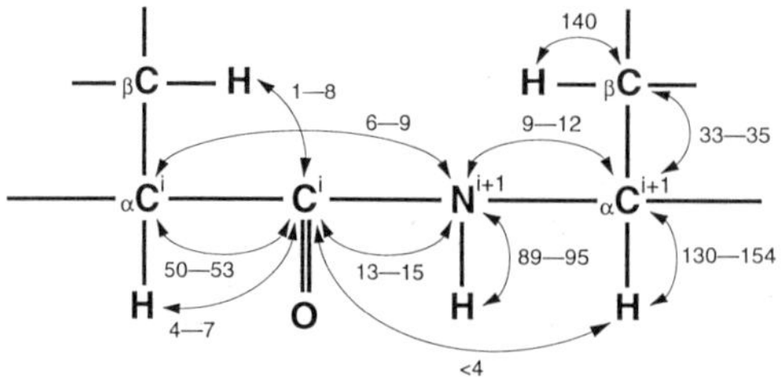

Figure 6. Segment of a polypeptide chain with indication of the scalar spin-spin couplings that provide the basis for obtaining sequential assignments by triple-resonance experiments with uniformly ^{13}C/^{15}N-doubly-labeled proteins. i and i + 1 identify backbone atoms of two sequentially neighboring residues.

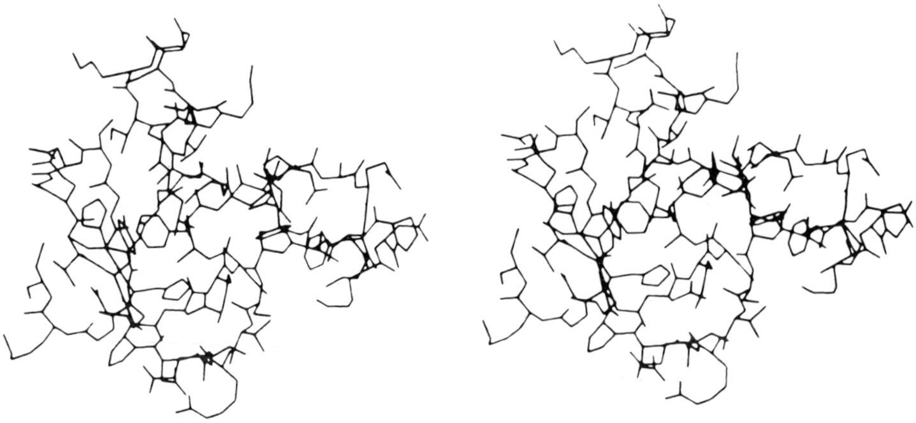

Figure 7. Stereo view of the NMR solution structure of the protein BUSI IIA (proteinase inhibitor IIA from bull seminal plasma). All heavy atoms of the 57-residue polypeptide chain are drawn. For the structure determination the protein had been dissolved in water. (Drawing prepared with the atomic coordinates from Williamson et al. (1985)).

cleotide, sequential and inter-strand ^{1}H-^{1}H NOEs are necessarily based on the assumption of a particular conformation type for the compound studied (Wüthrich, 1986). For B-type DNA, this approach is highly efficient for duplexes with up to about 20 base pairs. A conformation-independent approach is also available, which uses heteronuclear scalar ^{1}H-^{31}P couplings to establish sequential assignments. In unlabeled compounds these assignments cannot readily be extended to include the resonances of the bases, which limits their usefulness (Wüthrich, 1986). With suitably ^{15}N- and ^{13}C-labeled compounds, base-backbone connections can be established via heteronuclear scalar couplings, which may in the future enable complete assignments for oligonucleotide segments with non-regular conformations.

Collection of Conformational Constraints and Structure Calculation from NMR Data

The decisive information for macromolecular structure determination consists of NOE upper distance constraints, although some supplementary constraints may be derived from other NMR measurements (Wüthrich, 1986; 1989). For a high-quality structure determination, the maximum possible number of NOE conformational constraints must be collected as input for the calculation of the complete three-dimensional structure. This is accomplished by using the chemical shift lists obtained as a result of the sequence-specific resonance assignments to attribute the cross peaks in 2D (Fig. 2), 3D (Fig. 3) or 4D [^{1}H,^{1}H]-NOESY spectra to distinct pairs of hydrogen atoms. Since one has to anticipate some accidental chemical shift degeneracies, the data collection is achieved in several cycles (Fig. 1), where ambiguities in the NOESY cross peak assignments can usually be resolved by reference to preliminary structures calculated from incomplete input data sets (Güntert et al., 1993).

In present practice the NOE input data have the format of allowed distance ranges, which circumvents intrinsic difficulties that might arise from attempts at quantitative distance measurements. The lower limit is usually taken to correspond to the sum of two hydrogen atomic radii, i.e., 2.0 Å. For short-range connectivities between protons separated

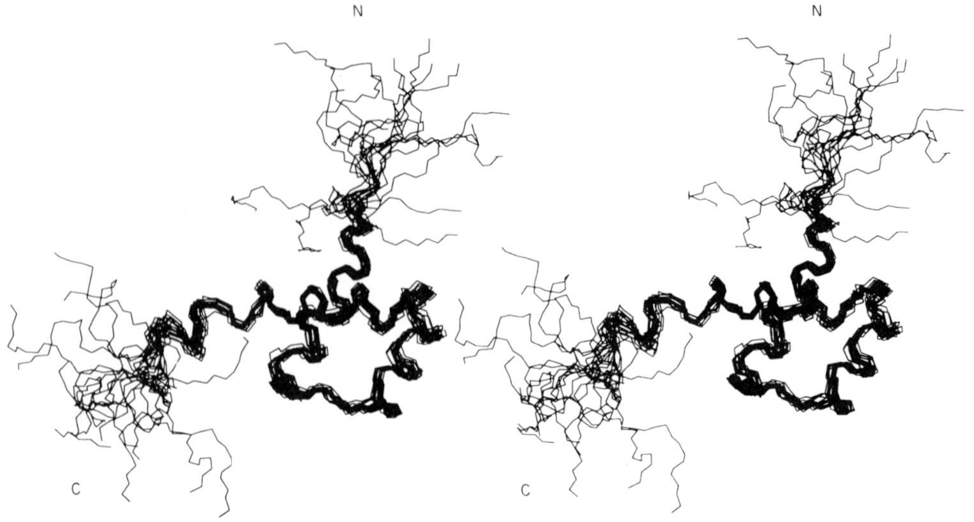

Figure 8. Stereo view of the polypeptide backbone of 19 energy-refined conformers selected to represent the NMR solution structure of the *Antennapedia* homeodomain. Conformers 2 to 19 were superimposed for pairwise minimum root mean square deviation (RMSD) of the backbone atoms N, C^{α} and C′ of residues 7- 59 with conformer 1. From residues 7 -59 (identified in the figure) the structure is well defined by the NMR data, as is manifested by the small dispersion among the group of conformers, but the chain-terminal segments 0-6 and 60-67 are disordered. (Drawing prepared with the atomic coordinates from Qian et al., 1989).

by not more than three single bonds in the chemical structure, that is, intraresidual and sequential distances (Wüthrich, 1986), the NOE intensities are translated into corresponding upper bounds, typically in steps of 2.5, 3.0, and 4.0 Å. For all longer-range connectivities, a predetermined upper limit is applied, usually 4.0 or 5.0 Å, depending on the protein and the experimental conditions used. Unless stereospecific, individual assignments for pairs of diastereotopic substituents were obtained, in particular for the β-methylene groups in amino acid side chains and the isopropyl groups of valine and leucine, correction factors for the use of *pseudoatoms* must be added to the observed upper limits on distances to hydrogen atoms of prochiral centers (Wüthrich et al., 1983). The resulting allowed distance ranges may then be as large as from 2.0 to about 10 Å. Supplementary conformational constraints, for example, from spin-spin coupling constants, are represented in the input by similar allowed ranges, which account for the limited accuracy of the individual experimental measurements. In computer-supported collection of the input data, the stepwise classifications may be replaced by a smoothed fit of upper distance bounds vs. NOE intensity, and the calibration may be updated in successive cycles of the structure calculation.

Distance geometry algorithms can be used to obtain the Cartesian coordinates of spatial molecular structures that are consistent with a predetermined set of intramolecular, interatomic distances (Crippen and Havel, 1988). To this end, such algorithms identify the conformations

$$\{r_i = (x_i, y_i, z_i);\ i = 1, 2, \ldots N\} \quad (1)$$

which are consistent with the inequality

$$L_{ij} \leq | r_i - r_j | \leq U_{ij}, \quad (2)$$

where L_{ij} and U_{ij} are lower and upper bounds, respectively, on the Euclidian distance $| r_i - r_j |$ between the points r_i and r_j, and N is the number of points (or atoms) in the system studied. The problem described by the inequality [2] may alternatively be expressed in the form of an error function, which can then be minimized. In the process of this conversion of distance information into Cartesian coordinates, one inevitably encounters the problem of local minima. Metric matrix methods (Havel and Wüthrich, 1984) and variable target functions in dihedral angle space (Braun and Gô, 1985; Güntert et al., 1991) have been implemented for the search for protein conformations obeying inequality [2], and the energy functions in molecular dynamics programs have been supplemented with terms representing the constraints expressed by inequality [2] (Brünger et al., 1986; Kaptein et al., 1985). Overall, corresponding results are obtained with these different computational approaches, although details may be subject to further analysis.

Model calculations show that the quality of a NMR structure determination depends critically on the density of NOE distance constraints, while the procedure is remarkably robust with regard to low precision of the individual distance constraints (Havel and Wüthrich, 1985). Multiple rounds of data collection and structure calculation (Fig. 1) are usually needed to obtain the desired, high-density input.

The result of a single distance geometry calculation (Fig. 7) represents one molecular geometry that is compatible with the NMR data, but it cannot tell whether this solution is unique. Therefore, the calculation is repeated with different boundary conditions. For each calculation, convergence is judged by the criterion that there should be only small residual constraint violations, and all satisfactory solutions are included in a group of conformers used to represent the NMR structure of the protein. The final result of a structure determination is then commonly presented as a superposition of a group of conformers for pairwise minimum root mean square deviation (RMSD) relative to a predetermined conformer, or relative to the mean atom coordinates. Fig. 8 shows a superposition of the polypeptide backbone in 19 conformers of the *Antennapedia* homeodomain that were selected to represent the solution structure (Qian et al., 1989). The precision of the structure determination, as reflected by the dispersion among this group of conformers, varies along the polypeptide chain. From residues 7-59, all conformers are similar, have nearly the same overall dimensions and contain the same secondary structure, showing that this part of the structure is well defined by the NMR data. In contrast, both chain ends are disordered, as is manifested by the large dispersion among the 19 conformers. Large variations are typically observed also in the precision of the structure determination of the amino acid side chains, with well defined interior side chains representing the protein core and flexibly disordered side chains located predominantly near the protein surface.

Structure determination with nucleic acids is pursued along similar lines as described here for proteins. As a general guide to an evaluation of the results obtained, one can assume that "global folds", for example, formation of duplexes, triplexes, quadruplexes, or loops can be well defined by NMR data. On the other hand, because of the short range and the low precision of individual NOE distance measurements, information on details in a given nucleic acid structure type, for example, bending of DNA duplexes or non-regular base stacking, may be more difficult to characterize.

Structure Refinement

The individual conformers obtained as the solutions of distance geometry calculations generally contain some close inter-atomic contacts and correspondingly high conformational energies. These are routinely removed in a separate final refinement step (Fig.1). Alternatively, in typical molecular dynamics calculations the energy minimization may be performed in concert with the fit to the distance constraints. Methods for direct refinement

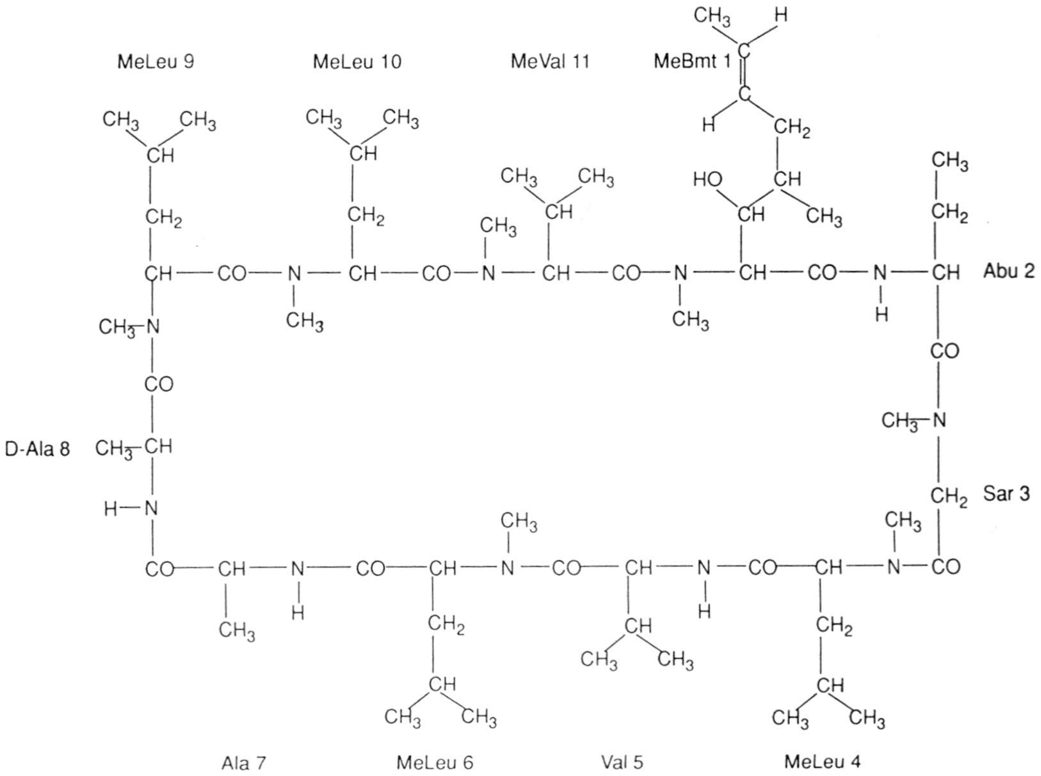

Figure 9. Chemical structure of cyclosporin A (CsA) with the standard numeration of the amino acid residues. Symbols used for residues other than common amino acids: MeBmt, (4R)-4-[(E)-2-butenyl]-4,N-dimethyl-L-threonine; Abu, L-α-aminobutyric acid; Sar, sarcosine; MeLeu, N-methylleucine; MeVal, N-methylvaline.

against the NOESY spectra by complete relaxation matrix calculations are in principle available (e.g. Boelens et al., 1989; Borgias et al., 1990; Mertz et al., 1991; Yip and Case, 1989), but because of a variety of fundamental and technical difficulties they have so far hardly been used in actual structure determinations.

THREE-DIMENSIONAL PROTEIN STRUCTURES, PROTEIN ENGINEERING AND DRUG DESIGN

Naturally occurring proteins perform a wide range of life-sustaining functions. Examples are mechanical protection by the proteins in the skin, hair or feathers and nails, catalysis by enzymes, metabolic regulation by hormones, transcriptional regulation by proteins that bind to deoxyribonucleic acid (DNA), electron transfer by cytochromes, oxygen transport by hemoglobin, or the reactions of the immune system. It has long been known that small alterations of the chemical polypeptide structure, even the replacement of a single amino acid residue in a given sequence position by a different residue type, can dramatically modify the functional properties of a protein and, for example, in genetic disease be the cause of severe pathological disturbance. Variation of the amino acid sequence of natural proteins is therefore a promising avenue to modified or possibly novel protein functions, and as an alternative, novel functionalities might also be obtained by *de novo* design of protein molecules. Today two methods are available which enable, in principle, the generation of an unlimited number of different polypeptide sequences: i.e., chemical synthesis of polypeptides for the production of smaller peptides, and DNA recombination in the genes that encode the polypeptide sequence.

The number of different amino acid sequences that can be generated by recombinant and chemical techniques is staggering. If one assumes that only the 20 common amino acid residues are used, it amounts to 20^n, where n is the number of residues in the chain (for n = 10, one obtains about 10^{13} different sequences). Therefore, a systematic approach of screening all possible amino acid sequences for desirable functional properties is hardly practical. Much rather, promising protein engineering relies on 3D structure-based, "rational" protein design. Instead of systematic variation of the amino acid sequence, one searches the 3D structure (Fig. 7) for patterns of spatially adjacent residues that can be related to protein functions, where residues forming such patterns may be far apart in the sequence. The most obvious examples are the groups of residues that form "active sites" in proteins. In search of improved or altered protein functions, amino acid substitutions are then performed exclusively within the limited number of sequence locations that are involved in the formation of a functionally important 3D array of amino acid residues.

Although one estimates that the amino acid sequences of over 80 000 natural proteins have been determined, still only about 1000 three-dimensional protein structures are presently known. The scarcity of 3D structures is clearly a bottleneck in today's protein design and protein engineering. Fortunately, progress has been rapid (Hendrickson and Wüthrich, 1991; 1992; 1993; 1994), and the outlook is that this pace will be maintained also in the foreseeable future.

AN ILLUSTRATIVE EXAMPLE: THE THREE-DIMENSIONAL STRUCTURES OF FREE AND RECEPTOR-BOUND CYCLOSPORIN A

Cyclosporin A (CsA) is a member of a group of immunosuppressive agents which bind with high affinity to their respective immunophilins (immunosuppressant-binding proteins) and act as inhibitors of specific signal transduction pathways that lead to T lymphocyte activation (Borel, 1986; Schreiber, 1991). CsA is currently the favoured therapeutic agent for prevention of graft rejection in clinical organ transplantations. In addition to the fundamental interest in the molecular basis of immunosuppressive action, this fact has further supported the generation of vigorous research activity on the structural basis of the function of CsA and related compounds. In this context, cyclophilin (Cyp), which is a cytosolic binding protein with molecular weight 17900, was identified as a cellular receptor of CsA (Handschumacher et al., 1984). The protein Cyp has been found in diverse organisms from E. coli to man (Koletsky et al., 1986), and is present at high abundance in all tissues studied.

In its chemical structure, CsA is a cyclic undecapeptide with pronounced hydrophobic character (Fig. 9). Besides hydrocarbon moieties there is only a single -OH group in the amino acid side chains, and along the polypeptide backbone seven of the eleven peptide groups are N-methylated. The 3D structure of CsA was determined by X-ray diffraction in single crystals (Loosli et al., 1985) and by NMR in solution (Loosli et al., 1985; Kessler et al., 1990). As free CsA is very little soluble in aqueous media, the NMR structure was determined in chloroform solution, and the CsA crystals were grown from a solution in a nonpolar solvent. The NMR structure in chloroform at -20 °C is virtually identical to the X-ray crystal structure (Fig.10). In both environments the polypeptide backbone forms a twisted β-sheet involving the residues 1, 2, 5, 6, 7 and 11, and a type II′ turn at Sar 3 and MeLeu 4 (see Fig. 9 for the numeration). The loop formed by the remaining amino acid residues 8-10 includes a *cis*-peptide bond between MeLeu 9 and MeLeu 10. The four amide protons (Fig. 9) are involved in the three transannular hydrogen bonds Abu 2 NH-Val 5 C′=O, Val 5 NH-Abu 2 C′=O and Ala 7 NH-MeVal 11 C′=O, and the hydrogen bond D-Ala 8 NH-MeLeu 6 C′=O. In Fig. 10 the three transannular hydrogen bonds are in central locations in the upper half of the molecule, and the fourth hydrogen bond is on the right edge just below the center. The near-identity of the CsA structures in the crystal and in chloroform solution extends also to the amino acid side chains, which form two hydrophobic clusters. One of these includes Abu 2, Val 5, Ala 7 and MeVal 11 and is in front of the plane of the cyclic polypeptide backbone in the orientation of Fig. 9, and the other is made up of MeBmt 1, MeLeu 4, MeLeu 6 and MeLeu 10 and is behind the ring plane (Fig. 10).

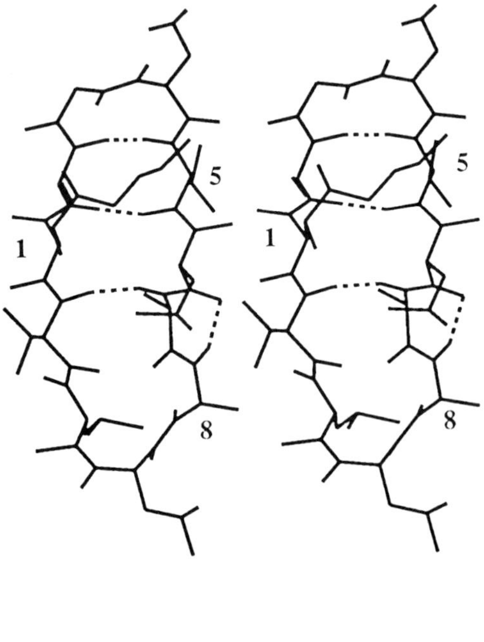

Figure 10. Stereo view of the X-ray crystal structure of free CsA. The molecule has been oriented so that Sar 3 is at the top, MeLeu 9 at the bottom, MeBmt 1 in the center on the left and MeLeu 6 in the center on the right (see Fig. 9 for the residue numeration). (The drawing was prepared from the atomic coordinates of Loosli et al. (1985)).

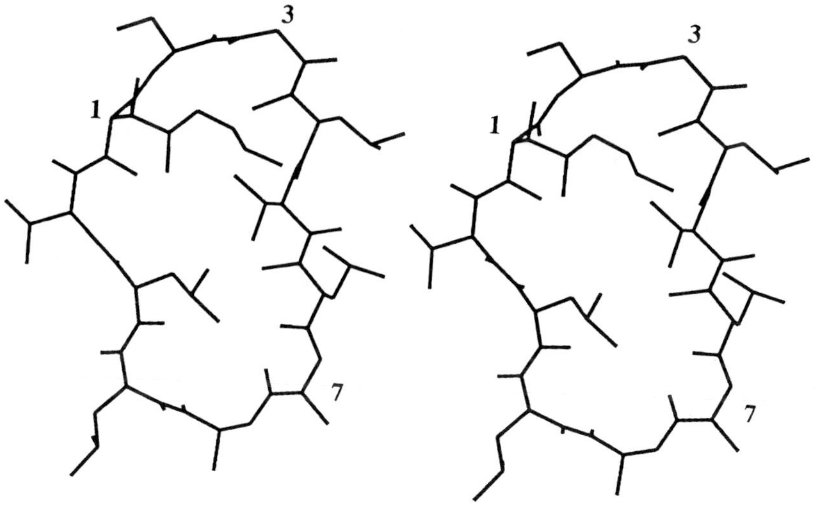

Figure 11. Stereo view of the NMR structure of cyclophilin-bound CsA. The same atoms are shown as in Fig. 10. The orientation of CsA is as in Fig. 10, except that the residues are shifted clockwise by about one unit, so that Abu 2 is at the top and D-Ala 8 at the bottom. (Drawing prepared with the atomic coordinates from Weber et al. (1991)).

In stark contrast to free CsA, CsA bound to Cyp (Fig. 11) contains no regular secondary structure and no intramolecular hydrogen bonds (Fesik et al., 1991; Weber et al., 1991). The two backbone segments from residues 4 - 7 and 9 -2 (Fig. 6), which form the antiparallel β-structure in free CsA, have their planar peptide bonds rotated out of the plane defined by the cyclic polypeptide backbone, which contrasts with regular β-structures. Four of the seven N-methyl groups line the inside of the cyclosporin ring, i.e., those of residues 1, 4, 6 and 10, and all four amide protons are exposed to the outside (Fig.11). All peptide bonds are in the *trans* conformation. A side chain cluster of MeBmt 1, MeLeu 4, MeLeu 6 and MeLeu 10 is located in front of the ring plane in the orientation of the molecule in Fig.9 and a side chain cluster of Abu 2, Val 5, Ala 7 and MeVal 11 is behind the ring plane. Overall, the conformation of CsA bound to Cyp thus includes the following differences relative to that of free CsA in a nonpolar solvent: The peptide bond between residues 9 and 10 is *trans* rather than *cis*, all four amide protons are exposed along the edges of the ring rather than hydrogen-bonded in the central region of the ring plane, and the two side chain hydrophobic clusters are located on opposite sides of the ring plane. The structural rearrangement of this cyclic polypeptide when going from chloroform solution to the receptor binding site is reminiscent of the inversion of a glove, whereby the hydrophobic exterior edges formed by the N-methyl groups in free CsA (Fig. 10) are replaced by a polar surface of amide protons and carbonyl oxygens (Fig. 11). For specific recognition by the receptor protein, the "bound" form of CsA can participate in inter-molecular hydrogen bonds both as a proton donor and a proton acceptor (Fig.11), whereas the structure of free CsA offers only a limited choice of proton acceptor sites (Fig. 10). The recent determinations of the complete structure of the Cyp-CsA complex (Pflügl et al.,1993; Spitzfaden et al., 1992; 1994; Thériault et al., 1993) showed indeed that a hydrogen bond involving an amide proton of CsA is prominent among the intermolecular interactions in the complex.

CONCLUSIONS

Three-dimensional protein structures have a key role in biomedical research, where they provide a basis for investigating correlations between molecular structure and physiological functions in natural systems, and a platform for protein engineering as well as for rational drug design supported by knowledge on the receptor structure. Ten years ago NMR spectroscopy in solution was introduced as a second method for protein structure determination, besides X-ray diffraction in protein crystals. In the meantime, about 200 NMR solution structures of proteins have been determined, which is an important contribution toward easing the bottleneck in protein design caused by the scarcity of data on 3D structures. More importantly, perhaps, the structural data collected by NMR in solution complement the information obtained from crystallography, in particular by additional characterization of dynamic molecular properties (Wüthrich, 1990). Combined use of information from the two techniques can therefore establish a more reliable structural foundation for modern biomedical research. In the previous section this was illustrated with the cyclophilin-cyclosporin A system, which is of keen interest in the biomedical field. It is particularly worth noting that although NMR observations had indicated the formation of multiple conformers upon addition of polar solvents to chloroform solutions of CsA (Ko and Dalvit, 1992), the large extent of the conformational differences observed when comparing this cyclic peptide (Fig. 9) in crystals or in chloroform solution (Fig. 10) and in the receptor-bound state (Fig. 11) was entirely unexpected. For example, modeling studies using molecular dynamics calculations with the crystal structure of CsA (Fig. 10) gave no indication of a propensity for major conformation changes (Lautz et al., 1987). CsA thus convincingly illustrates the importance of experimental studies with both the free and the receptor-bound states of effector molecules for obtaining a reliable guide to improved molecular design. In practice, one is quite regularly faced with situations such as that with the thrombin-hirudin system (Szyperski et al., 1992): on the one hand, the complex could be crystallized but would be too big for a NMR structure determination, while on the other hand the free, individual components could not be crystallized so that NMR was the only applicable method. Similar to the Cyp-CsA system, the thrombin-hirudin studies illustrate that experimental structure determinations should, if at all possible, be repeated in different milieus (different crystal forms, different solvents) and with different techniques, especially since one often has to compromise in selecting rather unphysiological conditions in order to obtain crystals that are suitable for X-ray diffraction studies, or to obtain solutions that can be used for NMR measurements. With regard to future applications in similar drug-receptor systems, it should also be added that the NMR structure of receptor-bound CsA (Fig. 11) was solved using a novel strategy of isotope labeling with ^{13}C and ^{15}N, and heteronuclear editing of the two-dimensional ^{1}H NMR spectra recorded with the labeled compounds. Binary complexes are ideally suited for efficient isotope labeling schemes, since the two components can be labeled separately with ^{13}C or ^{15}N before complex formation and subsequently combined with unlabeled partner molecules. Heteronuclear editing (Griffey and Redfield, 1987; Fesik, 1988) can then be used to separate the ^{1}H NMR lines of the two molecules in the complex. In particular, use of heteronuclear half-filters (Otting et al., 1986; Otting and Wüthrich, 1989; 1990; Wider et al., 1990) in two-dimensional ^{1}H NMR spectra represents a valid alternative to using three-dimensional or even higher-dimensional experiments, making NMR studies of receptor-bound low molecular weight effector molecules highly efficient and comparatively inexpensive.

Quite generally, these techniques thus represent an attractive avenue for use of NMR studies in conjunction with projects on drug design.

ACKNOWLEDGMENTS

Financial support by the ETH Zürich and by the Schweizerischer Nationalfonds (project 31.32033.91) is gratefully acknowledged. I thank Mrs. R. Hug for the careful processing of the manuscrcipt.

REFERENCES

Bax, A., and Grzesiek, S., 1993, *Acc. Chem. Res.* 26:131-138.

Billeter, M., Braun, W., and Wüthrich, K., 1982, *J. Mol. Biol.* 155:321-346.

Boelens, R., Koning, T. M. G., van der Marel, G. A., van Boom, J. H., and Kaptein, R., 1989, *J. Magn. Reson.* 82:290-308.

Borel, J. F., 1986, Ciclosporin, Karger, Basel.

Borgias, B. A., Gochin, M., Kerwood, O. S., and James, T. L., 1990, *Progr. NMR Spectrosc.* 22:83-100.

Braun, W., and Gô, N., 1985, *J. Mol. Biol.* 186:611-626.

Brünger, A. T., Clore, G. M., Gronenborn, A. M., and Karplus, M., 1986, *Proc. Natl. Acad. Sci. USA* 83:3801-3805.

Clore, G. M., and Gronenborn, A. M., 1991, Science 252:1390-1399.

Crippen, G. M., and Havel, T.F., 1988, Distance Geometry and Molecular Conformation. Wiley, New York.

Dubs, A., Wagner, G., and Wüthrich, K., 1979, *Biochim. Biophys. Acta* 577:177-194.

Ernst, R. R., Bodenhausen, G., and Wokaun, A., 1987, Principles of Nuclear Magnetic Resonance in One and Two Dimensions. Clarendon, Oxford.

Fesik, S. W., 1988, *Nature* 332:865-866.

Fesik, S. W., Gampe, R. T., Eaton, H. L., Gemmecker, G., Olejniczak, E. T., Neri, P., Holzman, T. F., Egan, D. A., Edalji, R., Simmer, R., Helfrich, R., Hochlowski, J., and Jackson, M., 1991, *Biochemistry* 30:6574-6583.

Griffey, R. H., and Redfield, A. G., 1987, *Q. Rev. Biophys.* 19:51-82.

Güntert, P., Braun, W., and Wüthrich, K., 1991, *J. Mol. Biol.* 217:517-530.

Güntert, P., Berndt, K. D., and Wüthrich, K., 1993, *J. Biomol. NMR* 3:601-606.

Handschumacher, R. E., Harding, M. W., Rice, J., and Drugge, R. J., 1984, *Science* 226:544-547.

Havel, T. F., and Wüthrich, K., 1984, *Bull. Math. Biol.* 46:673-698.

Havel , T. F., and Wüthrich, K., 1985, *J. Mol. Biol.* 182:281-294.

Hendrickson, W. A., and Wüthrich, K., 1991, Macromolecular Structures 1991, Current Biology, London.

Hendrickson, W. A., and Wüthrich, K., 1992, Macromolecular Structures 1992, Current Biology, London.

Hendrickson, W. A., and Wüthrich, K., 1993, Macromolecular Structures 1993, Current Biology, London.

Hendrickson, W.A., and Wüthrich K., 1994, Macromolecular Structures 1994, Current Biology, London.

Kaptein, R., Zuiderweg, E. R. P., Scheek, R. M., Boelens, S. R., and van Gunsteren, W.F., 1985, *J. Mol. Biol.* 182:179-182.

Kessler, H., Köck, M., Wein, T., and Gehrke, M., 1990, *Helv. Chim. Acta* 73:1818-1832

Ko, S.Y., and Dalvit, C., 1992, *Int. J. Peptide Protein Res.* 40:380-382.

Koletsky, A.S., Harding, M.W., and Handschuhmacher, R.E., 1986, *J. Immunol.* 137:1054-1059.

Lautz, J., Kessler, H., Kaptein, R., and van Gunsteren, W. F., 1987, *J. Computer-Aided Mol. Design* 1:219-241.

Loosli, H. R., Kessler, H., Oschkinat, H., Weber, H. P., Petcher, T., and Widmer, A., 1985, *Helv. Chim. Acta* 68:682-704.

Mertz, J. E., Güntert, P., Wüthrich, K., and Braun, W., 1991, *J. Biomol. NMR* 1:257-269.

Otting, G., and Wüthrich, K., 1989, *J. Magn. Reson.* 85:586-594.

Otting, G., and Wüthrich, K., 1990, *Q. Rev. Biophys.* 23:39-96.

Otting, G., Senn, H., Wagner G., and Wüthrich, K., 1986, *J. Magn. Reson.* 70:500-505.

Pflügl, G., Kallen, J., Schirmer, T., Jansonius, J. N., Zurini, M. G. M., and Walkinshaw, M. D., 1993, *Nature* 361:91-94.

Qian, Y. Q., Billeter, M., Otting, G., Müller, M., Gehring, W. J., and Wüthrich, K., 1989, *Cell* 59:573-580.

Schreiber, S. L., 1991, *Science* 251:283-287.

Sodano, P., Chary, K. V. R., Björnberg, O., Holmgren, A., Kren, B., Fuchs, J. A., and Wüthrich, K., 1991, *Eur. J. Biochem.* 200:369-377.
Spitzfaden, C., Weber, H. P., Braun, W., Kallen, J., Wider, G., Widmer, H., Walkinshaw, M.D., and Wüthrich, K., 1992, *FEBS Lett.* 300:291-300.
Spitzfaden, C., Braun, W., Wider, G., Widmer, H., and Wüthrich, K., 1994, *J. Biomol. NMR* 4:463-462.
Szyperski, T., Wider, G. Bushweller, J. H., and Wüthrich, K., 1993, *J. Am. Chem. Soc.* 115:9307-9308.
Szyperski, T., Güntert, P., Stone, S. R., Tulinsky, A., Bode, W., Huber, R., and Wüthrich, K., 1992, *J. Mol. Biol.* 228:1206-1211
Thériault, Y., Logan, T. M., Meadows, R., Yu, L., Olejniczak, E. T., Holzman, T. F., Simmer, R. L., and Fesik, S. W., 1993, *Nature* 361:88-91.
Wagner, G., and Wüthrich K., 1982, *J. Mol. Biol.* 155:347-366.
Weber, C., Wider, G., von Freyberg, B., Traber, R., Braun, W., Widmer, H., and Wüthrich, K., 1991, *Biochemistry* 30:6563-6574.
Wider, G., Weber, C., Traber, R., Widmer, H., and Wüthrich, K., 1990, *J. Am. Chem. Soc.* 112:9015-9016.
Williamson, M. P., Havel, T. F., and Wüthrich, K., 1985, *J. Mol. Biol.* 182:295-315.
Wüthrich, K., 1986, NMR of Proteins and Nucleic Acids. Wiley, New York.
Wüthrich, K., 1989, *Science* 243:45-50.
Wüthrich, K., 1990, *J. Biol. Chem.* 265:22059-22062.
Wüthrich, K., Billeter, M., and Braun, W., 1983, *J. Mol. Biol.* 169:949-961.
Yip, P., and Case D. A., 1989, *J. Magn. Reson.* 83:643-648.
Zuiderweg, E. R. P., Kaptein, R., and Wüthrich, K., 1983, *Eur. J. Biochem.* 137:279-292.

8

REDOX PROPERTIES OF CYTOCHROMES AND BIOELECTRONICS

Geoffrey R. Moore

Centre for Metalloprotein Spectroscopy and Biology
School of Chemical Sciences
University of East Anglia
Norwich, Norway 4 7TJ

The structural basis of metalloprotein action has been under investigation for many years by numerous researchers. The relationships they have uncovered should inform the discussion concerning the use of metalloproteins in synthetic electronic devices. The present paper is a contribution to this discussion. It concentrates on properties of cytochromes: these are haem-containing proteins in which the haem iron undergoes a reversible redox state change during their normal biological function (Belka et al., 1979). The paper will summarise some of the work currently going on in the author's laboratory and review certain topics relevant to the use of cytochromes in electronic devices. These include factors influencing rates of interprotein electron transfer and redox potentials of cytochromes *c*, and redox properties of non-haem-iron-containing cytochromes *b* which act as iron-storage proteins, bacterioferritins, in bacteria.

There are a number of important points that need to be made before we consider cytochromes further. Firstly, proteins have evolved to perform a biological function and not to create ideal components for synthetic electronic devices. Though obvious this needs to be stressed: most redox proteins are not chemically perfect in the sense of having, for example, rapid rates of electron transfer. Secondly, proteins themselves, without cofactors, are insulators and even with cofactors generally have only poor conductivity. Thirdly, though much has been learnt, our understanding of the physical basis of redox protein action, and our ability to change it in a predictable and controlled manner, remains poor compared to what is known, and achievable, with, for example, inorganic semiconductors and conductive organic polymers.

TYPES OF CYTOCHROME

The different classes of cytochrome, denoted by *a, b, c* and *d*, are distinguished by the nature of their haem groups (Belka et al., 1979). Haem *c* is one of the most common. This is covalently bound to the polypeptide chain via thioether bonds that are formed by condensation of the haem vinyl group substituents with the thiol side chains of cysteine

Molecular Manufacturing, edited by C. Nicolini
Plenum Press, New York, 1996

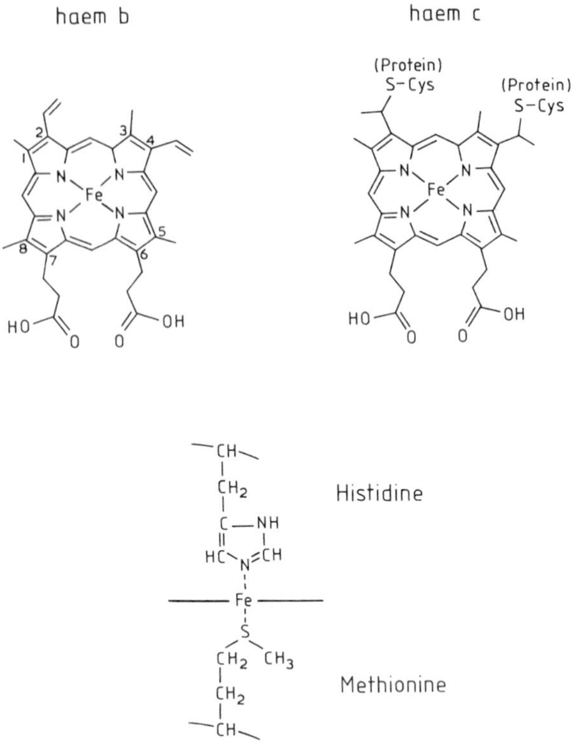

Figure 1. Common haem groups and axial ligands.

residues (Fig.1). In some cytochromes *c* only one such thioether bond is formed (Pettigrew and Moore, 1987). Cytochromes in which the haem vinyl substituents remain intact are termed cytochromes *b*. This is the type of haem that is also found in globins. Haem *a* has only been found in membrane bound cytochromes, and haem *d* only in bacterial cytochromes. Haems *b* and *c* are widespread, being found in both eukaryotic and prokaryotic organisms, and in both membraneous and water-soluble cytochromes (Pettigrew and Moore, 1987; Lemberg and Barrett, 1973; Meyer and Kamen, 1982).

When bound to the protein the haem iron has one or two axial ligands (Moore and Pettigrew, 1990). Usually these are from amino acid side chains, but the N-terminal amine is a ligand in some cytochromes *c* (Martinez et al., 1994) and H_2O is a ligand in some haem enzymes (Moore and Pettigrew, 1990). The most common axial ligands are histidine and methionine (Fig.1). Other amino acid side chains that may bind to the haem iron are those of tyrosine, cysteine, lysine, glutamate and aspartate. Generally electron transferring cytochromes have two axial ligands, though the five-coordinate Class II cytochromes *c* may be an exception (Moore and Pettigrew, 1990), and haem enzymes generally have at least one haem that will become five-coordinate at some stage of its reaction cycle.

In different cytochromes the haem iron can exist in high, low or mixed spin states. The spectroscopic and magnetic properties of these states differ and thus such measurements yield useful information for characterising cytochromes (Moore and Pettigrew, 1990). However, whether the spin state has a direct functional significance in its own right, or merely reflects other functionally important parameters such as the number and identity of the axial ligands, is not clear.

Table 1. Types of cytochromes *c*

A)	MONOHAEM ELECTRON TRANSFER (Fe^{2+}/Fe^{3+}) *e.g.Cytochromes c* & b_5 (2,5,7)
B)	MULTIHAEM ELECTRON TRANSFER (Fe^{2+}/Fe^{3+}) *e.g.Cytochromes* c_3 &c_7 (2,5,8) (trihaem,tetrahaem, octahaem)
C)	ENZYMATIC *e.g.* Bacterial cytochrome *c* peroxidase(dihaem enzyme; $Fe^{2+}/Fe^{3+}/Fe^{4+}$= 0; &)
D)	MULTIHEADED *e.g.* Cytochrome b_2 (haem *b* ; flavin)OH

e.g. Non-haem Fe containing cytochrome b_{557} or bacterioferritin (12,14) (24 subunits; 12 haem *b* ; up to 4500 non-haem Fe)

There are many kinds of cytochromes: it is probably the largest family of metallo-redox proteins. Table 1 describes briefly some of the better characterised cytochromes.

RATES OF INTERPROTEIN ELECTRON TRANSFER

The simplest reaction scheme for a bimolecular redox reaction is that shown in Fig.2 (Moore and Pettigrew, 1990; Marcus and Sutin, 1985). To react, oxidised protein A and reduced protein B come together to form an interprotein complex [1]. Though this complex may have only a transient existence it is essential for reaction to occur. Within the complex the electron transfer donor and acceptor energy levels become equalised [2] so that radiationless electron transfer can take place [3]. This is followed by relaxation [4] and dissociation [5]. In many cases the overall bimolecular rate constant is determined by protein association/dissociation work terms, though if substantial conformational rearrangements are required for step [2] then the resulting high reorganisation energy may lead to a low overall rate of reaction. Generally, however, with monohaem electron transferring cytochromes the reorganisation energy barriers are rather low (Moore and Pettigrew, 1990; Marcus and Sutin, 1985).

The effect of the protein association/dissociation work terms can be seen in the reaction of mitochondrial cytochrome *c* with its cytochrome *c* oxidase. *In vitro* assays show there are two binding sites for cytochrome *c* on the oxidase (Pettigrew and Moore, 1987); a high affinity site with a K_d of 0.1 μM and a k_{off} of 40 s^{-1} and a lower affinity site with a K_d

Association:	Ao + Br	→	(Ao --- Br)
Equalization of energy levels:	(Ao --- Br)	→	(Ao --- Br)*
Electron transfer:	(Ao --- Br)*	→	(Ar --- Bo)*
Relaxation:	(Ar --- Bo)*	→	(Ar --- Bo)
Dissociation:	(Ar --- Bo)	→	Ar + Bo

Figure 2. Reaction scheme for interprotein electron transfer. **o** and **r** indicate oxidised and reduced species respectively. When proteins **A** and **B** are the same the reaction is self-exchange; where they are different it is a cros-reaction.

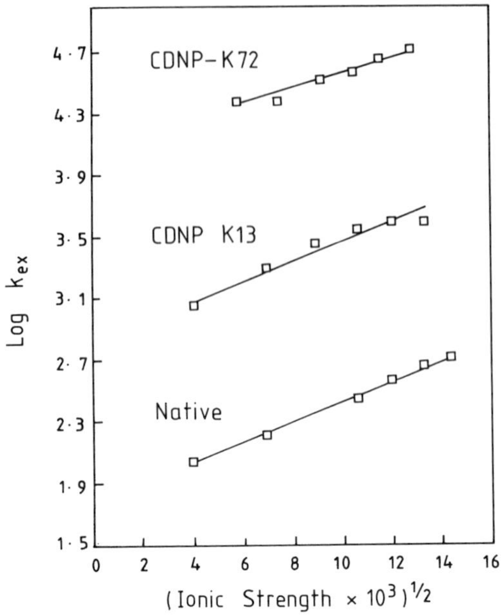

Figure 3. Self-exchange electron transfer rates of horse cytochrome *c*. The ionic strength dependence of the self-exchange rate was measured at 300 K and pH7 in sodium cacodylate buffer. Ionic strengths were calculated from the concentrations of sodium cacodylate. CDNP indicates the carboxy-dinitrophenol derivative of lysines 72 and 13 (Concar et al., 1986).

of 10 μM and a k_{off} of 200 s^{-1} (all at pH 6 and I = 0.1 M) (Ferguson-Miller et al., 1979; Wilms et al., 1981). These dissociation rates are comparable to the forward rates of electron transfer between cytochrome *c* and its oxidase *in vivo* in state 3 mitochondria (Chance, 1974) consistent with an overall rate through cytochrome *c* limited by its dissociation from its oxidase rather than by the electron transfer step [3] itself. The *in vivo* data also suggest that since electron transfer through cytochrome *c* is not the rate determining step of mitochondrial respiration, it is unlikely to have been optimised via evolutionary pressures.

The cytochrome *c* electron self exchange rates also reveal the influence of association work terms (Moore and Pettigrew, 1990; Marcus and Sutin, 1985; Gupta et al., 1972). Here the reactive complex is formed between molecules of similar charge and thus there are unfavourable electrostatic work terms to overcome before electron transfer can take place. For this reason the rates of self-exchange reactions are usually dependent on both ionic strength and pH. When the pH equals the isoelectric point of the protein, electrostatic repulsions are minimised and the rate thereby enhanced. Similarly increasing ionic strength increases the rate (Fig.3). An alternative way to enhance the rate is to chemically alter the charge on the protein (Concar et al., 1986). This can be done by synthetic chemistry or by site-directed mutagenesis. Fig.3 shows that the carboxydinitrophenol derivatives of horse cytochrome *c* modified at Lys 13 or Lys 72, both of which have a negatively charged group in place of the positively charged lysine, have higher rates of self-exchange than does the native protein, in one case by a factor of 10^2. Note that the two derivatised forms of cytochrome *c* have different rates; this shows that it is not just the net charge that is important, it is the charge distribution as well (Concar et al., 1986).

One reason many cytochromes have high net charges, or local areas of high charge density, is to provide some guidance for the formation of productive complexes (Pettigrew and Moore, 1987; Matthew, 1983; Koppenol and Margoliash, 1982). In the case of horse cytochrome *c* only about 0.6% of the protein surface is occupied by the haem and if there were no orienteering of the cytochrome and its reaction partner prior to reaction then random collisions would produce a low rate of formation of productive complexes. Thus the physiological partners of mitochondrial cytochrome *c*, which is highly positively charged

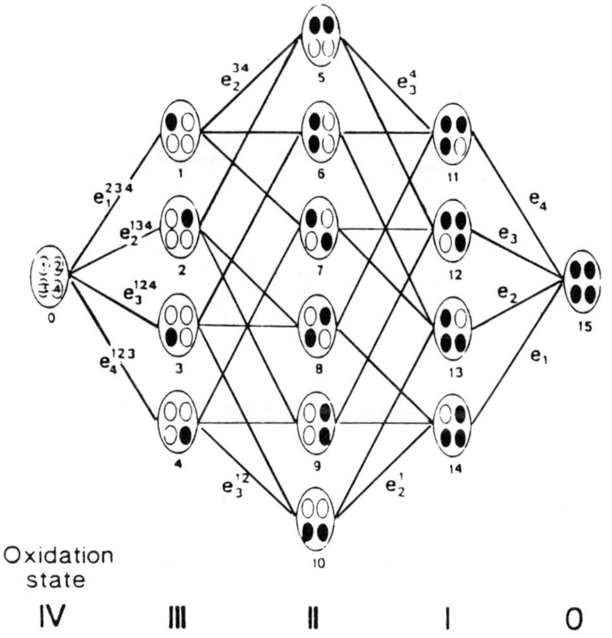

Figure 4. Redox states of tetrahaem cytochromes. Electron distribution scheme for a tetrahaem cytochrome. Oxidation state IV indicates the fully oxidised state and oxidation state 0 indicates the fully reduced state. States I, II and III are the partially reduced intermediate states. e_i^{jkl} is the microscopic midpoint potential of haem i when haems j, k and l are oxidised (Moura et al., 1982; Santos et al., 1984).

with a pI ~10, have negatively charged reaction domains for interacting with cytochrome *c* (Pettigrew and Moore, 1987).

Multihaem cytochromes, such as the tetrahaem cytochromes c_3 and photosynthetic reaction centre cytochrome *c*, exhibit a variety of electron transfer processes (Moore and Pettigrew, 1990; Xavier, 1986; Moura et al., 1982; Santos et al., 1984; Nitschke and Rutherford, 1994). In addition to intermolecular electron exchange, which is affected by interaction work terms in a similar fashion to that described above, there are intramolecular processes. Clearly there are sixteen possible redox states for a tetrahaem protein (Fig.4), and to fully describe the redox properties of such a protein the microscopic redox potentials and interstate electron transfer rates need to be determined. This is a formidable task, especially if haem-haem cooperativity occurs to remove degeneracy between some of the states. Nevertheless, considerable progress has been made in characterising both types of tetrahaem cytochrome. Not all the intramolecular rate constants are rapid; some for cytochrome c_3 may be as low as $10^3 s^{-1}$ (Moura et al., 1978). Since the haems of cytochrome c_3 are relatively close together [inter-iron distances are 10.9 - 17.3 Å (Haser et al., 1979; Higuchi et al., 1984)] such low rates indicate that some of the electron transmission pathways through the protein are not as effective as they could be or that there are redox linked conformation changes (Moore and Pettigrew, 1990).

REDOX POTENTIALS OF CLASS I CYTOCHROMES *C*

Class I cytochromes *c* are the most widespread type of water-soluble cytochrome (Pettigrew and Moore, 1987). Their Cys-Xxx-Yyy-Cys-His haem attachment sites are close to the N-terminus with their methionine sixth ligand located approximately 3/4 of the way along the sequence toward the C-terminus. The Class I group includes the mitochondrial cytochromes *c* and a wide variety of bacterial respiratory cytochromes *c*, including the cytochromes c_2 and *Pseudomonas aeruginosa* cytochrome c_{551}.

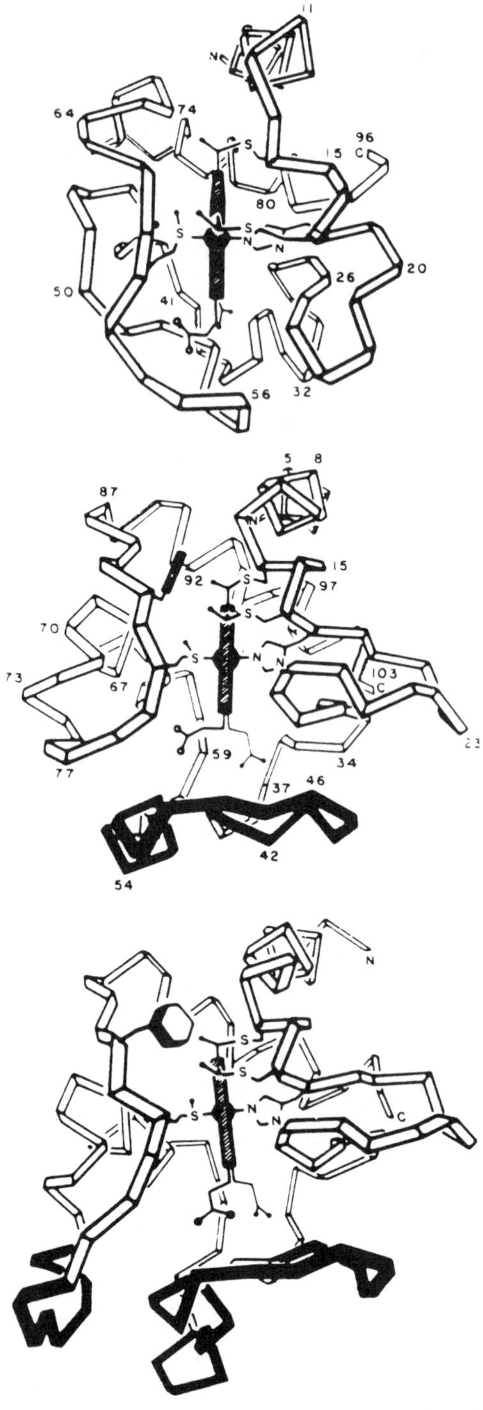

Figure 5. The class I cytochrome *c* protein fold. The ribbon represents the polypeptide chain. The shaded portions indicate additional residues compared to the top structure. The additional residues are located in surface loops. The haem groups are in the centre of each structure viewed edge-on. **Top:** *Chlorobium limicola* cytochrome *c*-555; **Middle:** Tuna cytochrome *c*; **Bottom**: Rhodospirillum rubrum cytochrome c_2. For further details see Meyer and Kamen, 1982; Moore and Pettigrew, 1990 and Dickerson, 1980; Salemme, 1977; Timkovich, 1979; Louie et al., 1988; Matsuura, 1982.

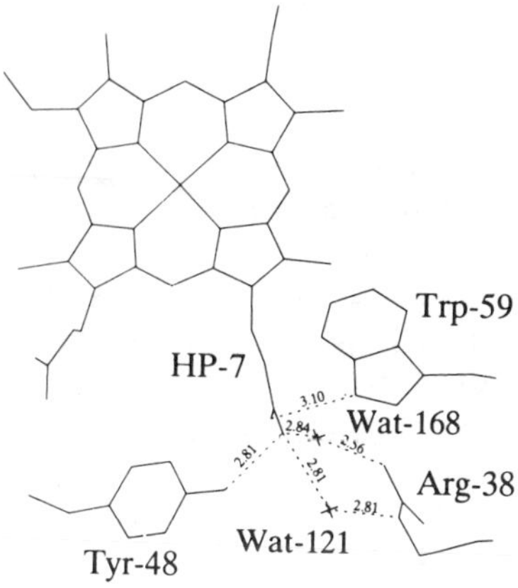

Figure 6. The environment of haem propionate-7 of yeast iso-1 ferrocytochrome *c* * indicate water molecules; residues replaced by site-directed mutagenesis are indicated (Louie et al., 1988; Davies et al., 1993).

Mitochondrial cytochromes *c* consists of 103-116 amino acids. It is an evolutionarily conservative class of proteins in that there are a large number of evolutionarily unvaried residues and a high proportion of conservatively substituted amino acids. In the compilation of 95 sequences analysed by Moore and Pettigrew (Moore and Pettigrew, 1990), that covers animals, plants and eukaryotic microorganisms, twenty-six amino acids were identified as evolutionarily unvaried. The tertiary structure of cytochrome *c* shows that the haem is embedded within the protein so that both the iron and its propionate substituents are shielded from the bulk solvent (Fig.5) (Moore and Pettigrew, 1990; Dickerson, 1980; Salemme, 1977;

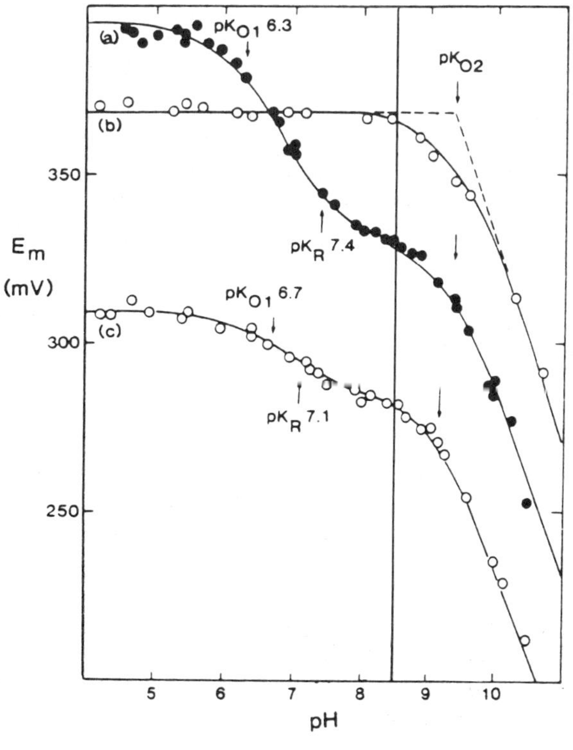

Figure 7. pH dependence of redox potential for some class I cytochromes *c*. (a) *Rhodomicrobium vannielii* cytochrome c_2; (b) *Rhodobacter capsulatus* cytochrome c_2; (c) *Rhodopseudomonas viridis* cytochrome c_2. Solid and open circles are experimental points. Solid lines are theoretical curves calculated as described by Pettigrew(Moore and Pettigrew, 1990). pK_{o2} is the alkaline isomerisation involving replacement of the methionine ligand by lysine, and pK_{o1} and pK_r are the ionisation of a haem propionate in (a) and a non-ligand histidine in (c) (Moore and Pettigrew, 1990; Moore et al., 1984).

Timkovich, 1979). Both propionate substituents interact with a variety of functional groups via hydrogen-bonding interactions. Those for haem propionate-7 of yeast iso-1 cytochrome *c* are illustrated in Fig.6 (Louie et al., 1988).

The bacterial class I cytochromes *c* have the same basic three-dimensional structure, even though in some cases there is minimal amino acid sequence identity and the proteins are of different size (Meyer and Kamen, 1982; Moore and Pettigrew, 1990; Higuchi et al., 1984; Dickerson, 1980). The structures illustrated in Fig.5 reveal that the size differences are largely accommodated by the insertion or deletion of loops at the surface of the molecule. The location and orientation of the haem and its axial ligands, together with the distribution of α-helices, is the same for all three structures.

Bacterial cytochromes c_2 tend to be larger than mitochondrial cytochromes *c*, ranging from 107 to 140 residues (Meyer and Kamen, 1982; Moore and Pettigrew, 1990). The cytochromes c_{551} are smaller. The size difference is reflected by the haem exposure to bulk solvent: only one of the haem propionates of *Pseudomonas aeruginosa* cytochrome *c*551 is completely shielded (Matsuura, 1982).

A comparison of the redox potentials of various Class I cytochromes *c* is instructive. Some are strongly pH dependent over the range pH 5-8 as a result of the ionisation of a haem propionic acid substituent while in others the haem propionic acid substituents do not ionise over this pH range and the haem redox potential is pH independent (Fig.7) (Moore and Pettigrew, 1990). The shift in E_M with varying pH results from the redox state dependence of the pK_a value of the ionising haem propionate, itself a consequence of an electrostatic interaction between the carboxylate negative charge and the net positive charge on the ferrihaem of the oxidised protein (Moore, 1983). Thus there are three reactions depending upon the pH:

$$H{-}Cyt^{III} + e^- \rightarrow H{-}Cyt^{II} \qquad (i)$$

$$H^+ + {}^-Cyt^{III} + e^- \rightarrow H{-}Cyt^{II} \qquad (ii)$$

$${}^-Cyt^{III} + e^- \rightarrow {}^-Cyt^{II} \qquad (iii)$$

The haem propionic acid ionisations are relatively fast with proton dissociation rates of $\sim 10^5$ s^{-1} (Moore and Pettigrew, 1990; Leitch et al., 1984).

The change in E_M at high pH for both cytochromes of Fig.7, indicated by pK_{695}, is a result of a relatively slow conformational change of the oxidised protein in which the axial methionine ligand is replaced by a lysine (Moore and Pettigrew, 1990; Moore et al., 1984). Not only do some class I cytochromes *c* exhibit the Type A behaviour of Fig.7 and others the Type B, but also some have redox potentials in the range of 240-260 mV at pH 7, including the mitochondrial cytochromes *c*, whilst others have redox potentials spanning from ~100 mV to 500 mV. The structural origins of these differences have been extensively investigated but there are no clear, quantitative relationships between structural differences and redox potential differences established.

It is clear that the nature of the solvent is important, at least in influencing the haem propionate pK_a, as predicted by theoretical considerations of electrostatic interactions in proteins (Rogers et al., 1985) and by experimental results with *P. aeruginosa* cytochrome c_{551} in various solvent systems (Table 2) (Rigby et al., 1991). The solvent too is expected to be influential in determining E_M (Kassner, 1973). The protein is clearly important but whether it is the combined effects of all the protein dipoles and charges that control the level of the E_M, or the effect of a few residues that dominate is unclear.

Since it was observed that some of the evolutionarily unvaried residues of mitochondrial cytochrome *c* were haem propionate contact residues (Fig.6) (Moore and Pettigrew,

Table 2. Apparent pK values for haem propionate-7 of *P. aeruginosa ferricytochrome* c_{551} in mixed aqueous/organic solvents (Rigby et al., 1991)

HAEM PROPIONATE IONISATION

	solvent dielectric constant	pK_a
H_2O/D_2O	78.5	6.2
D_2O/glycerol	65.0	6.9
D_2O/propylene glycol	58.3	6.8

1990), and that in cytochromes c_2 lacking an arginine equivalent to Arg 38 the haem propionate ionised with pK_a values in the range 6 to 8 (Moore et al., 1984), a mutagenesis study was undertaken to replace the contact residues Arg 38, Tyr 48 and Trp 59 (Cutler et al., 1989; Thurgood et al., 1991; Davies et al., 1993). The results of this study are summarised in Table 3.

In none of the variants does a haem propionate ionise in the pH range 5-7, showing that the local environment is not the dominant factor. However, the redox potentials of the variants were modified. 2D-NMR studies showed that the amino acid substitutions did not produce substantial accompanying conformational changes and thus the redox potential decreases were ascribed to increasing solvent accessibility of the haem or to variations in dipole interactions with the ferrihaem charge resulting from an altered polarity of haem contact residues.

REDOX CHARACTERISTICS OF FERRITIN AND BACTERIOFERRITIN

Ferritins and bacterioferritins are iron storage proteins of animals and bacteria respectively (Steifel and Watt, 1979; Moore, 1994; Harrison et al., 1991, Theil, 1991). Though not generally regarded as redox proteins, this group of molecules have a range of properties that may make them good candidates for use in electronic systems. Both are composed of twenty-four subunits, of mass about 20 kD, that aggregate into an approxi-

Table 3. Redox potentials of *S. cerevisiae* cytochrome *c* and variants of it produced by site-directed mutagenesis (Cutler et al., 1989; Davies et al., 1993)

cytochrome	E_m at pH 5 (mV)	pK_{695}
control: R38,Y48,W59	285	8.0
H38	259	8.3
N38	248	8.2
Q38	245	8.4
A38	237	8.3
F48	263	8.6
F48,F59	215	7.6
A38,F48,F59	220	7.0

[pK_{695} is an indicator of the stability of the ferricytochrome. It is the pK_a for loss of the axial Met ligation to the Fe^{3+}]

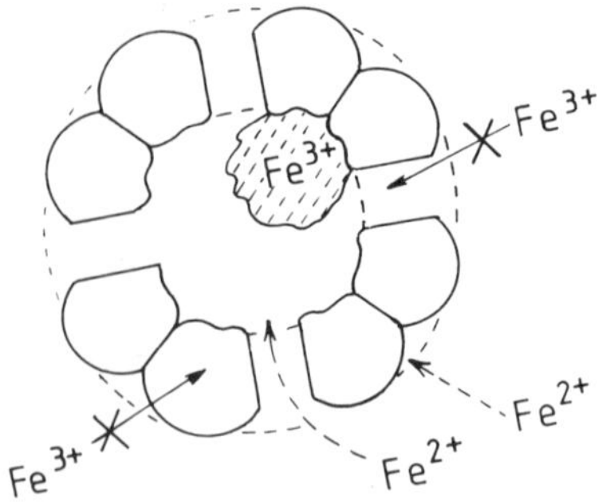

Figure 8. Schematic representation of the structure of ferritin and bacterioferritin in cross-section. The sub-unit dimers forming the protein coat are shown as solid lines, and the non-haem iron mineral in the core is shaded (Harrison et al., 1991; Ford et al., 1987). Iron entry is indicated to be possible at appreciable rates only for Fe^{2+} with entry being via the channels between sub-units or through a protein sub-unit (Harrison et al., 1991; Theil, 1991; Ford et al., 1987; Rohrer et al., 1987). In the latter case Fe^{2+} entry is generally associated with the ferrooxidase reaction (Treffry et al., 1992; Xu and Chasteen, 1991). Oxidation of Fe^{2+} can occur on the growing mineral in the cavity and this reaction has a different oxidation pathway to the ferrooxidase reaction:

Ferrooxidase site [low Fe loading] $2Fe^{2+}+O_2 + 4H_2O \rightarrow 2FeOOH + H_2O_2 + 4H+$

Cavity site [high Fe loading] $4Fe^{2+}+O_2 + 6H_2O \rightarrow 4FeOOH + 8H^+$

mately spherical oligomer of 120 Å diameter. The protein subunits form a coat of about 20Å thickness that encapsulates a central cavity of 80 Å diameter (Ford et al., 1984). Into this cavity a non-haem iron mineral can be formed (Fig.8). Some of the characteristics of this mineral are:

1. Up to 4500 Fe^{3+} ions per molecule can be stored as crystalline ferrihydrite, $Fe_2O_3.9H_2O$. Crystalline cores can be laid down both *in vitro* and *in vivo* (Ford et al., 1984; Rohrer et al., 1987; Mann et al., 1987)
2. A less dense amorphous core of iron-oxyhydroxide-phosphate can be formed, both *in vitro* and *in vivo* (Steifel and Watt, 1979; Moore et al., 1986; Rohrer et al., 1987; Mann et al., 1987)
3. Antiferromagnetic coupling between the Fe^{3+} ions of both amorphous and crystalline cores produces a relatively low magnetic susceptibility (Granick, 1946).
4. Crystalline cores exhibit superparamagnetic behaviour at low temperatures (St.Piere et al., 1989).
5. Fe^{2+} cores are stable under non-oxidising conditions and in the absence of Fe^{2+} chelators (Watt et al., 1985; Moore et al., 1992).
6. The redox potential of the core of ferritin is strongly pH dependent. At pH 7 the crystalline core of animal ferritin has an E_M of -190 mV whilst that of the amorphous core of bacterioferritin an E_M of -420 mV (Watt et al., 1985; Watt et al., 1986).

As isolated, bacterioferritins contain between 1 and 12 haems *b* per molecule each axially coordinated to two methionine sidechains (Cheesman et al., 1990). Animal ferritins do not contain haem and nor can they be induced to bind haem *in vitro* until they have undergone an, as yet, undefined ageing process. Aged animal ferritin has been shown to bind sixteen haem groups per molecule each at bis-histidine coordination sites (Kadir and Moore, 1990; Moore et al., 1992,). Aged bacterioferritin from *P. aeruginosa* also binds additional haem *in vitro*, in addition to its intrinsic haem, to give a final haem loading of twenty-four haems per molecule (Moore et al., 1992). Should these ageing processes be defined, protein engineering may allow the construction of stable variant ferritins capable of binding haem.

Both the mechanisms of iron uptake and those for iron release are unclear. *In vitro* reductive iron release assays show that haem bound to ferritin and the haem of bacterioferritin promote iron release to chelators, probably by enhancing electron transfer through the protein coat into the core Fe^{3+} (Moore et al., 1992; Kadir et al., 1992,). Rapid iron uptake into ferritin and bacterioferritin in *in vitro* assays only occurs with Fe^{2+} (Harrison et al., 1991; Theil, 1991; Treffry et al., 1992; Le Brun et al., 1993; Xu and Chasteen, 1991) (Fig.8). Under aerobic conditions the Fe^{2+} becomes oxidised and an Fe^{3+} core laid down. Recent work has identified several stages in the iron uptake pathway into bacterioferritin (Le Brun et al., 1993) that may involve a catalytic centre, termed the ferrooxidase centre, proposed to be a common feature of all ferritins (Treffry et al., 1992). Work is underway in several laboratories to investigate this.

PROTEIN AGEING AND PROTEIN STABILITY

Protein stability is a widely used term that has a variety of accepted definitions. It may refer to the temperature or chaotroph concentration, at which a protein denatures, as monitored by spectroscopic methods, or experimental conditions under which the protein loses 50% of its activity in standard assays (Creighton, 1993). It is usually assumed that attempts to increase by protein engineering, thermal structural stability will also lead to an enhancement of functional stability and stability to a variety of chaotrophs. However, this may not always be the case.

Protein stability is not always a matter of zero activity or 100% activity. Often individual molecules age and their intrinsic activity become modified. This is particularly so with metalloproteins that take part in redox chemistry. Ferritin and bacterioferritin provide a good example of this. With age the properties of such molecules alter so that, for example, they are capable of binding haem and have altered rates of iron uptake. Also aged ferritins and bacterioferritins contain quinones which may be the result of the oxidation of tyrosine residues (Al-Massad et al., 1992). Whether these two phenomena are related has not yet been established.

The point needs to be stressed, however, that the performance and stability of proteins may not have the reproducibility and stability desirable for their incorporation into electronic devices.

PROTEIN ENGINEERING OR *DE NOVO* PROTEIN DESIGN?

In the attempt to find a protein suitable for an electronic device what approach should be pursued, protein engineering or *de novo* protein design? The former assumes that by using techniques such as site-directed mutagenesis an existing protein, perhaps a cytochrome *c*, can be altered to a form with acceptable properties; and the latter approach assumes it will be more productive to use techniques such as peptide synthesis or gene synthesis followed by protein expression. While it remains the case that most metalloprotein studies are focused on naturally occurring proteins, generally because such studies are carried out by biochemists aiming to define biological mechanisms, a variety of semi-synthetic and synthetic studies have been carried out. The work of Dutton *et al.* (Choma et al., 1994; Robertson et al., 1994) is a good example of the synthetic approach. These workers have designed a number of molecular maquettes based on a synthetic 62 residue protein that fold into 2 helices and binds

2 haem groups. The aim of this work is to synthesise such maquettes to reproduce the redox characteristics of proteins such as cytochrome *bc*$_1$, cytochrome *aa*$_3$ and the photosynthetic reaction centre. It seems probable that of all the approaches possible, this synthetic approach will be the one most likely to provide useful metalloproteins for electronic applications.

ACKNOWLEDGEMENTS

The author thanks the Biology and Biotechnology Science Research Council and the Engineering and Physical Sciences Research Council, through their joint Biomolecular Sciences Committee, and the Wellcome Trust for supporting his work on ferritins and bacterioferritins; NATO and the CEC for supporting collaborative projects on cytochromes; and Professors M.T. Wilson (Essex), A.G. Mauk (UBC Vancouver) and A.J. Thomson (UEA) and Dr. N. Le Brun (UEA) for helpful discussions.

REFERENCES

Al-Massad, F. K., Kadir, F. H. A., and Moore, G. R., 1992, *Biochem. J.* 283:177-180.

Bielka, H., Horecker, B. L., Jacoby, W. B., Karlson, P., Keil, B., Liebecq, C., Lindberg, B., and Webb, E. C., 1979, in Recommendations of The Nomenclature Commission of The IUB. Academic Press, New York.

Chance, B., 1974, *Ann. N.Y.Acad.Sci.* 227:613-626.

Chapman, S. K., White, S. A., and Reid, G. A., 1991, *Adv. Inorg. Chem.* 36:257-301.

Cheesman, M. R., Thomson, A. J., Greenwood, C., Moore, G. R., and Kadir, F. H. A., 1990, *Nature* 346:771-773.

Choma, C. T., Lear, J. D., Nelson, M. J., Dutton, P. L., Robertson D. E., and DeGrado, W. F., 1994, *J. Am. Chem. Soc.* 116:856-865.

Concar, D. W., Hill, H. A. O., Moore, G. R., Whitford, D., and Williams, R. J. P., 1986, *FEBS Letters* 206:15-19.

Creighton, T. E.,1993, in Proteins 2nd edition, 507pp. W.H.Freeman, New York.

Cutler, R. L., Davies, A. M., Creighton, S., Warshel, A., Moore, G. R., Smith, M., and Mauk, A. G., 1989, *Biochemistry* 28:3188-3197.

Davies, A. M., Thurgood, A. G. P., Greenwood, C, Guillemette, J. G., Smith, M., Mauk, A. G., and Moore, G. R., 1993, *Biochemistry* 32:5431-5455.

Dickerson, R. E., 1980, *Sci Amer* 242:136-153.

Ellfolk, N., Ronnberg, M., Aasa, R., Andreasson L-E., and Vanngard, T., 1983, *Biochim Biophys Acta* 743:23-30.

Ferguson-Miller, S., Brautigan, D. L., and Margoliash, E., 1979, in The Porphyrins 7 (Ed. D. Dolphin), Academic Press, London New York 149-240.

Foote, N., Peterson, J., Gadsby, P. M. A., Greenwood C., and Thomson, A. J., 1985, *Biochem J* 230:227-237

Ford, G. C., Harrison, P. M., Rice, D. W., Smith, J. M. A., Treffry, A., White, J. L., and Yariv, J.,1984, *Phil. Trans. R. Soc. Lond.* B304:551-565.

Granick, S., 1946, *Chemical Revs.* 38:379-403.

Gupta, R. K., Koenig, S. H., and Redfield, A. G., 1972, *J Mag Res* 7:66-73.

Harrison, P. M., Andrews, S. C., Artymiuk, P. J., Ford, G. C., Guest, J. R., Hirsmann, J., Lawson, D. M., Livingstone, J. C., Smith, J. M. A., Treffry, A., and Yewdall, S. J., 1991, *Adv. Inorg. Chem.* 36:449-486.

Haser, R., Pierrot, M., Frey, M., Payan, F., Astier, J. P., Bruschi, M., and Le Gall, J., 1979, *Nature* 282:806-810.

Higuchi, Y., Kusunoki, M., Matsuura, Y., Yasuoka, N., and Kakudo, M., 1984 *J Mol Biol* 172:109-139.

Kadir, F. H. A., Al-Massad, F. K., and Moore, G. R., 1992, *Biochem. J.* 282:867-870.

Kadir, F. H. A.,and Moore, G. R., 1990, *FEBS Lett.* 276:81-84.

Kassner, R. J., 1973, *J Am Chem Soc* 95:2674-2677.

Koppenol, W. H., and Margoliash, E., 1982, *J Biol Chem* 257:4426-4437.

Le Brun, N. E., Wilson, M.,T., Andrews, S., Guest, J. R., Harrison, P. M., Thomson, A. J., and Moore, G. R., 1993, *FEBS Letters* 333:197-202.

Leitch, F. A., Moore, G. R., and Pettigrew, G.W., 1984, *Biochemistry* 23:1831-1838.

Lemberg, R., and Barrett, J., 1973, in The Cytochromes, Academic Press, London New York.

Louie, G. V., Hutcheon, W. L. B., and Brayer, G. D., 1988, *J Mol Biol* 199:95-314.
Mann, S., Williams, J. M., Treffry, A., and Harrison, P. M., 1987, *J. Mol. Biol.* 198:405-416.
Marcus, R. A., and Sutin, N., 1985, *Biochim Biophys Acta* 811:265-322.
Martinez, S. E., Huang, D., Szczepaniak, A., Cramer W. A., and Smith J. L., 1994, *Structure* 2:95-105.
Mathews, F. S., 1984, *Prog. Biophys. Mol. Biol.* 45:1-56.
Matsuura,Y., Takano, T., and Dickerson, R. E., 1982, *J Mol Biol* 156:389-409.
Matthew, J. B., Weber, P. C., Salemme, F. R., and Richards, F. M., 1983, *Nature* 301:169-171.
Meyer, T. E., and Kamen, M. D., 1982, *Adv Prot Chem* 35:105-212.
Moore, G. R., 1983, *FEBS Letters* 161:171-175.
Moore, G. R., 1994, in Encyclopedia of Inorganic Chemistry *in press,* John Wiley publishers.
Moore, G. R., and Pettigrew, G. W., 1990, in Cytochromes *c* : Evolutionarily, Structural and Physicochemical Aspects, Springer-Verlag. 478.
Moore, G. R., Cheesman, M. R., Kadir, F. H. A., Thomson, A. J., Yewdell, S.J., and Harrison, P.M., 1992, *Biochem. J.* 287:457-460.
Moore, G. R., Harris, D. E., Leitch, F. A., and Pettigrew, G. W., 1984, *Biochim. Biophys. Acta.* 764:331-342.
Moore, G. R., Kadir, F. H. A., and Al-Massad, F. K., 1992, *J. Inorg. Biochem.* 47:175-181.
Moore, G. R., Mann, S., and Bannister, J. V., 1986, *J.Inorg.Biochem.* 28:329-336.
Moura, J. J. G., Santos, H., Moura, I., LeGall, J., Moore, G. R., Williams, R. J. P., and Xavier, A. V., 1982, *Eur.J.Biochem.* 127:151-155.
Moura, J. J. G., Xavier, A. V., Cookson, D. J., Moore, G. R., Williams, R. J. P., Faque, Bruschi, M.G., and Le Gall, J., 1978, *Biochem. Soc. Trans.* 6:1285-1287.
Nitschke, W., and Rutherford, A. W., 1994, *Biochem. Soc. Trans. in press.*
Pettigrew, G. W., and Moore, G. R., 1987, in Cytochromes c : Biological Aspects, Springer-Verlag. 282.
Rigby, S. E. J., Burch, A. M., and Moore, G. R., 1991, *Mag. Reson. Chem.* 29:1036-1039.
Robertson, D. E., Farid, R. S., Moser, C. C., Urbauer, J. L., Mulholland, S. E., Pidikiti, R., Lear, J. D., Wand, A. J., DeGrado, W. F., and Dutton, P. L., 1994, *Nature* 368:425-432.
Rogers, N. K., Moore, G. R., and Sternberg, M. J., 1985, *J.Mol.Biol.* 182:613-616.
Rohrer, J. S., Frankel, R. B., Papaefthymiou, G. C., and Theil, E. C., 1987, *Inorg. Chem.* 28:3393-3395.
Salemme, F. R., 1977, *Ann Rev Biochem* 46:299-329.
Santos, H., Moura, J. J. G., Moura, I., LeGall, J., and Xavier A. V., 1984, *Eur.J.Biochem.* 141:283-296.
St. Pierre, T. G., Webb, J., and Mann, S., in Biomineralization, (Eds Mann, S., Webb, J., and Williams, R. J., P.), VCH Publ., Weinheim, 1989, 295-344.
Steifel, E. I., and Watt, G. D., 1979, *Nature* 81-83.
Theil, E. C., 1991, *Adv. Enzymol.* 63:421-449.
Thurgood, A. G. P., Davies, A. M., Greenwood, C., Guillemette, J. G., Smith, M., Mauk, A. G., and Moore, G. R., 1991, *Eur. J. Biochem.* 202:339-347.
Timkovich, R., 1979, in: The Porphyrins 7 (Ed. D. Dolphin) Academic Press, London New York, 241 - 294.
Treffry, A., Hirzmann, J., Yewdall, S. J., and Harrison, P. M., 1992, *FEBS Lett* 302:108-112.
Watt, G. D., Frankel, R. B., and Papaefthymiou, G. C., 1985, *Proc. Natl. Acad. Sci. USA* 82:3640-3643.
Watt, G. D., Frankel, R. B., Papaefthymiou, G. C., Spartalian, K., and Steifel, E. I., 1986, *Biochemistry* 25:4330-4336.
Wilms, J., Veerman, E. C. I., Konig, B. W., Dekker, H. L., van Gelder, B. F., 1981, *Biochim Biophys Acta* 637:168-176.
Xavier, A. V., 1986, in Frontiers in Bioinorganic Chemistry (Ed.A.V.Xavier) VCH Publishers, 722-725.
Xu, B., and Chasteen, N. D., 1991, *J. Biol. Chem. 266:19965-19970.*

9

HIGH-YIELD RECOMBINANT CYTOCHROME P450 SCC AND ITS OPTIMIZATION FOR BIOELECTRONICS FROM AB INITIO CONSIDERATIONS

P. Paschkevitch,[2] V. Sivozhelezov,[2] S. Vakula,[2] V. Sidorovich, M. Eldarov, F. Nozza,[1] and C. Nicolini[1]

Center "Bioengineering", Russian Academy of Sciences
Prosp. 60-let Oktyabrya, 7/1, 117234 Moscow, Russia
[1] Institute of Biophysics, University of Genoa
Via Giotto 2, 16153 Genoa, Italy
[2] EL.B.A. Foundation, Via A. Moro 17
57033 Marciana Marina (LI), Italy

INTRODUCTION

Metalloproteins play diverse and important roles in various vital bioenergetic reactions and metabolic pathways, including respiration, photosynthesis, monooxygenation reactions and many others [Harison P.M., 1985]. They have been chosen as the most plausible materials for application in bioelectronics on the basis of several of their well-documented properties discussed in [Nicolini, 1996, Sivozhelezov et al., in preparation] of which the most important are, on the one hand, the ability to use electron as information carrier in the designed molecular electronic devices, and, on the other hand, the pronounced anisotropy of physical properties allowing to consider these proteins as cellular automata [Sivozhelezov et al., in preparation] characterized by finite numbers of functional states separated by energy barriers high enough to be overcome by thermal fluctuations. Those anisotropic properties include the presence of highly selective pathways for electron transfer and highly asymmetric distributions of electrostatic fields and hydrophilic/hydrophobic amino acid residues over their surfaces, and can also be used for controlled association of those proteins into supramolecular structures like thin films suitable for macroscopic characterization and/or direct industrial applications [Nicolini, 1996].

The biophysical, biochemical and enzymatic properties of different classes of metalloproteins, like haemoglobins, peroxidases, blue-copper proteins, cytochromes P450 and other cytochromes have been studied extensively and, as a result of genetic engineering effort, the genes for most of these systems are available. Genetic engineering is therefore envisaged as a powerful and most important tool for designing or improving stabilities and

Molecular Manufacturing, edited by C. Nicolini
Plenum Press, New York, 1996

Table 1. Relative expression levels of recombinant metalloproteins in bacterial and yeast systems

Representive metalloprotein	Heterologous host	Expression level (mg/l)	Reference
Azurin	E.coli	10	(Germanas et al., 1993)
Cytochrome b5	E.coli	50	(Holmans et al.,1994)
Cytochrome C	S.cerevisiae	0,2-2	(Lu et al., 1993)
Cytochrome P450E1	E.coli	25	(Imai et al., 1993)
Cytochrome P450A1	yeast	20	(Truan et al., 1993)

electron-transfer characteristics of those proteins, the three-dimensional structures known for many of these proteins providing good starting points for this strategy.

In designing the biocomputer architecture, metalloproteins have the advantage over the conventional devices both in spatial dimensions and in speed, that consequently results in final advantage in performance per unit of volume of as much as 10^7-10^9 which provides a significant challenge for material designers [Nicolini, 1996, Sivozhelezov et al., 1995].

The use of recombinant DNA technology had revolutionarized the study of electron-transfer mechanisms in metalloproteins and their complexes. The most widespread application of this approach is the introduction of precise structural changes in the protein of interest by site-directed mutagenesis. This has resulted in dramatic broadening of our understanding of relations between the structure and the mechanistic, spectroscopic, electrochemical, kinetic and other parameters of redox-active metalloproteins [Mauk, 1991, Canters and Gillardi, 1993, Caffrey and Cusanovich, 1994].

As compared with the method of extraction from natural source or chemical synthesis, heterologous synthesis of recombinant metalloproteins is cheaper and advantegeous for obtaining highly purified and precisely modified protein of interest in large amounts. With the development of powerful bacterial and eukaryotic expression systems [Sambrook et al., 1989, Ausubel et al., 1991, Studier et al., 1990] the achievement of protein production in sufficient quantity became rather sraightforward. *Escherichia coli* and *Saccharomyces cerevisiae* appear to be the vehicles of choice for the expression of different classes of metalloproteins, like cupredoxins, cytochromes C and b5, haemoglobins, cytochromes P450 and others [Mauk, 1991]. As summarized in Table 1, the production level in several cases range from 1 to 50 mg per liter of cell culture of recombinant strain, and yields up to 30% of the total cellular protein accounting to the produced target metalloprotein are not uncommon.

Genetic engineering also provides several convenient tools for the subsequent purification of recombinant metalloproteins [Nygren et. al., 1994]. This can be achieved by introduction of small affinity tags and tails to the N- or C—terminal parts of the expressed metalloproteins, and these modifications usually do not influence the intrinsic metalloprotein properties. Other useful approaches for structural modification of metalloproteins in the ways applicable to bioelectronics, include the engineering of unique attachment sites that can be used for controlled orientation on planar surfaces and chromatography supports [Stayton et al., 1990, McLean et al., 1993], and increasing their thermal stability by introduction of additional disulphur bonds and metal-chelating sites [Arnold and Zhang, 1994].

ELECTRON TRANSFER IN CYTOCHROME P450 SYSTEMS

One of the most promising electron-transfer systems that can be used as a source of new materials for bioelectronics, for the design of new biosensors and multistage enzyme processing systems, is that based on cytochrome P450-dependent monooxygenase reactions. The fact that electron transfer in those systems is coupled to an enzymatic reaction makes it possible to use the energy of the reaction for switching between the redox states of their metal sites. In view of the cellular automata concept, this makes the cytochrome P450-containing systems a particularly attractive object for bioelectronic applications. One other advantage over the small metalloproteins functioning as mere electron carriers is the size of P450-containing systems considerably lowering the probability of random electron transfer and thus favoring the above-mentioned specific electron-transfer pathways.

Cytochromes P450 enzymes comprise a large group of heme-containing proteins which catalyze monooxygenation reactions involving the insertion of molecular oxygen into the substrate. Such reactions are responsible for a wide variety of important biological functions, for example the biosynthesis of all steroid hormones and the metabolism of almost all lipophilic drugs and environmental toxins and carcinogens [Gonzales, 1990, Guengerich, 1992].

Recent advances in recombinant DNA technology have revealed the multiplicity of cytochromes P450 at the molecular level. More than 200 cDNAs and genomic DNAs coding for P450s have so far been isolated, sequenced, and classified into 36 gene families on the basis of their aminoacid sequence similarities [Degtyarenko and Archakov, 1993]

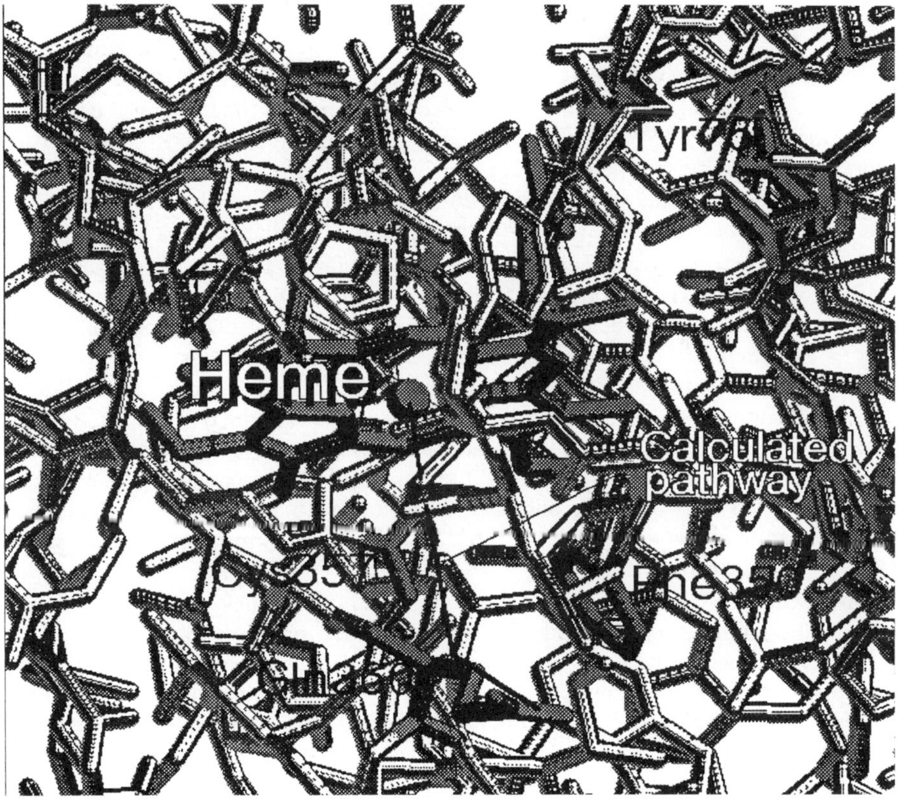

Figure 1. The 3-D structure of bacterial cytochrome P450cam, showing the residues involved in electron transfer pathway - Cyt357 and Gln360 [Bernhardt et al., 1994], the nearby aromatic residues Tyr75 and Phe350, and the heme which is imbedded deep within the protein structure.

All P450 enzymes have about 500 aminoacids and function as terminal oxidases in an electron-transport chain from NADPH. Cytochromes P450 enzymes fall into two broad groups: the microsomal ("type II") enzymes with a NADPH reducible flavoprotein containing FAD/FMN as electron donor, and mitochondrial and most procaryotic monooxygenases, requiring two redox proteins: a flavoprotein containing a single FAD and an iron-sulfur protein redoxin, that connects it to the heme protein [for recent review see Degtyarenko and Archakov, 1993].

The understanding of the detailed mechanisms of electron transfer and dioxygen activation in this rather complex enzymatic systems is important for the fundamental studies with this important class of heme oxygenases, and also promising for the design of novel biocatalysts with improved reactivity and stability.

The "paradigmatic" experimental model in this studies is represented by the bacterial cytochrome P450cam which 3D-structure became available nearly a decade ago [Poulos et al., 1987].

From the X-ray crystallogrphic data it is evident that the heme in P450cam is deeply embedded into the protein being not accessible from the surface. Thus, no direct pathway from the prosthetic group of the electron donor to the prosthetic group of P450cam seems to be available for the electron to be transferred, but a pathway via protein groups is necessary (Figure 1).

The role of charged and aromatic amino acid residues in mediating inter and intra-protein electron transfer within different P450 systems have been assessed using site-directed mutagenesis and chemical modifications [Shimada et al., 1994, and refs. therein]. For P450cam, this was accompanied by the predictions made by using molecular modeling, and, particularly, the electron transfer pathways analysis technique developed by Beratan and coworkers [Regan et al., 1993]. Promising results in studying the intraprotein electron transfer in the P450cam molecule were obtained using the approach based on the incorporation via genetic engineering of a specific attachment site on the surface of P450cam and the derivatization of this site with ruthenium bipyridal chelate which allows rapid input of reducing equivalents into the iron porphyrin system. [Bernhardt et al., 1994]

In cytochrome P450scc, however, the most important task at the moment seems to be verification of the 3D structure obtained by sequence homology, basing on the 3D structure of cytochrome P450cam [Vijayakumar & Salerno, 1992]. One feature of this structure is that the helix-containing amino acids 270-300 which, basing on the sequence homology, should be the substrate-binding helix, is located quite close to the heme. Particularly, the amino acid Thr291 is within a few Angstrom of the oxygen-binding pocket. Mutagenesis of amino acids located in this area can influence oxygen or CO binding as well as electron transfer properties by altering the heme iron's redox potential. Accordingly, mutagenesis of Ala286, Asn290, and Thr291 residues to Ile or Phe was carried out. Preliminary results show that the mutation of Asn290 to Phe prevents the protein from binding CO [in preparation].

Site-directed mutagenesis of Lys339 and Lys343 residues has shown them to be involved in the adrenodoxin binding site through which electron transfer occurs during interaction of the two proteins [Wada & Waterman, 1992]. On the model Cyt P450scc structure, an aromatic residue Phe410 is located nearby which may participate in the electron transfer process. It is therefore a possible target for further mutagenesis studies aiming to confirm the model structure.

The sucessful application of both theoretical and experimental methods for investigating electron transfer in different P450 systems critically depends on the progress in the development of methods for the production of these proteins in microbial and eukaryotic hosts.

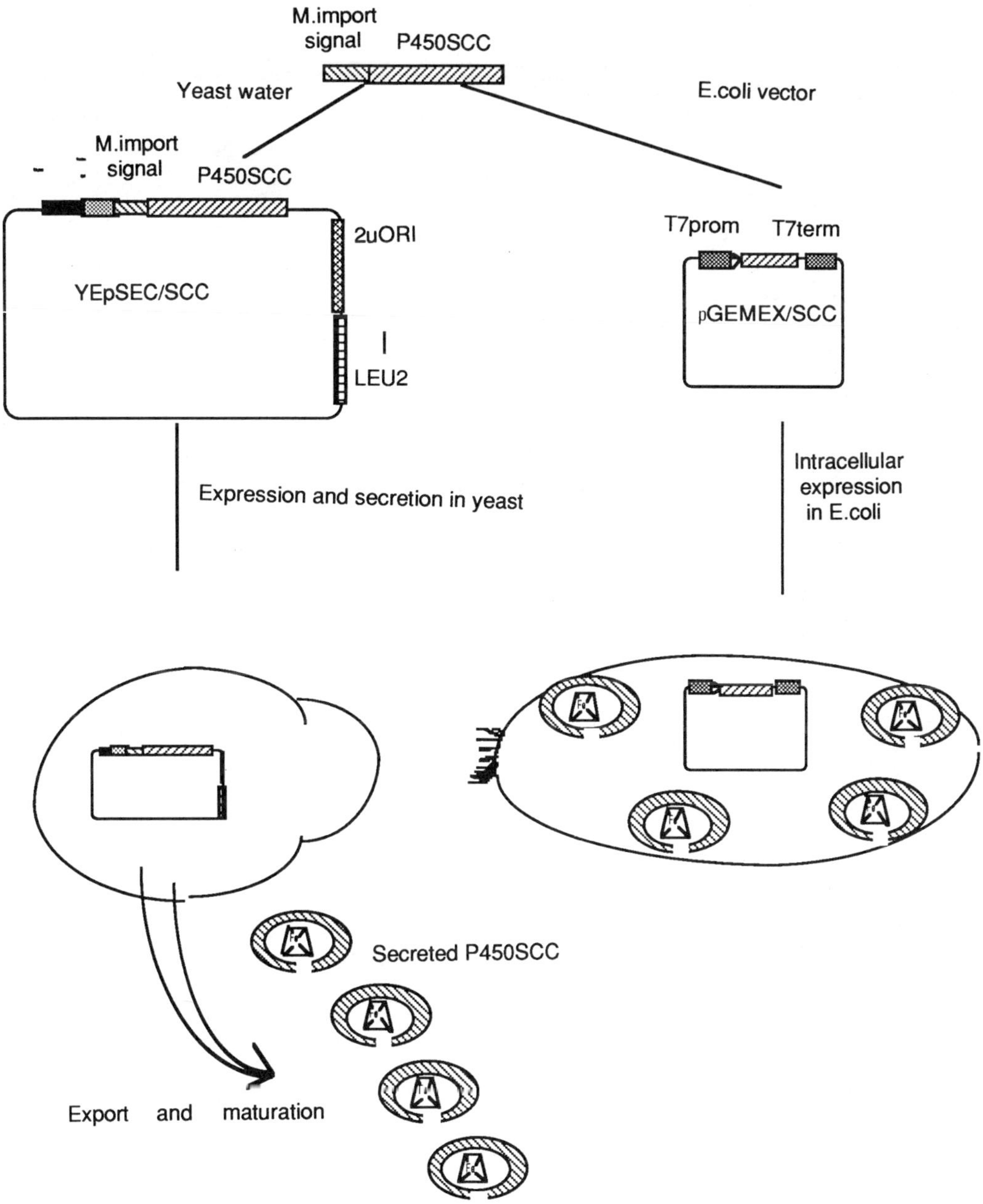

Figure 2. Scheme of electron flow from NADPH to cytochrome P450SCC and the catalyzed reaction. Abbreviations are: AR, adrenodoxin reductase; ADX, adrenodoxin; Two electrons are sequentially delivered to the P450SCC cytochrome to support the classical scheme of cytochrome-P450-catalyzed oxygen activation: substrate binding -> initial electron transfer->oxygen binding -> second electron transfer -> oxygen activation -> substrate oxidation.

Heterologous expression of mammalian cytochromes P450 has become an area of major interest for a variety of reasons and is being applied to study the role of individual isozymes in drug metabolism, carcinogen activation or deactivation, to study structure-activity relationships, in the generation of cell lines capable of synthesizing steroid hormones from cholesterol, and in the development of biocatalysts etc. [Waterman, 1994]. Several heterologous expression systems for P450 enzymes have been developed during last decade, including bacteria, yeast, mammalian and insect cell lines and plants. However, despite of significant progress in this field, succesful expression of P450s remains largely empirical at present.

DESIGN OF EXPRESSION SYSTEMS FOR THE PRODUCTION OF CYTOCHROME P450SCC

We have compared the efficiency of production of a mammalian mitochondrial cytochrome P450scc in the two most popular expression systems - *Escherichia coli* and *Saccharomyces cerevisiae*. This cytochrome plays a critical role in steroid metabolism in various steroidogenic tissues, where it functions within the mitochondrial "mini-electron transfer chain", including also adrenodoxin and adrenodoxin reductase [Usanov et al., 1990]. The cholesterol side chain-cleavage reaction to yield pregnenolone, catalyzed by this cytochrome (Figure 2) is known to be regulated by specific protein-protein interactions between the electron transfer partners in this system, and also dependent on substrate-induced enhancement of electron transfer [Lambeth, 1989]. The fundamental studies of electron transfer within this system, as well as its application for the creation of novel biocatalysts and biosensors require elevated amount of this protein.

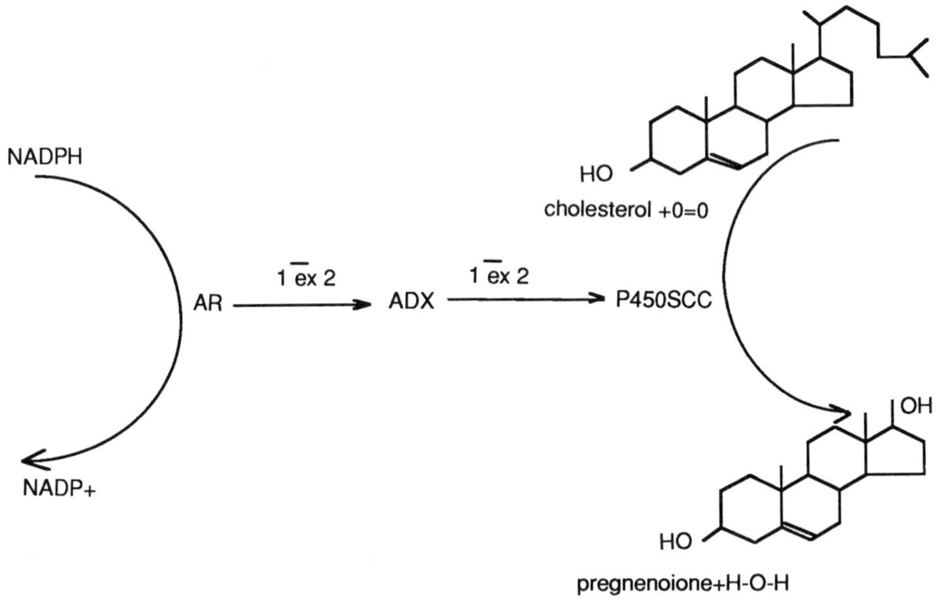

Figure 3. Essential steps in the development of bacterial and yeast expression systems for cytochrome P450scc starting from the gene-precursor and ending with the creation of recipient strains, transformed by the constructed plasmids.

~~~
A.
  Nco I
CCATGGTCACAAAGACCCCTCGCCCCTACAGTGAGATCCCCTCCCCTGGTGACAATGGCTGGCTTAACCTCTACCATTTC
   Met ValThrLysThrProArgProTyrSerGluIleProSerProGlyAspAsnGlyTrpLeuAsnLeuTyrHisPhe
``````````````````````````` mature P450scc ````````````````````````````

B.

---------------GAGCTC   ATG AAT ATA TTT TAC ATA TTT TTG TTT TTG CTG TCA TTC GTT CAA GGT------
-----------------------Met Asn Ile Phe Tyr Ile Phe Leu Phe Leu Leu Ser Phe Val Gln Gly------
----------------------|--------K.lactis secretion signal-------------------------------

  Sph I
GGCATGCTAGCAAGGGGGCTTCCCCTCCGCTCAGCCCTGGTCAAAGCCTGCCCACCCATCCTGAGCACAGTGGGG
GlyMetLeuAlaArgGlyLeuProLeuArgSerAlaLeuValLysAlaCysProProIleLeuSerThrValGly
|---------------------cytochrome--P450 scc--signal-for mitochondrial ----------

GAGGGCTGGGGCCACCACAGGGTGGGCACTGGAGAGGGAGCTGGCATCTCCACAAAGACCCCTCGCCCCTACAGT
GluGlyTrpGlyHisHisArgValGlyThrGlyGluGlyAlaGlyIleSerThrLysThrProArgProTyrSer
-import------------------------------------|   mature P450scc ```````````````````````
~~~

Figure 4. Nucleotide and aminoacid sequences of the 5′-coding part of the modified cytochrome P450scc gene, contained within the bacterial expression vector pGEMEX/SCC (A) and yeast secretion vector YEpSEC/SCC (B). Vertical bars indicate the sites of cleavage of N-terminal extension peptides by the signal protease. Modifications introduced due to the subcloning procedures are indicated in bold, together with the restriction sites used during the construction of the "precursor" and "mature" genes. The K.lactis killer toxin secretion signal is indicated in italic.

EXPRESSION OF MATURE P450SCC IN *E. COLI*

The two different strategies that we have employed for the production of bovine cytochrome P450scc in heterologous hosts are depicted on Figure 3.

For the bacterial expression of bovine P450SCC we have used the powerful T7 promoter/T7 polymerase system [Studier et al., 1990]. The complete cDNA gene for P450scc was structurally modified to code for a mature protein by eliminating the N-terminal mitochondrial targeting sequence as shown on Figure 4(A). The gene was cloned in the pGEMEX1-vector and the level of the expressed protein in obtained recombinant strains was assessed by means of electrophoretic separation of cell lysates, Western blotting and measurement of CO-difference spectra in the solubilized membranes of producing strain, as well as of purified recombinant protein after affinity chromatography on cholate-sepharose and adrenodoxine-sepharose.

Induction of P450SCC synthesis in the BL21(DE3) recipient strain, bearing the pGEMEX/SCC plasmid leads to rapid accumulation of a 56 kDa protein, that is found predominantly in inclusion bodies (Figure 5). As estimated from the densiometric scanning of the electrophoregrams of total cellular proteins, separated in SDS PAAG, the production level of rP450scc accounted to 60-80 mg/l bacterial culture. However, only a limited portion of the expressed protein was correctly folded and assembled with heme, as evidenced from the spectroscopic estimation of P450 content in solubilized membranes of the recombinant strain.

HIGH YIELD EXPRESSION OF CYTOCHROME P450SCC IN SECRETED FORM IN YEAST

In order to overcome the problems with correct folding and proper heme assembly of the bacterially expressed P450scc, we have designed a system for its secretion from

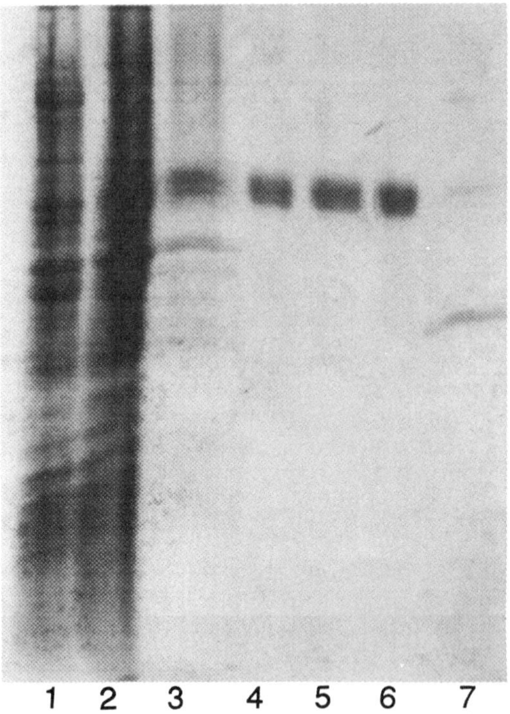

Figure 5. Expression and purification of cytochrome P450SCC with the developed bacterial and yeast systems. SDS-PAGE analysis of total cell lysates, culture medium concentrates and purified recombinant proteins.
Lane 1. Molecular weight markers (Pharmacia low, 14, 45, 66 kDa)
Lane 2. Total cell lysates of the control plasmid free bacterial recipient strain BL21 (DE3) , growth in TB medium to OD600=1, 50 μl of pelleted and lysed bacterial cell culture per lane.
Lane 3. The same for the strain transformed with the pGEMEX/SCC plasmid, 5h after induction with IPTG.
Lane 4. Inclusion bodies from the same strain.
Lane 5. Purified bacterial P450scc.
Lane 6. Purified yeast P450scc.
Lane 7. Purified native P450scc.

S.cerevisiae cells. The production of the heterologously expressed proteins in secreted form has several advantages. The proteins are protected from the cellular proteases, their subsequent purification is simplified and correct folding and formation of disulfur bonds is facilitated. As compared with mammalian cell culture, yeast expression has the advantage of both higher production levels and cost efficiency. Though the P450 cytochromes have no identifiable secreted homologues in nature, several examples justify that non-covalent heme attachment and folding may not interfere with protein secretion. Many fungal peroxidases are naturally produced as secreted proteins. The eukaryotic cytochrome b5 was shown to be efficiently translocated across bacterial inner membrane and processsed to yield authentic mature protein within the periplasm, when appended by a secretory signal for bacterial secretion [Karim et al., 1993]. And, finally, experiments involving the manipulation with both the microsomal and mitochondrial cytochrome P450 N-terminal signals, have shown that the structural and functional integrity of these proteins is tolerable to the changes in their subcellular localization [Waterman, 1994].

We have utilized the efficient yeast secretion vector YepSEC1 to direct sceretion of mature P450scc to the extracellular medium of recombinant yeast strains. The vector contains the secretion signal of *K.lactis* killer toxin and the tightly regulated GAL UAS-CYC1 hybrid promoter, that is advantageous to overcome possible problems with the toxicity of the overexpressed heterologous cytochrome P450scc [Baldari et al., 1987].

Figure 3 depicts the scheme employed to create the construction suitable for expression and secretion of cytochrome P450scc. The first step was to clone the gene, coding for the P450scc with intact N-terminal mitochondrial targeting signal "in frame " with the secretory signal of the *K.lactis* killer toxin, modified by insertion of specific adaptor. This was achieved after several subcloning steps. Figure 4 (B) shows the nucleotide and respective N-terminal amino acid sequence of the P450scc derivative, coded by the YepSEC/SCC vector. This clone was selected after restriction endonuclease analysis and sequencing of the *K.lactis* signal peptide/P450scc junction.

The transformants of the recipient *S.cerevisiae* strain, obtained with the constructed plasmid, were grown under conditions inducing for the GAL UAS-CYC promoter, and the presence of the expressed P450SCC protein in the culture medium supernatants, cell lysates and mitochondrial fractions was analyzed using Western-blotting with the anti-P450scc specific antibodies. This analysis showed, that the majority of the expressed P450scc have been exported to the culture medium. The accumulated secreted P450scc visually transformed culture medium from the yellowish to red color, testifying the very high secretion level.

The produced and secreted recombinant P450 cytochrome was concentrated from the culture medium supernatants using the "Amicon" ultrafiltration unit with the 30 kDa

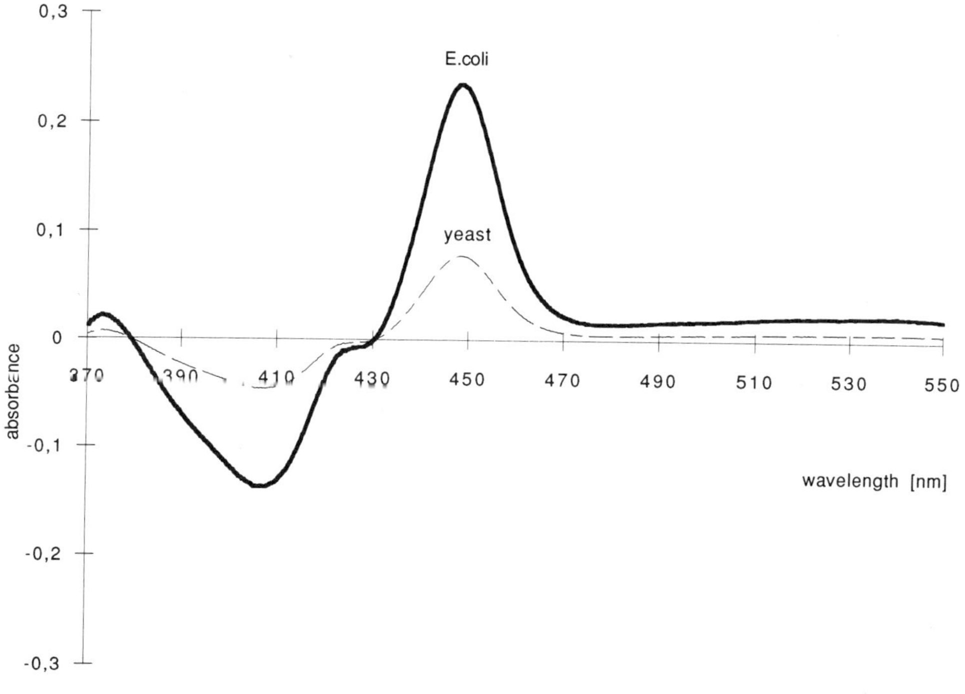

Figure 6. The CO-difference spectra of reduced recombinant cytochrome P450scc expressed in *E.coli* (thick line), and yeast (thin line).

"cut-off" threshold and further purified using the affinity chromatography on cholate sepharose and adrenodoxine-sepharose columns.

To determine the essential steps of the processing and maturation of the expressed secreted P450scc we have performed the spectroscopic measurement of P450 content in the cell lysates and concentrated culture media supernatants and sequencing of the N-terminal part of the purified protein. The obtained characteristic CO-difference spectrum confirms the structural integrity and correct assembly of P450scc holoenzyme following its processing and export from the producing yeast strain (Figure 6).

Quantification of the amount of the purified secreted rP450scc, done using the extinction coefficient 91 $mM^{-1} \times cm^{-1}$ for the absorbance difference between 450 nm and 490 nm indicates the production level accunting to 10mg of rP450scc per liter of culture medium supernatants. Determination of the N-terminal aminoacid sequence of the obtained rP450scc showed its identity to those of the cytochrome from the natural source (data not shown) and this fact demonstrates that the processing of the fusion secretory signal -precursor protein in the yeast cell secretory pathway reproduces the essential feautures of the processing of the proteins, naturally evolved to be translocated to mitochondria. The absence of the holo-form of this protein inside the cell indicates, that maturation and heme assembly take place at some terminal stage of processing and export, unfortunately not identified by us.

However, after repeated experiments to scale up the production of yeast-secreted P450scc, we observed, that the initial success with efficient processing and export of P450scc from recombinant yeast tended to be a very unstable process and the production levels dropped significantly. We attribute it to the specific feautures of the yeast translocation, protein sorting and export machinery and discuss the possible reasons for it in the concluding section.

DISCUSSION

Applying conventional rules of the signal hypothesis we have demonstrated that a eukaryotic protein of mitochondrial origin with intact mitochondrial-targeting signal is efficiently transported outside the yeast cell when addressed by the secretory signal sequence linked to gene precursor. The biochemical and biophysical analysis demonstrated, that yeast secreted cytochrome P450scc is structurally and enzymatically equivalent to its natural counterpart. The achieved level of productivity was very promising and far exceeded what was reported before concerning the production of eukaryotic cytochromes in yeast and higher eukaryotic cells.

This result opens the eciting new opportunities for the application of yeast secretion machinery for the production of metalloproteins, necessary for bioelectronics. It also demonstrates considerable flexibility of the protein sorting, processing and export machinery in yeast. This notion is supported by several recent studies. For instance, it was shown that overexpressed mammalian mitochondrial protein rhodanese can be efficienly secreted from mammalian cells by a nonclassical pathway [Sloan et al., 1994], and that this process does not require the amino-terminal mitochondrial import signal. The authors attribute it to the saturation of the mitochondrial import machinery by overexpressed protein and subsequent interaction of the cytosolic complex of rhodanese with chaperonins with the import sites on plasma membrane resulting in efficient export. This explanation can also be applicable in our case for the overexpressed P450scc. Another reason is the presence of the very strong secretion signal in our construct and inefficiency of the natural P450scc mitochondrial signal in mediating import in the yeast mitochondria, that results in the predominant "rerouting" of the expressed protein in the secretory pathway.

It remains unsettled, whether import into mitochondria *in vivo* occurs co-translationally or post-translationally [Verner, 1993]. However, some of the essential features of the mitochondrial import and secretion process are rather similar, involving the critical role of the hsp-proteins in mediating translocation-competent conformation during both secretion and protein import, and the homology between the apparata for maturation of secreted and imported proteins. From the evolutionary point of view the "prokaryotic " origin of mitochondria and plastids may reflect the utilization of bacterial secretion machinery for protein import, and explain our observation of the correct processing of the complex presequence of the yeast-expressed P450scc by the ER-localized signal peptidase.

The complexity of the eukaryotic sorting, translocation and export machinery makes the overexpressed foreign protein, directed by signals of co-or post-translational transfer, rather sensitive to subtle alterations in the conformation of interacting partner proteins and sophisticated membrane structures. Accumulation of incorrectly folded protein is recognized by the cell as a stress signal and activates the efficient ubiquitin-dependent degradation pathway. So, special precautions and strain improvement approaches are made to overcome this problems. However, biotechnologists working with yeast are well aware of the frequent problems of instability in heterologous expression and significant strain-to-strain and clone-to-clone variations. The recent demonstration of the operation of prion-like effects in yeast justifies that non-inheritable changes may be a common source of such dissapointing instability, as in our case with the P450scc secretion.

Bacterial expression is more stable and reproducible, and despite the accumulation of significant part of the expressed P450SCC in insoluble form, can be a cheap and efficient source of this protein for different needs.

REFERENCES

Arnold, F.H., and Zhang, J.H., 1994, Trends in Biotechnology 12:189-192.

Ausubel, F.M., Brent, R., Kingston, R.E., Moore, D.D., Seidman, J.D. Struhl, K. (eds) Current Protocols in Molecular biology. J.Wiley&Sons, New York.

Baldari, C. et al., 1987, EMBO J. 6:229-234.

Bernhardt, R., Beckert, V., Uhlmann, H., Sligar, S.G., 1994, In: Cytochrome P450, M.C. Lechner (ed.), John Libbey Eurotext, Montrouge, France, pp. 387-394.

Caffrey, M.S., Cusanovich, M.A., 1994, Biochim et Biophys Acta 1187:277-288.

Canters, G.W., Gilardi, G., 1993, FEBS Letters 325:39-48.

Degtyarenko, K.N., Archakov, A.I., 1993, FEBS Lett. 332:1-8.

Germanas, J.P., Di Bilio, A.J., Gray, H.B., Richards, J.H., 1993, Biochemistry 32:7698-702.

Gonzales, F.J., 1990, Trends Pharmacol. Sci. 13:346-352.

Guengerich, F.P., 1992, FASEB J. 6:667-668.

Holmans, P.L., Shet, M.S., Martin-Wixtrom, C.A., Fisher, C.W., Estabrool, R.W., 1994, Arch. Biochem. Biophys. 312:554-565.

Imai, T., Globerman, H., Gertner, J.H., Kawawa, N., Waterman, M.R., 1993, J. Biol. Chem. 268:19681-19689.

Karim, A., Kaderabhai, N., Evans, A., Harding, V., and Kaderabhai, M., 1993, Bio/technology 11:612-618.

Lambeth, J.D., 1990, In: Molecular mechanisms of adrenal steroidogenesis and aspects of regulation and application. Frontires in biotransformation. vol.3. Akademie-Verlag, Berlin, Ruckpaul K. and Rein H. (eds). p. 60-100.

Lu, Y., Casimiro, D.L., Bren, K.L., Richards, J.H., Gray, H.B., 1993, Pro. Natl. Acad. Sci USA 90:11456-11459.

Mauk, A.G., Electron transfer in Genetically Engineered Proteins. The cytochrome C paradigm. Structure and Bonding, 1991, Springer-Verlag. p. 132-156.

McLean, M.A., Stayton, P.S., Sligar, S.G., 1993, Anal. Chem. 65:2676-2678.

Nicolini, C. 1996, Molecular Bioelectronics, World Scientific Publishing, Singapore, in press.

Nygren, P-A., Stahl, S., and Uhlen, M., 1994, Trends in Biotechnology 12:184-189.

Poulos, T.L., Finzel, B.C., Howard, A.J., 1987, J.Mol.Biol. 195:687.

Regan, J.J., Risser, S.M., Beratan, D.N., Onuchic, J.N., 1993, J. Phys. Chem. 97:13083.

Sambrook, J., Fritsch, E.F., Maniatis, T., 1989, Molecular Cloning. A laboratory manual. 2 nd edn. Cold Spring Harbor Laboratory Press.
Shimada, H., Makino, R., Unno, M., Horiuchi, T., Ishimura, Y., 1994, In: Cytochrome P450, M.C. Lechner (ed.), John Libbey Eurotext, Montrouge, France, pp. 299-306.
Sloan, I.S., Horowitz, P.M., Chirgwin, J.M., 1994, J.Biol. Chem. 269:27625-27630.
Stayton, P.S., Bohn, P.W., Sligar, S.G,. 1990, J.Cell.Biochem. Suppl.14B:367.
Studier, F.W., Rosenberg, A.H., Dunn, J.J., Dubendorf, J.W., 1990, Methods Enzymol. 185:60-89.
Truan, G., Cullin, C., Reisdorf, P., Urban, P., Pompon, D., 1993, Gene 125:49-55.
Usanov, S.A., Chashchin, V.L., Akhrem, A.A., 1990, In: Molecular mechanisms of adrenal steroidogenesis and aspects of regulation and application. Frontiers in biotransformation. vol.3. Akademie-Verlag, Berlin, Ruckpaul K. and Rein H. (eds). p. 1-57.
Vijayakumar, S., Salerno, J.C., 1992, Biochim.Biophys.Acta 1160:281
Verner, K., 1993, Trends Bioch. Sci. 18:366-371.
Wada, A., Waterman, M.R., 1992, J. Biol. Chem. 267:22877.
Waterman, M.R., 1994, Advances Enzymol. Relat. Areas Mol.Biol. 68:37-66.

10

MODIFICATION, CHARACTERIZATION AND HANDLING OF PROTEIN MOLECULES AS THE FIRST STEP TO BIOELECTRONIC DEVICES

Hiroyuki Sasabe

Frontier Research Program
The Institute of Physical and Chemical Research (RIKEN)
2-1 Hirosawa, Wako, Saitama 351-01, Japan

INTRODUCTION

Molecular electronic devices (MED) have attracted many researchers in both academia and industries since Dr. Forrest Carter of Naval Research Laboratory (USA) organized 1st International Workshop on MED in 1981 (Carter, 1982). A number of international conferences, symposia, forums and/or workshops have been organized since then (Table 1), and various approaches to realize MED have been discussed from the viewpoint of molecular design, molecular assembly, characterization and device design. However, still there are a lot of problems to show a prototype device. When biological molecules including proteins and enzymes are used instead of synthetic organic molecules, MED can be called bioelectronic devices (BED). Compared to MED, BED have shown some outputs such as biosensors, but this is far from the real molecular electronics.

For the realization of MED, the following three items are key issues (Sasabe et al., 1988): (1) Molecules or Materials which show switching or memory function by themselves, (2) Molecular Assembly Techniques which control the higher order structure of functional molecules, and (3) Lithography Techniques to form the device circuits. Various molecules have been designed with relevant ideas from the early stage of MED researches worldwide, but many of them should be discarded because of thermal stability and reliability for the long term usage. In the case of BED, on the other hand, the genetic engineering of proteins may give a clue to solve these problems. Some examples will be shown in the later section in this paper. For molecular assembly techniques, there are two different approaches; the one based on wet process such as Langmuir-Blodgett (LB) technique and/or adsorption technique, and the other based on dry process such as organic molecular beam epitaxy (OMBE). In both cases the self-assembly of molecules is desired for the close packed regular array formation (so called two dimensional crystal). This indicates the importance of the feed-back of experimental results to the molecular design and modification. To form the designed pattern such as circuit pattern

Molecular Manufacturing, edited by C. Nicolini
Plenum Press, New York, 1996

Table 1. History of International Conference/ Symposium/ Forum on Molecular Electronics

International Workshop on Molecular Electronics Devices (MED)			
1981	1st	F. L. Carter (Naval Research Laboratory)	Washington, DC
1983	2nd	F. L. Carter (Naval Research Laboratory)	Washington, DC
1986	3rd	F. L. Carter (Naval Research Laboratory)	Washington, DC
International Conference on Molecular Electronics - Science & Technology			
1989	1st	A. Aviram (IBM Yorktown)	Hawaii
1991	2nd	A. Aviram (IBM Yorktown)	Virgin Island
1994	3rd	A. Aviram (IBM Yorktown)	Hawaii
International Symposium on Bioelectronic and Molecular Electronic Devices (BMED) (sponsored by Association for Future Electron Devices)			
1985	1st	M. Aizawa (Tokyo Institute of Technology)	Tokyo
1988	2nd	M. Aizawa (Tokyo Institute of Technology)	Fuji-Yoshida
1990	3rd	M. Aizawa (Tokyo Institute of Technology)	Kobe
1992	4th	M. Aizawa (Tokyo Institute of Technology)	Miyazaki
International Symposium on Molecular Electronics and Bio-Computing (MEBC)			
1988	1st	Gèza Biczò (Hungarian Academy of Science)	Budapest
1989	2nd	P. I. Lazarev (Academy of Science, USSR)	Moscow
1991	3rd	R. R. Birge (Syracuse University)	New York
1993	4th	R. S. Potember (Applied Physics Lab., JHU)	Gaithersburg
1994		G. Govil (Tata Inst. Fundamental Res.)	Goa, India
IEEE Engineerings in Medicines & Biology Society (EMBS): Molecular Electronis Tract			
1988		Felix Hong (Wayne State University)	New Orleans
1990		Felix Hong (Wayne State University)	
1992		Felix Hong (Wayne State University)	
International Conference on LB Films (LB#)			
1983	1st	G. G. Roberts (Durham University)	Durham
1985	2nd	G. L. Gains, Jr. (General Electric Co.)	Schenectady
1987	3rd	D. Möbius (Max-Planck Institute)	Göttingen
1989	4th	K. Fukuda (Saitama University)	Tsukuba
1991	5th	A. Barraud (Sacley)	Paris
International Conference on Organized Molecular Films			
1993	6th	R. M. Leblanc (Université du Québec)	Trois-Riviéres
European Conference on Molecular Electronics (ECME)			
1992	1st		Padova
1994	2nd	H. C. Wolf (Univ. Stuttgart)	Kloster Banz

in the molecular scale, the usual lithographic techniques of electron beam, ion beam, and/or X-ray can be applicable for the two dimensional crystal of organic molecular assembly, but these may cause the fatal damages to organic molecules. Therefore the promising technique is the combination of well patterned substrate with chemical modification and self-assembly of molecules on this pattern. A novel technique to handle the molecules is based on STM (scanning tunneling microscope) tip proposed by Hameroff (Hameroff, 1986) and successfully demonstrated by Eigler et al. (Stroscio and Eigler, 1991)

The neural network formation based on real neurons is quite attractive. Towards the development of "native" bioelectronic devices with neurons it is required to control the growth of nerve cells on a solid substrate, especially the growth direction of the nerve cell neurites and the formation of synaptic contact between two cells via neurites. The control of cell growth along a patterned solid substrate such as a silicon wafer or a polymer surface has been carried out by means of chemical (Kleinfeld et al., 1988, Stenger et al., 1992), photo-chemical (Matsuda and Inoue, 1990) and/or conventional lithographic (Hirono et al., 1988) techniques. We have also developed micro-patterning of cells by means of the ion implantation technique (Lee et al., 1993). Adhesion and proliferation of bovine aorta endothelial cells were drastically improved on the surface of a segmented polyurethane modified by ion implantation (Suzuki et al., 1992). It was suggested that a carbonized layer formed by ion implantation on the polymer substrate induced the remarkable improvement of cell adhesion and proliferation (Lee et al., 1994).

In this paper we will show the two dimensional crystallization of proteins characterized by SEM (scanning electron microscope) and STM or AFM (atomic force microscope) and patterning of cell growth on the polymeric substrate as a first step to BED.

TWO DIMENSIONAL CRYSTALLIZATION OF PROTEINS

A few proteins have been crystallized in two dimensions by means of a modified LB technique, where binding of proteins to a lipid layer incorporated with ligand molecules is exploited. However, this technique has disadvantage that proteins crystallize as patches not a uniform film and the nature of crystallization is capricious. For fabrication of electronic devices of two-dimensional protein crystals, we should get a uniform film where protein molecules are densely packed preserving their functions even though they are not crystallized in two dimensions. We aim in the first place at establishing an LB technique to gain such uniform films with ease. Macroscopically monodomain two-dimensional crystal will be an ultimate structure for bioelectronic devices.

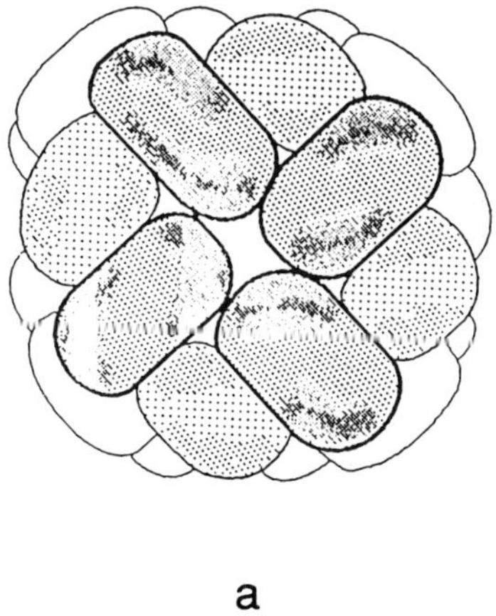

a

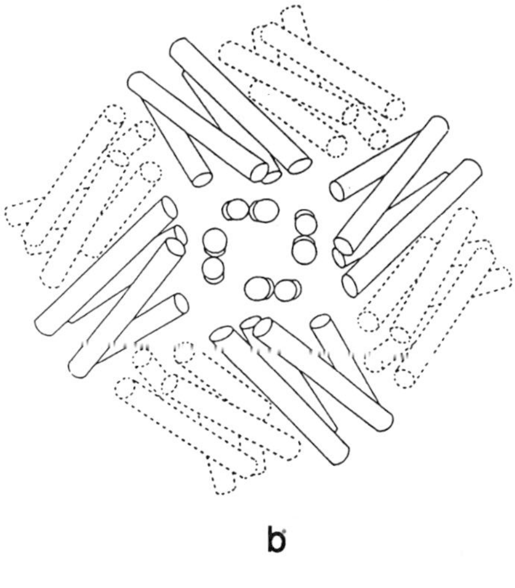

b

Figure 1. Schematic representation of the subunit arrangement in an apoferritin molecule viewed down a fourfold axis. (a) Complete molecular structure with a spherical shell of 24 polypeptide subunits. Centers of the quaternary structure are considered to be channels passing through the polypeptide shell which provide access route for iron atoms forming an iron oxide core. (b) Subunits related by a fourfold axis following the electron density map.

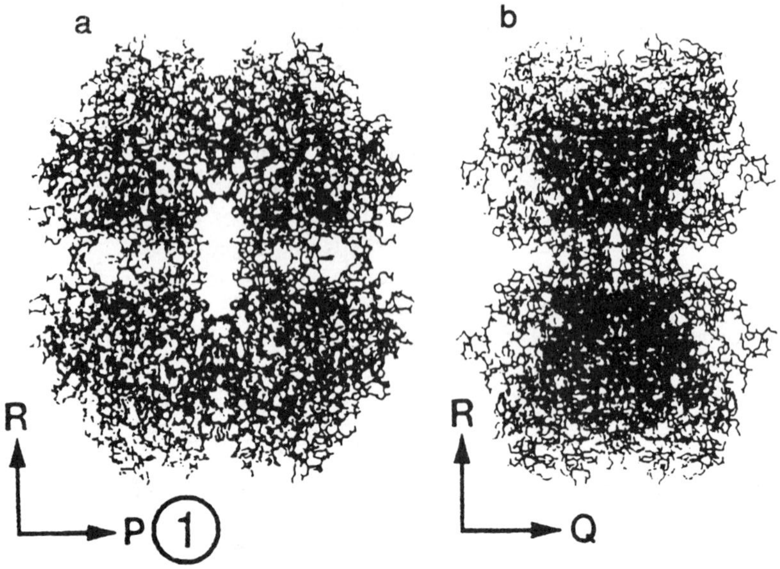

Figure 2. Wore-frame model of catalase projected to P-R (a) and Q-R (b) planes. The tetramer model was constructed by a symmetry operation for the three-dimensional coordinates of the dimer in Brookheaven data base using the program QUANTA on the GS1000 computer (Stellar).

For the development of BED, critical steps are the selection of proper proteins/enzymes/cells and the development of general methods for controlling the spatial arrangement and orientation of them to form highly-ordered, macroscopic materials. For structural analyses, the availability of two-dimensional crystals of biological macromolecules also offers many advantages. In our investigations, three kinds of protein, *i.e.*, ferritin, catalase and cytochrome b_{562} (Cyt-b_{562}) have been chosen as a model protein for the fabrication of two-dimensional crystals. The shape of ferritin is a spherical shell with outer and inner diameters of 13 and 7 nm, respectively (Figure 1) (Banyard et al., 1978). A core of iron oxide composed maximally of 400 iron atoms is held inside the protein shell. The second protein is beef liver catalase with molecular weight of 240 k, which is composed of 4 identical subunits (Figure 2) (Murthy et al., 1981). The approximate size is 10x11 nm^2 when viewed to direction which gives maximum dimensions of the molecule. The third protein Cyt-b_{562} is a small (M_w= 12,000) water-soluble protein found in *Escherichia coli*, consisting of 106 amino acid residues, and classified as an electron transfer protein. The structure of the *Escherichia coli* ferricytochrome b_{562} has been defined by X-ray crystallography to a resolution of 2.5 Å. X-ray crystallographic studies revealed that ferricyt-b_{562} consists of four α-helices (Figure 3) (Mathews et al., 1976). The heme, protoporphyrin IX, is noncovalently bound and is ligated to the polypeptide chain through methionine 7 on the N-terminal helix and histidine 102 on the C-terminal helix. Three approaches are presented in this paper in making two-dimensional crystals of cyt-b_{562}.

Binding of Ferritin and Catalase to PBLH Films

Among the two-dimensional crystallization studies conducted at the air/water interface, the preformed film had two kinds of role: the first is to bind proteins dissolved in the buffer, and the second is to suppress the interfacial denaturation of the bound proteins. In

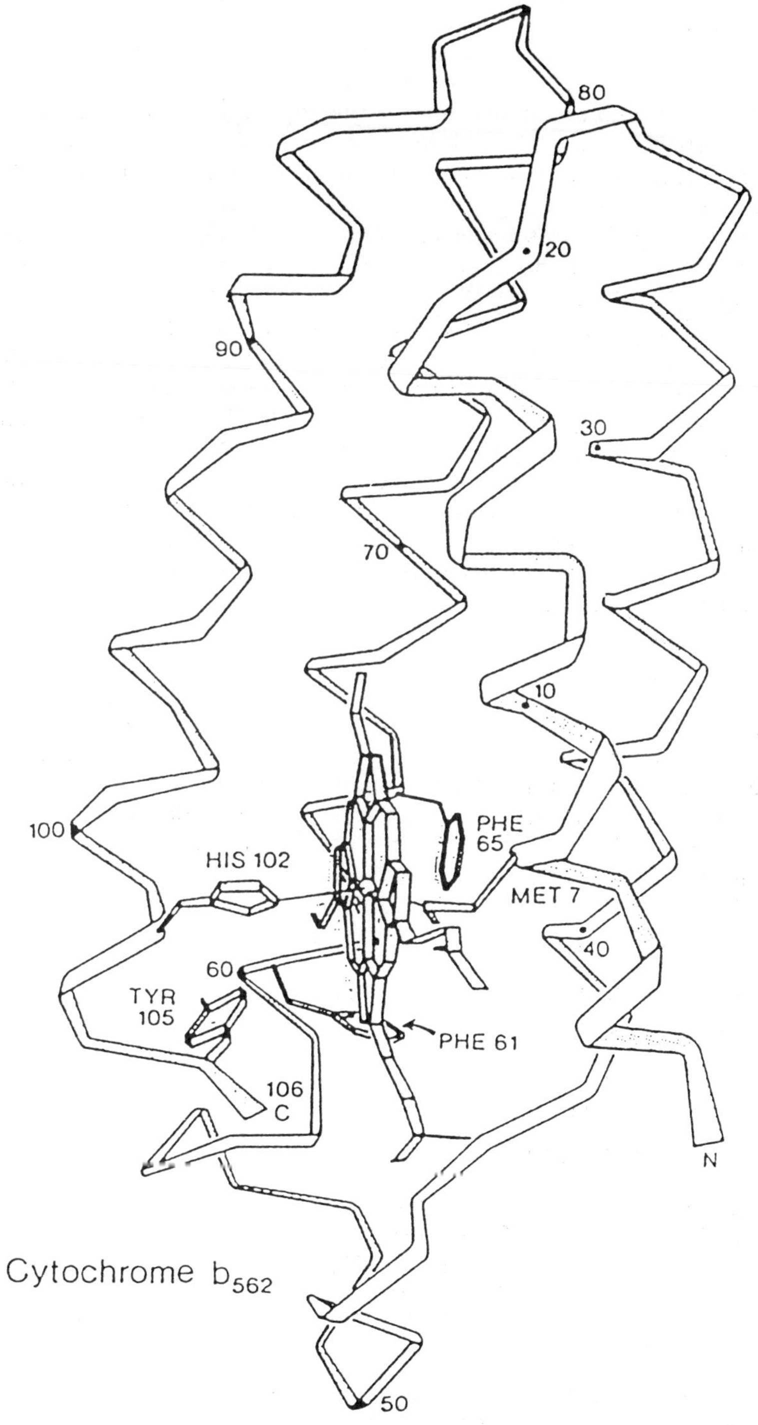

Figure 3. Native structure of cytochrome b_{562} and its heme group [P. C. Weber *et al.*, J. Biol. Chem., **256**, 7702 (1981)]. The heme iron is bonded on opposite sides of the heme plane to a histidine nitrogen and methionine sulfur.

our approaches, a polypeptide film of poly-1-benzyl-L-histidine (PBLH) was spread over the protein solution (Figure 4) (Furuno et al., 1989). The PBLH film was supposed to have the following features suitable for binding proteins, and probably for two dimensional crystallization: (1) Its imidazole rings, positively charged at acidic conditions, are electrostatically interactive with acidic proteins. (2) Due to its polymeric structure, some of the hydrophhobic benzyl groups are facing water, allowing us to assume hydrophobic interactions between PBLH and proteins, which might stabilize the binding further. (3) The film is flexible and slightly fluid.

The film of PBLH was spread over the protein solution. The film was incubated on a solution of horse spleen ferritin (purchased from Sigma) for 1-3 h around pH 6 (concentration of ferritin 10-40 μg/ml). Beef liver catalase was also bound to PBLH film by the similar method. The subphase pH was adjusted to 4.8-6.4 with phosphate buffers. The incubation time was 1-6 h. Following preliminary studies, thermal treatment of repeated heating and cooling between 23 and 35-37 °C was applied to the catalase/PBLH film *in situ* on the water surface. The crystal growth was activated with this treatment, and larger crystals than those obtained at room temperature were formed (Furuno et al., 1992).

The interfacial films were transferred onto a hydrophobic silicon surface by a horizontal transfer method. The transferred film which emerged wet was rinsed with pure water or diluted subphase buffer, and then stained. To quickly dry up a stained film, the wafer plate wetted with the solution was placed in a microcentrifuge tube and the excess solution was blown off by centrifugation (at 6000 rpm, 2000 g) for 10-20 s.

Ferritin

The ordered monolayer array of ferritin was negatively stained with ammonium molybdate (0.2 % w/v, pH 4.9) containing 0.2 % glucose. The film was dried naturally by keeping the wet substrate vertically. The monolayer in Figure 5(a) is a uniform dense packing of ferritin molecules, as confirmed by inspecting the unstained film prepared simultaneously (Furuno et al., 1989). However, as can be seen in the figure, there are two kinds of region: one region is the assembly of brightly seen ferritin particles, and in the other corpuscular

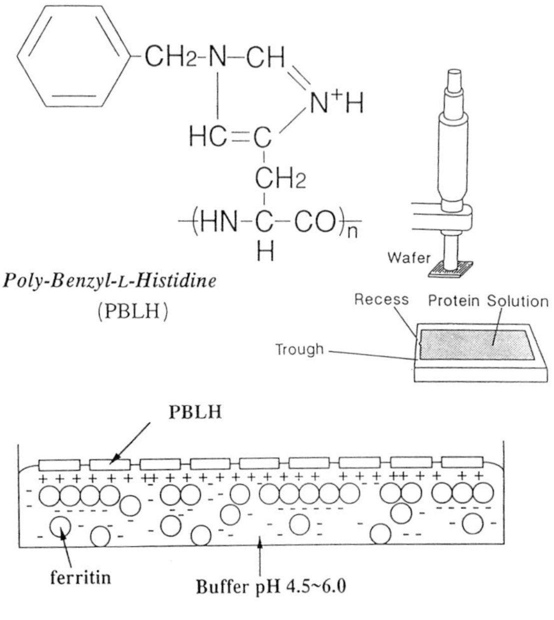

Figure 4. Chemical structure of poly-1-benzyl-L-histidine (PBLH) and the formation of PBLH monolayer at the air/water interface.

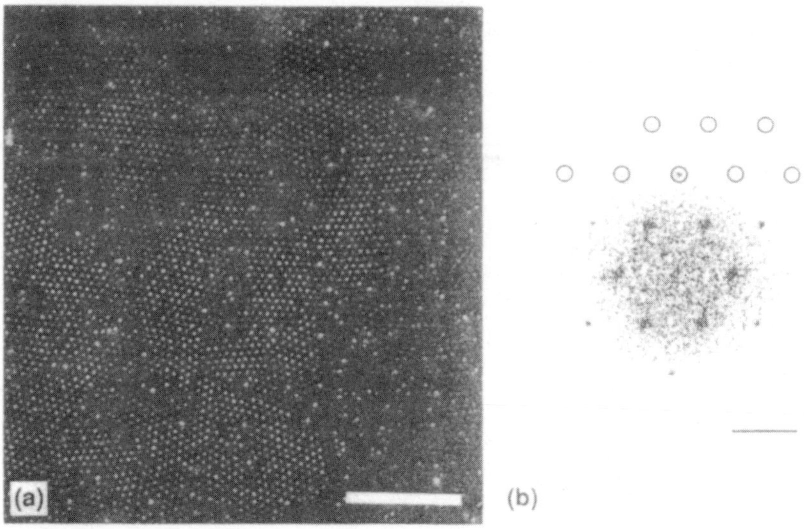

Figure 5. (a) Scanning electron micrograph of monolayer arrays of ferritin stained with ammonium molybdate containing glucose. The arrays of ferritin were covered inhomogeneously with a layer of the stain. Scale bar = 200 nm. (b) Computed diffraction pattern of a lightly stained well-ordered array of ferritin. The lattice undending was applied. Diffraction bar = 0.1 nm^{-1}.

images are almost invisible or faintly visible. The secondary electrons from iron cores of ferritin are blocked if the stain is accumulated too thick over the ferritin arrays. This result shows clearly that in the scanning electron microscopy (SEM) of stained protein monolayers the image is seriously dependent on staining, especially on the thickness. Figure 5(b) is a Fourier transform of a well ordered ferritin array. The spacing of the highest diffraction spot is 3.8 nm.

Catalase

The catalase-bound PBLH film was heated between 23 and 35 °C after 1h incubation at room temperature (pH 4.9, 10 mM phosphate, 10 mM NaCl, 10 μg/ml concentration of catalase). The molecular shape is visible due to stain accumulation at the spaces between neighboring molecules (Figure 6) (Furuno et al., 1992). This kind of view would be common to a dense packing of large proteins as catalase, when successfully stained. The binding of catalase to PBLH film is not explained simply by an electrostatic force between the oppositely charged molecules, since from the pI value, beef liver catalase should have net positive charge at pH 4.9 as PBLH. However, it was shown that the binding of catalase to PBLH film to form a dense packing was far more rapid at pH 6.0 than at pH 4.9 (data not shown), indicating that the net charge polarity was nevertheless effective in the binding process.

In the Fourier transform of the well-ordered area in Figure 6 the diffraction spots extended to 3.4 nm spacing. It is possible to find out some molecules which exhibit a tetrameric habit typical for the catalase molecule. However, the Fourier averaged image does not present such characteristics. The artifact or noise coming from an inhomogeneous staining would be the reason for this, since SEM images the variation of the secondary electron yield which is sensitive to the surface structure.

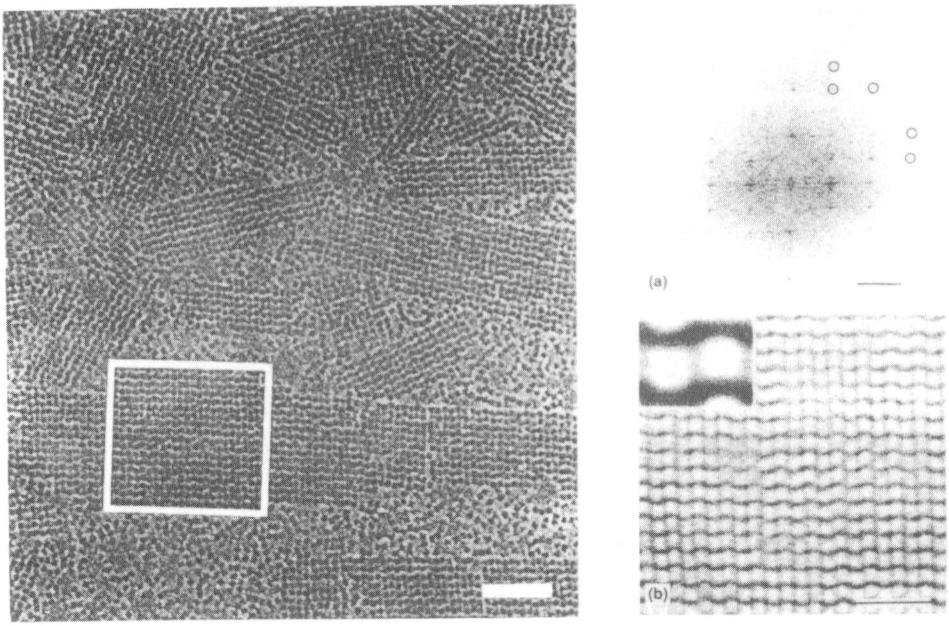

Figure 6. Scanning electron micrograph of two-dimensional crystal of catalase bound to PBLH film. A thermal treatment was applied to the film at the air/water interface. Scale bar = 100 nm. The boxed region was subjected to image processing: (a) Computed diffraction pattern after lattice unbending. Diffraction bar = 0.1 nm^{-1}. (b) Fourier-filtered image and correlation averaged one (inset). Scale bar = 50 nm.

Alkylated Cytochrome b_{562}

The first is the formation of monomolecular layer of cyt-b_{562} reconstituted with alkylated heme. If the protein molecules have amphiphilic property at air/water interface, the protein solution can be spread directly onto the water surface. The heme prosthetic group of cyt-b_{562} is not covalently bound to the protein, and can therefore be removed and replaced *in vitro* with other prosthetic groups such as modified porphyrin. As a model case, we synthesized protoheme IX having long alkyl chain, and reconstituted apocyt-b_{562} with the modified protoheme IX. This would give cyt-b_{562} an amphiphilic property. Figure 7 shows the π-A isotherms of stearylated holocyt-b_{562} spread on 0.25 mM $CdCl_2$, pH 7.7, at 21°C (Kobayashi et al., 1993a). Stearylated holocyt-b_{562} was prepared under the conditions of dimethyl-sulfoxide content 63%, temperature 10°C, reconstitution time 40 minutes. In Figure 7, curve A shows the isotherm of native b_{562} and curve B shows that of stearylated holocyt-b_{562}. It is clear that the curve B shows surface-active property. Similar results were obtained from the curves C and D with the addition of dimethylformamide and isopropyl alcohol, respectively. Thus, it is concluded that the presence of organic solvent is not necessary in spreading stearylated holocyt-b_{562} onto a buffer surface. The surface pressure of *ca.* 26 dyn cm^{-1} given in Figure 7 is thought to be the value inherent to stearylated holocyt-b_{562}. After the compression, the surface pressure decreased gradually and showed a steady value of *ca.* 20 dyn cm^{-1} (data not shown), which indicates that the stearylated holocyt-b_{562} forms a stable monomolecular layer on the buffer surface.

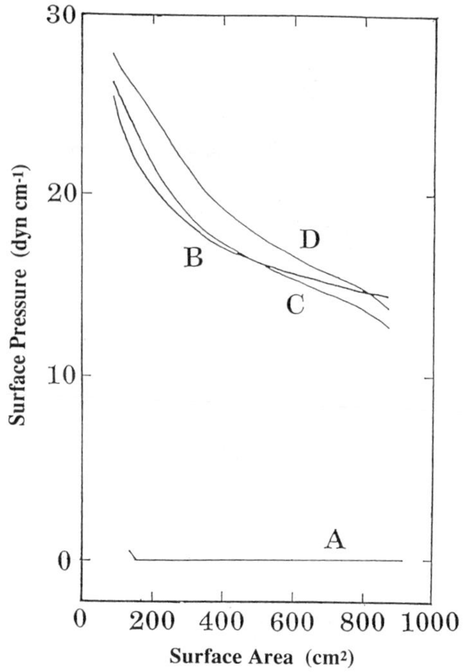

Figure 7. Surface pressure - area (π-A) isotherms spread on 0.25 mM $CdCl_2$, pH 7.7. A: Native Cyt-b_{562}, B: stearylated Cyt-b_{562} in pure water, C: stearylated Cyt-b_{562} in pure water and DMF, D: stearylated Cyt-b_{562} in pure water and isopropyl alcohol.

Enhancement of Stability by Site-Directed Mutagenesis

The second approach is to increase the stability of proteins by protein engineering. When crystallizing proteins are spread at interfaces, most proteins denature due to surface pressure effects. The enhancement of structural stability of the protein by site-directed mutagenesis would be an approach to prevent denaturation. An effective method of making a thermally stable mutant protein would be to introduce a disulfide bridge formed by two cysteine residues. Therefore, we made a mutant protein having both asparagine 22 and glycine 82 changed to cysteines (A22C/G82C). These positions were calculated to be suitable for disulfide bridge formation by computer simulation (Figure 8) (Kobayashi et al., 1993b). Another effective method would be to introduce an increased hydrophobic interaction between the α-helices of the protein or to replace a poor helix-forming residue within an α-helix with a good helix former. Thus, we made another mutant protein having glycine 64 changed to an alanine (G64A). The thermal stabilities of these mutants with oxidized heme were examined and were compared with that of the wild-type protein with oxidized heme. Thermally induced denaturation of cyt-b_{562} was monitored by the UV-Vis (Soret band at 418 nm) and CD (at 222 nm) spectral changes, and the microDSC measurement. Each can follow the dissociation of the heme moiety, the transition of α-helix to random coil and the thermal transition of heat capacity, respectively. The results of UV-Vis, CD, and DSC measurements indicate that the dissociation of the heme moiety initiates the denaturation of α-helices. It is clear from the DSC data (Figure 9) that the denaturation temperature of G64A mutant is almost 78°C and is 2 degrees higher than that of wild-type, whereas that of A22C/G82C mutant is about 10 degrees lower. It is suggested that the unexpected instability in A22C/G82C mutant is due to the weak binding of the heme moiety to the peptide chain which was induced by structural distortion of apo-mutant protein.

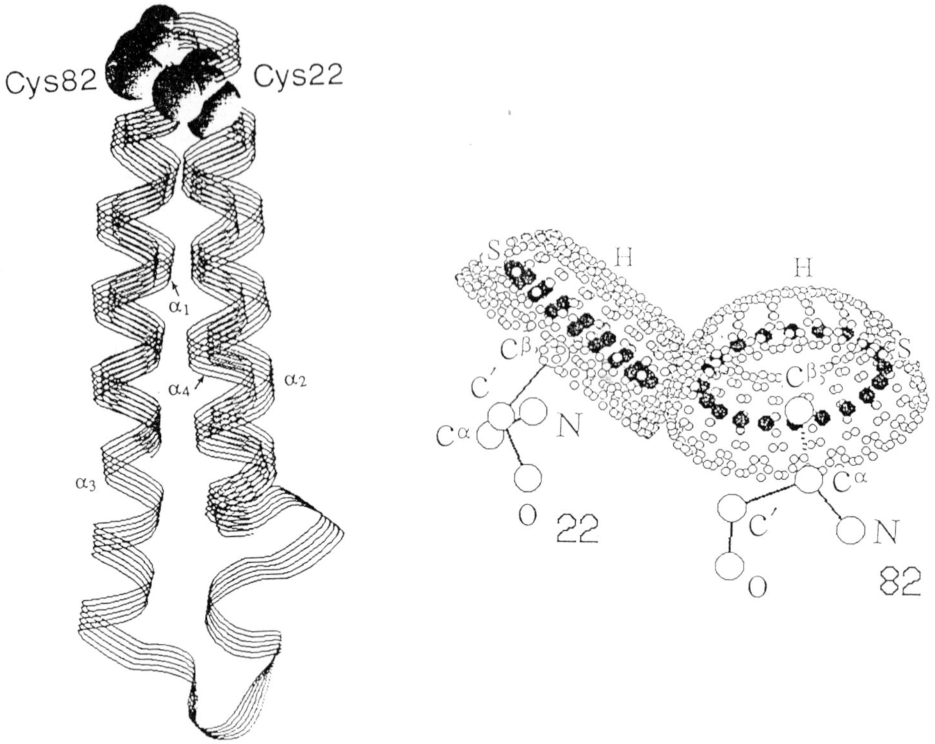

Figure 8. Structure of the mutated Cyt-b_{562} (N22C/G82C), and the lampshades of Cys22 and Cys82 in the mutated Cyt-b_{562} (N22C/G82C).

Control of Surface Charge Distribution by Site-Directed Mutagenesis

The third approach is the control of surface charge distribution of protein molecule and to utilize the electrostatic interaction between molecules in the subphase and the charged monolayer apread at the interface. From the computer simulation of surface charge distribution of cyt-b_{562}, it is suggested that the single mutation of aspartic acid 73 to lysine (Asp73L) gives a good distribution of positive and negative charges separately on the surface. We are now preparing this mutant for monolayer formation due to the LB/adsorption method.

AFM IMAGINGS OF FERRITIN MOLECULES

AFM has been attracting our attention increasingly as an entirely new approach to the study of surface topology of biological specimens. We reported the AFM images of ordered arrays of ferritin molecules bound to LB films of PBLH as a function of film forming conditions such as pH and surface pressure (Ohnishi et al., 1992). It was suggested that PBLH films are useful for AFM imagings as a general substrate for fixing water-soluble protein molecules without denaturation.

The AFM system used in our study was a commercially available NanoScope II (Digital Instruments, Inc., Santa Barbara, CA, USA). After slowly circulating pure water around the sample in a fluid cell, a Si_3N_4 cantilever, which is V-shaped and 200 μm long

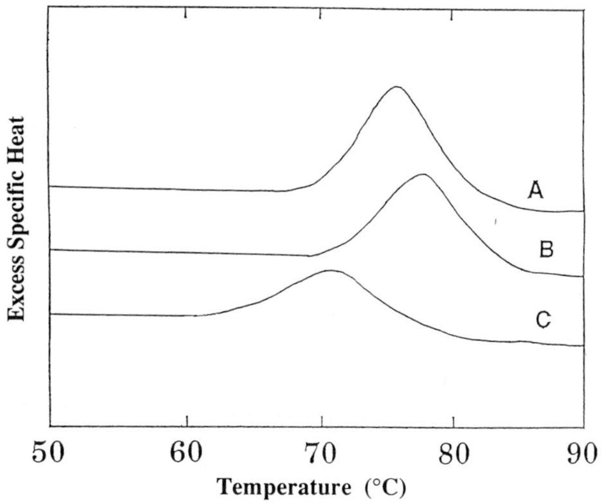

Figure 9. Differential scanning calorimetry (DSC) data for wild-type Cyt-b_{562}, G64A mutant and 22C/82C mutant. These results were obtained in 10 mM phosphate buffer. The temperature was increased at a rate of 0.5 °C/min from 5 to 90 °C. A: wild-type Cyt-b_{562}, B: G64A, C: 22C/82C.

with a spring constant of 0.12 N/m (reported by Digital Instrument, Inc.) was positioned and scanned over the sample in pure water at room temperature. AFM images (400x400 pixels) were obtained using the 'height mode', which kept the force constant. The software provided with the AFM from the manufacturer was used to measure force *vs.* distance curves.

pH Dependence of Forces Between Tip and Surface of PBLH Layer

Horse spleen ferritin has an isoelectric point of about 4.5 and is negatively charged in a solution of phosphate buffer of pH 5 or above (Arbuthnott and Beeley, 1975). The imidazole ring of PBLH, on the other hand, should be positively charged when PBLH is spread on a subphase at pH lower than the pKa of the histidyl residue (in the range of 6-7). Therefore it is expected that ferritin in the subphase would bind electrostatically to a PBLH film at a slightly acidic pH range of 4.5-6.5 and form only a monolayer, even if there exist excess ferritin molecules near the interface. In this case, electrostatic charges at the ferritin-PBLH interface are cancelled, thereby forming a more efficient screening of the surface charge of ferritin monolayers, which we called the 'self-screening effect' in our previous paper (Ohnishi et al., 1992).

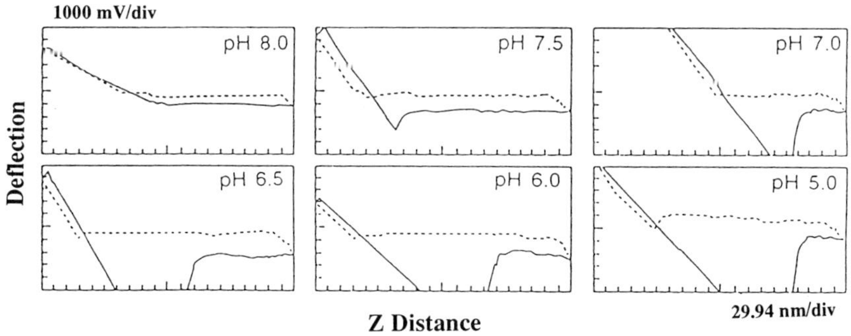

Figure 10. A series of force vs. distance curves between the Si_3N_4 tip and the PBLH surfaces in different pH solutions. The dashed and solid curves are for approach and withdrawal of the tip, respectively. pH 8.0, 7.5, 7.0: 10 mM tris-HCl buffer, pH 6.5, 6.0, 5.5: 10 mM phosphate buffer.

To elucidate the surface properties of PBLH films in the buffer solution, we investigated the forces of the tip interacting with a PBLH surface at various pH conditions. Figure 10 shows a series of force *vs.* distance curves between the Si_3N_4 tip and the PBLH surfaces in different pH solutions. We used 10 mM tris-HCl buffer for pH 8.0, 7.5 and 7.0, and 10 mM phosphate buffer for pH 7.0, 6.5, 6.0 and 5.0. At pH 8.0, a repulsive force was observed upon approach and withdrawal of the AFM tip from the sample. When lowering the pH, the repulsive force decreased upon approach of tip, while an attractive force increased upon withdrawal. Since there is no large difference between the force curves at pH 7.0 in tris-buffer and phosphate buffer (data is not shown), the attractive force is independent of the species of electrolyte. From pH 6.5, as the pH value decreased, attractive forces increased upon approach and withdrawal of tip. Actually, we did not get any unambiguous images of the PBLH layers, owing to these repulsion and attraction.

This variation of the forces with changing a pH condition can be explained by electrostatic and hydrophobic interactions. The attractive force can be attributed mainly to hydrophobic interaction, since the PBLH film on the silicon substrate has a hydrophobic surface. Moreover, the surface of the tip is negatively charged, because silicon nitride bears a slight negative surface charge in water. Therefore the interaction between negatively charged tip surface and positively charged PBLH may also enhance the electrostatic attraction below pH 6.5. Above pH 7.0, a repulsive force is expected because both surfaces bear negative charges. From these results, it is clear that the applied forces between tip and PBLH film in the solution are controllable by changing the pH. In addition, these results support our previous data with regard to the effect of pH of the buffer on the stability of the biological sample and the AFM imagings (Ohnishi et al., 1992). In this system, the PBLH film plays a critical role as an electrostatically charged substrate to fix feritin molecules.

AFM Imagings of Ferritin Molecules

Figure 11(a) shows an AFM image of the ferritin sample on a silicon surface, which was prepared following the 'self-screening effect' with the charged PBLH layer. Small spheres covering over the surface can be observed in an area of at least 738x738 nm^2, approaching the limiting scanning area of this AFM. In Figure 11(b), we show the typical unfiltered AFM image of the molecular packing of ferritin on a silicon surface taken in pure water at room temperature. Individually distinguishable spherical patterns and regular alignment are observed.

Figure 11(c) shows an AFM image in a small scanning area. It is evident that seven spheres form a hexagonally packed arrangement. The hexagonally packed structure is in good agreement with the two-dimensional array of ferritin molecules observed by high resolution SEM as shown in Figure 5. In addition, the spacing between these spheres in this pattern is about 14 nm with a diameter of about 12 nm, which agrees well with that of a ferritin crystal deduced from X-ray analysis (Harrison, 1963). From these results, each spherical image obtained with AFM is identified as an individual ferritin molecule. These imagings were carried out for the ferritin monolayers prepared with phosphate buffer at pH 5.3. The applied forces operated in these images were estimated from the force curves to be 10^{-10}-10^{-11} N. We have discussed in our prevous paper (Ohnishi et al., 1992) about the suitable pH condition for the AFM imagings during the sample preparation.

In the AFM image of the sample prepared at the pH value of around 5.0, we observed some domains in which ferritin molecules were bound to PBLH film in ordered arrays (Figure 11(d)). Ferritin molecules is likely to form some domains at pH 5.0, because the negative charge density of ferritin molecule becomes smaller as the pH increases. From these results, we conclude that the sample preparation with electrostatic binding is one of the suitable methods to achieve the nondestructive low-force AFM imagings for

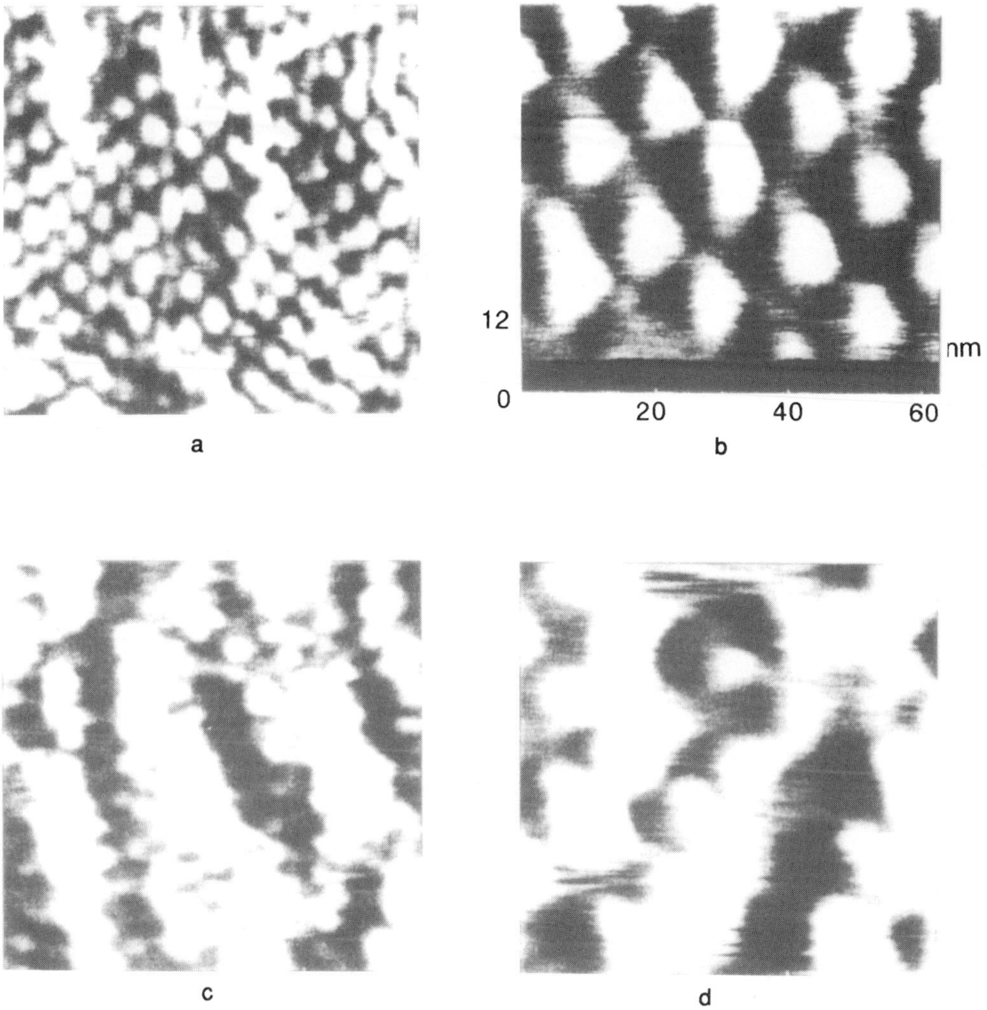

Figure 11. Unfiltered AFM images of ferritin samples prepared at pH: (a-c) 5.3 and (d) 5.0. The image area is: (a) 738x738 nm^2, (b) 200x200 nm^2, (c) 45x45 nm^2, and (d) 200x200 nm^2. The imaging was carried out in pure water.

biomacromolecules. In addition, controlling of the binding condition of protein molecules is also an important factor to get unambiguous AFM imagings.

CELL PATTERNING USING AN EXCIMER LASER

The excimer laser ablation is one of the most important techniques in the polymer surface modification (Niino and Yabe, 1993, Phillips et al., 1991). Here the cell adhesion and proliferation on the oxygen-plasma-treated polystyrene (O-PSt) film, the ablated O-PSt film by an excimer laser and the polystyrene (PSt) are presented. The laser ablation and modification techniques are also applied to micro-patterning of cultured cells.

O-PSt dishes (Falcon 3001, Becton-Dickinson, USA) provide a good substrate for the adhesion and proliferation of NG108-15 neuroblastoma x glioma hybrid cells (NG108-15 cells) (Nirenberg et al., 1983). This substrate was ablated by irradiation with a KrF excimer laser of 248 nm wavelength and 34 ns pulse width. The laser beam was incident in air nearly perpendicular to the sample surface through a cylindrical lens and a homogenizer which provides almost spatial uniformity better than 95 % with an irradiation area of 5 x 5 mm^2. The laser fluence at the sample surface was adjusted to about 200 mJ/cm^2 using an attenuator. The number of laser pulses and the repetition rate were kept at 10 and 1 Hz, respectively. For micro-patterning of cultured cells, a stainless-steel stencil mask placed in contact with the sample surface was used to define the ablated regions. It should be noted here that the cell adhesion on PSt (Falcon 1008, Becton-Dickinson, USA) is quite poor compared to O-PSt.

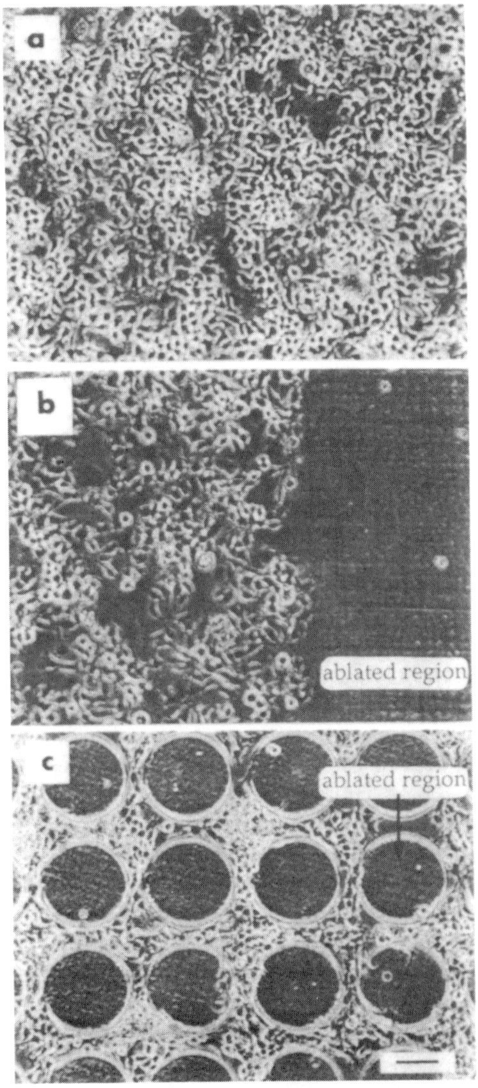

Figure 12. Phase contrast micrographs of cell adhesion and proliferation after NG108-15 cell seeding (cell number; 5 x 10^4 cells/ml): (a) on O-PSt surface at 5 days, (b) at the interface between O-PSt ablated by KrF laser at 200 mJ/cm^2 and five pulses at 3 days, (c) on a surface patterned by a KrF laser at 200 mJ/cm^2 and 5 pulses at 3 days. The dark region on which no cells attached corresponds to the ablated region. In both (b) and (c), the dark region on which no cells attached corresponds to the ablated region. (bar=100 μm).

Surface Roughness

Firstly, the adhesion and proliferation of cells on the O-PSt and ablated O-PSt surfaces was examined. The surface profile was obtained by a surface profiler (DekTak II) after the excimer laser ablation at 200 mJ/cm^2 and 5 pulses with a mask. The ablated region shows smooth surface within wild microsteps at the border of ablated/non-ablated regions. There are no effects on the cell adhesion and proliferation at a laser fluence below 50 mJ/cm^2 even with more than 100 laser pulses, since the laser ablation does not occur at the laser fluence lower than the threshold. The exposed regions appear to be roughened as shown in Figures 12(c) and 13(a) which are observed by transmission type phase contrast microscope. However, the ablated region (unmasked region) and non-ablated regions have microscopically smooth surface as observed by scanning electron microscope and surface profiler. The reason is not clear so far, but the roughened structures observed by phase contrast microscope may come out from difference in surface density of bulk polystyrene melted by laser ablation.

Chemical Structures

The O-PSt and ablated O-PSt surfaces as well as PSt were characterized by X-ray photoelectron spectroscopy (XPS) and contact angle measurements. The XPS spectra of O-PSt surface present two peaks corresponding to the binding energies of carbon and oxygen. There was, however, only one peak corresponding to the carbon energy in the spectra of ablated O-PSt. The spectra of the ablated surfaces were the same as those of PSt surface. The contact angle of water on PSt, O-PSt and ablated O-PSt surfaces were 87, 60, and 87, respectively. Thus, the modified layer with oxygen is completely removed by the laser ablation and the PSt surface reappears.

Cell Adhesion

Among all surfaces tested, the adhesion strength of the cells on PSt was the weakest. The cells adhered and proliferated well on O-PSt surface. Adhesion and proliferation of NG108-15 cells are observed on pristine O-PSt surface (left half of Figure 12(b)), but not on the laser-ablated surface (right half of Figure 12(b)). This is attributable to the reformation of a PSt surface, which is repulsive for cell adhesion, after the surface of O-PSt was ablated by a excimer laser. These results indicate that adhes ion and proliferation of the cells do not take place on the ablated surface. As shown in Figure 12(c), cultured cell patterning was realized on the surface patterned by the laser ablation using a contact mask with circular holes (200 mm in diameter). This cell pattern results from selective cell adhesion and also from the migration of cells from repulsive to attractive surfaces. Therefore, the repulsive surface is important for cell patterning.

Neurite Outgrowth

We checked the neurite outgrowth of NG-108-15 cell on an O-PSt surface. The surface is clearly patterned by excimer laser ablation with a mask as shown in Figure 13(a). Circular holes of 200 mm diameter are ablated on the O-PSt. Cells are distributed on both domains (non-ablated/ablated) immediately after seeding. After a period of 6 hrs, all cells have migrated from ablated domains to non-ablated domains, and then adhered to the non-ablated surfaces. Such cell migration was reported in the previous paper (Lee et al., 1993). In that case the cell migration occurred from a poor surface to a good surface for cell adhesion within about 2 hrs after seeding. For comparison, a chemical treatment with a solution of prostaglandin E1 (PGE1) and isobutylmethyl-xanthine (IBMX) (Docherty et al.,

1991) was carried out to promote the neurite growth. Figure 13(b) is a phase contrast micrograph taken 3 days after treatment with PGE1 and IBMX. The neurites are formed and connected between cells by treating of PGE1 and IBMX. The neurites are also spread only on the O-PSt surface.

Micro-Patterning of Cells

From these results, the cell adhesion on O-PSt surface became better than on PSt surface. The cell attractive surfaces (O-PSt) were obtained by oxygen-plasma-modification

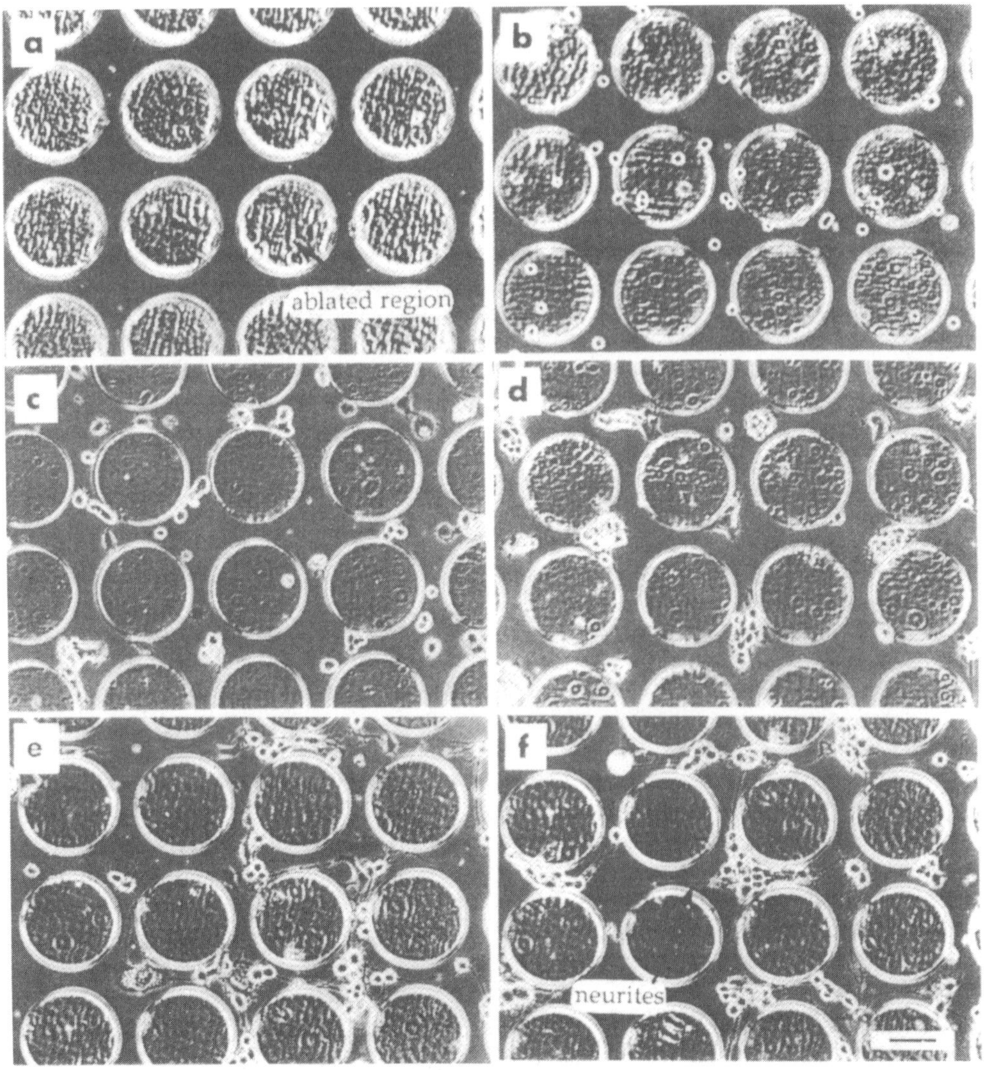

Figure 13. Phase contrast micrographs of patterns of cells and neurites: (a) surface ablated by a KrF laser ablation at 200 mJ/cm^2 and 5 pulses, (b) distribution of cells immediately after seeding (cell number 2x104 cells/ml), (c) at 6 h after seeding, (d) at one day after seeding, (e) at two days after PGE1 and IBMX treatment and (f) at three days after PGE1 and IBMX treatment. (bar=100 μm).

of repulsive surfaces (PSt). Therefore it is concluded that the surface chemical structure is one of important factors in adhesion and proliferation of cells. Repulsive and attractive domains on the surface appeared simultaneously by the surface processing technique such as plasma modification and laser ablation. It should be noted that the excimer laser ablation and laser modification are dry processes and operated in air which give advantages to fabricate a micro-pattern for cell culture.

Cell adhesion and proliferation did not occur on the laser-ablated domains which is a so-called negative type cell pattern, but were controlled in patterning. In contrast, those on PSt domains modified by a ArF excimer laser with a fluence of 2 mJ/cm^2 and 3000 laser pulses in air show positive type cell patterns. The details of positive type patterning will be presented elsewhere.

ACKNOWLEDGMENT

This work has been done under the Frontier Research Program, RIKEN. The author would like to express his sincere thanks to members of FRP, especially to Drs. T. Furuno, A. Sato for ferritin and catalase LB films, Drs. K. Kobayashi, Y. Kobayashi and Prof. T. Nagamune for Cyt-b_{562}, Drs. M. Hara and S. Ohnishi for STM/AFM, and Dr. J.-S. Lee for cell patterning.

REFERENCES

Arbuthnott, J. P., and Beeley, J. A., 1975, 'Isoelectric Focusing', Butterworths, London, p. 125.

Banyard, S. H., Stammers, D. K., and Harrison, P. M., 1978, *Nature (London)* 271:282.

Carter, F. L. (Ed.), 1982,'Molecular Electronic Devices', Dekker, New York.

Docherty, R. J., Robbins, J., and Brown, D. A., 1991, *in* 'Cellular Neurobiology', J. Chad and H. Wheal (Eds.), IRL, Oxford, pp75-95.

Furuno, T., Sasabe, H., and Ulmer, K. M., 1989, *Thin Solid Films* 180:23.

Furuno, T., Sato, A., and Sasabe, H., 1992, *in* 'Synthetic Microstructures in Biological Research', J. M. Schnur and M. Peckerar (Eds.), Plenum Press, New York, pp. 121-129.

Hameroff, S. R., 1986, Abstract of 3rd International Workshop on Molecular Electronics, October, Arlington, USA.

Harrison, P. M., 1963, *J. Mol. Biol.* 6:404.

Hirono, T., Torimitsu, K., Kawana, A., and Hukuda, J., 1988, *Brain Research* 446:189.

Kleinfeld, D., Kahler, K. H., and Hockberger, P. E., 1988, *J. Neurosci.* 8:4098.

Kobayashi, K., Kim, K.-H., Saito, K., Nagamune, T., Otomo, J., Kobayashi, Y., Furuno, T., and Sasabe, H., 1993a, *in* 'Electricity and Magnetism in Biology and Medicine', M. Blank (Ed.), San Francisco Press, San Francisco, pp 191-194,

Kobayashi, Y., Sasabe, H., and Saito, N., 1993b, *J. Protein Chem.* 12.121.

Lee, J.-S., Kaibara, M., Iwaki, M., Sasabe, H., Suzuki, Y., and Kusakabe, M., 1993, *Biomaterials* 14:958.

Lee, J.-S., Kaibara, M., Iwaki, M., Sasabe, H., Suzuki, Y., and Kusakabe, M., 1994, *Mol. Cryst. Liq. Cryst.* 247:365.

Mathews, F. S., Bethge, P. H., and Czerwinski, E. W., 1976, *J. Biol. Chem.* 254:1699.

Matsuda, T., and Inoue, K., 1990, *ASAIO Trans.* 36:M161.

Murthy, M. R. N., Reid, T. J. III, Sicignano, A., Tanaka, N., and Rossman, M. G., 1981, *J. Mol. Biol.* 152:465.

Niino, H., and Yabe, A., 1993, *Appl. Phys. Lett.* 63:3527.

Nirenberg, M., Wilson, S., Higashida, H., Rotter, A., Krueger, K., Busis, N., Ray, R., Kenimer, J. G., and Adler, M., 1983, *Science* 222:794.

Ohnishi, S., Hara, M., Furuno, T., and Sasabe, H., 1992, *Biophys. J.* 63:1425 .

Phillips, H. M., Callahan, D. L., Sauerbrey, R., Szabo, G., and Bor, Z., *Appl. Phys. Lett.* 58:2761.

Sasabe, H., Wada, T., Hara, M., Furuno, T., Aoyagi, Y. and Yamada, A., 1988, *in* 'Molecular Electronic Devices', F. L. Carter, R. E. Siatkowski and H. Wohltjen (Eds.), Elsevier Sci. Publ., New York, pp. 57-67.

Stenger, D. A., Georger, J. H., Dulcey, C. S., Hickman, J.J., Rudolph, A. S., Nielsen, T. B., McCort, S. M., and Calvert, J. M., 1992, *J. Am. Chem. Soc.* 114:8435.
Stroscio, J. A., and Eigler, D. M., 1991, *Science* 254:1319.
Suzuki, Y., Kusakabe, M., Lee, J.-S., Kaibara, M., Iwaki, M., and Sasabe, H., 1992, *Nucl. Instr. Meth.* B65:142.

11

BACTERIORHODOPSIN APPLICATIONS IN OPTICAL INFORMATION PROCESSING - GENETIC ENGINEERING AS A TOOL IN MATERIAL SCIENCE

D. Zeisel and N. Hampp*

Institute for Physical Chemistry, University of Munich
Sophienstr. 11, D-80333 München, Germany
Consortium für elektrochemische Industrie GmbH
Zielstattstr. 20, D-81379 München, Germany

Over the last few decades the use of biological materials in different fields of modern technology has grown rapidly. Examples are the widespread use of enzymes for stereospecific biosynthesis of materials and gene technological production of biologically important macromolecules.

In the field of optical information storage and processing the use of only synthetic, photochromic and photorefractive materials as reversible recording media has been discussed and investigated until recently. So far, however, no suitable medium has been found for the wide range of possible applications, due mainly to limitations arising through different photochemical side reactions or the complicated and expensive crystal growth procedure.

The retinal protein bacteriorhodopsin (BR) from *Halobacterium salinarium*, which is related to the human visual pigment (Oesterhelt and Stoeckenius, 1971) has been optimized by nature during a long period of evolution and may be used as an alternative dynamic recording material instead of conventional synthetic materials. Its outstanding properties include its surprising stability, excellent reversibility and different physiochemical features, which allow the production of bulky, but optically homogenous thin films (Hampp et al., 1992a).

In addition, genetic modification of the halobacterial strain allows the isolation of variants with optimized photochromic properties, adapted to the specific demands required of the material in optical information processing (Hampp et al., 1990).

This article will give a brief discussion of the physiochemical properties of the molecule and explain the possibilities of chemical and genetic variation of the properties of

* Present address for both: Phillips University Marburg, Fachbereich Physikalische Chemie, Hans-Meerwein-Str. Geb. H, D-35032 Marburg, Germany.

Molecular Manufacturing, edited by C. Nicolini
Plenum Press, New York, 1996

BR. This will be followed by the characterization of BR-films as a holographic recording material, which forms the basis for applications in optical information storage and processing. Finally, different examples of applications with the genetically modified variant D96N

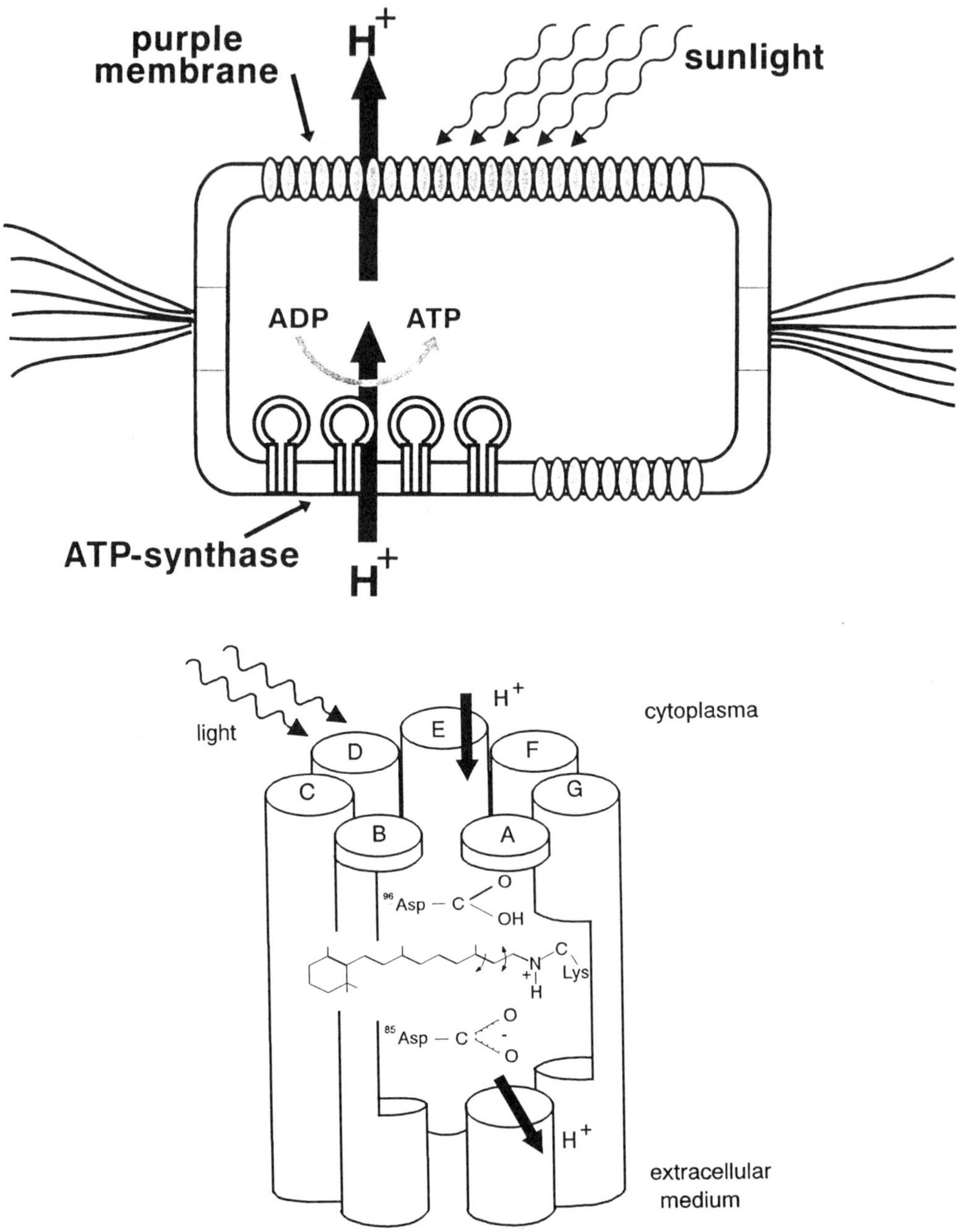

Figure 1. Scheme of a) the halobacterial cell and b) the bacteriorhodopsin molecule in the cell membrane with the proton pore and the main functional groups (Asp85, Asp96 and Schiff base) participating in proton translocation. The encaged retinylidene residue is bound to helix G, it separates the cytoplasmic from the extracellular part of the proton pore.

Table 1. Biochemical and photophysical properties of bacteriorhodopsin

Biochemical properties		Photophysical properties	
molecular weight	26,000 Dalton	initial absorption	$\varepsilon_{B568} = 63000\ lmol^{-1}cm^{-1}$
structure			$\varepsilon_{M412} = 43000\ lmol^{-1}cm^{-1}$
primary	248 amino acids		$\varepsilon_{O640} = 67000\ lmol^{-1}cm^{-1}$
secondary	7α-helical domains	quantum efficiency	$\phi\ (B \rightarrow K) = 64\ \%$
tertiary	cage with proton pore		$\phi\ (M \rightarrow B) \leq 64\ \%$
quarternary	trimers, 2-D crystal		$0.02\ \% \leq \phi\ (O \rightarrow P) \leq 1\ \%$
lipid bilayer	transmembrane, 10 lipid molecules per BR	refractory periodafter relaxation	none
chromophoric group	retinal	photochromism	trans/cis isomerization
biological function	light-driven proton pump		
stability	- constant illumination - oxygen + light - temperatures > 80° C - pH 3-11 - most proteases	instabilities	- UV - light - organic solvents

in real time interferometry and pattern recognition demonstrate the new and successful genetic approach in material sciences.

THE BIOLOGICAL FUNCTION AND THE STRUCTURE OF BACTERIORHODOPSIN

BR is the key protein in the photosynthetic system of the bacterium *Halobacterium salinarium* (formerly *Halobacterium halobium*) [reviews see for example (Kouyama et al., 1988; Birge, 1990; Oesterhelt, et al., 1991). It is found as a two-dimensional crystalline lattice in sheets of the purple membrane, which is part of the halobacterial cell membrane (fig. 1a) and is typically about 1 μm in diameter and 5 nm thick. Its biological function is that of a light-driven proton pump (Mathies et al., 1991). Under light exposure BR transports protons from the inner of the halobacterial cell (cytoplasm) to the extracellular medium and thereby converts light energy into chemical energy (see fig.1 - the proton pump direction is indicated with thick black arrows). The proton gradient generated over the cell membrane is utilized by the membrane-bound enzyme ATP-synthase for the regeneration of ATP (adenosine-tri-phosphate) from ADP (adenosine-di-phosphate).

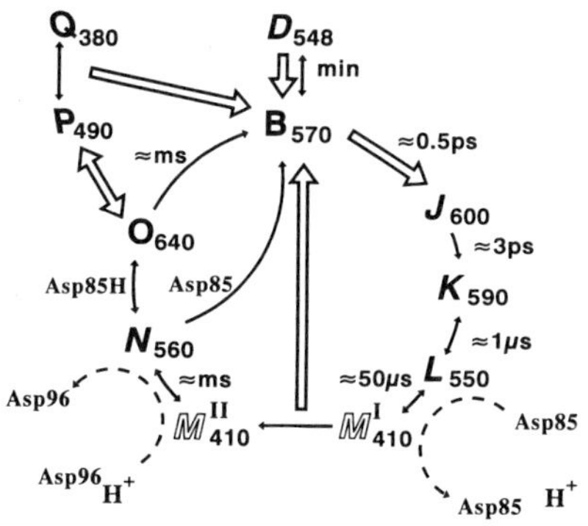

Figure 2. Photocycle of bacteriorhodopsin. The different intermediate states are represented by the common single letter code and can be distinguished by their absorption maxima, given as subscripts, the retinal configuration (upright / italic letters right / left slanted letters correspond to all-*trans* / 13-*cis* / 9-*cis*) and the protonation state of the Schiff base. Thin black arrows indicate thermal, outlined arrows photochemical transitions.

Each BR molecule consists of a single chain of 248 amino acids which is arranged as seven transmembrane α-helices (A-G) in the lipid bilayer of the cell membrane (fig. 1b). Inside the cage formed by the protein moiety the retinylidene residue is bound in the form of an initially protonated Schiff base linkage at the ε-amino group of lysine-216 to the protein moiety. The retinylidene residue together with an inner shell of amino acids forms the chromophoric system of BR (Fischer and Oesterhelt, 1979). The proton transport occurs through the proton pore which is divided by the chromophoric system into an inner (cytoplasmic) and outer (extracellular) half channel. Intermolecular proton transport takes place via the Schiff base, some water molecules and different donor and acceptor groups (Oesterhelt et al., 1992). Two of them have been identified as the aspartic acids at position 85 and 96 and are labeled in fig.1. Further important biophysical and biochemical parameters are summarized in table 1.

Absorption of a photon leads to a *trans - cis* photo-isomerization of the retinylidene residue around the 13,14 double bond within approximately 500 fs (Dobler et al., 1988; Mathies et al., 1988) with a quantum yield of 0.64 (Xie, 1990; Tittor and Oesterhelt, 1990). The following reversible thermal relaxation steps of protonation and deprotonation of different amino acids and the Schiff base linkage as well as the conformational changes of the protein backbone are intimately linked with the proton transport and are usually summarized in the photocycle scheme (fig. 2). The different photointermediates are represented by the common single-letter code and can be distinguished by the protonation state of the Schiff base, the retinal configuration and their absorption maxima, the latter being given as subscripts. Outlined thick arrows symbolize photochemical transitions, whereas thin black arrows stand for thermal steps of relaxation. Broken lines refer to the release and uptake of protons from the Schiff base linkage (Varo and Lanyi, 1990). The retinal configuration is indicated by upright letters for all-*trans*, right slanted italics for 13-*cis* and left slanted italics for 9-*cis*.

The initial B-state is characterized by an all-*trans* configuration of its retinylidene group. In the dark BR equilibrates with the D-state (dark-adapted BR) which has a 13-*cis* configuration. Illumination reverts it back to the so-called light-adapted B-state with a pure all-*trans* retinal population. After photoisomerization BR relaxes thermally, as a result of continued protein conformation changes, via the intermediate K_{590} to the L_{550} - state. In the step $L_{550} \rightarrow M_{410}^{I}$ the Schiff base releases its proton to the proton acceptor group aspartic acid at position 85 (Asp85 or D85). Asp85 is located in the outer part of the proton pore and is part of a complex counter ion in the environment of the protonated Schiff base.

Simultaneously, a proton is released to the outer medium from a group still unidentified, but probably Arg82 (Heberle and Dencher, 1990). The reprotonation of the Schiff base occurs through the cytoplasmic half-channel and aspartic acid 96 (Asp96 or D96) acts as a proton donor in the step $M_{410}^{II} \rightarrow N_{560}$. Finally, the reprotonation/deprotonation steps of Asp96/Asp85 take place during the steps $N_{560} \rightarrow O_{640}$ and $O_{640} \rightarrow B_{570}$, respectively. In addition, the O_{640} state can be photochemically converted into the P_{490} state, which is characterized by a 9-*cis* retinal configuration. The P_{490} thermally relaxes to the Q_{380} state where the retinylidene residue is released from the Schiff base linkage but remains in the binding pocket. The spectrally identical M-intermediates $M_{410}^{I,II}$ can be photochemically reconverted to the initial B-state by the use of actinic light with a proper wavelength inside the absorption band of $M_{410}^{I,II}$.

These related changes of the absorption spectrum of the BR-molecule (photochromism) are the most interesting properties of the BR-molecule which can be used for optical storage and information processing. The most important feature is the number of possible write/erase cycles which determines the lifetime of such a recording media. Since the BR system was developed during evolution to enable the halobacterial cell to survive under extreme conditions, such as low oxygen content and highly concentrated salt solutions in the desert, the stability of the protein has to be very high otherwise the energy consumed for its synthesis would be higher than the chemical energy transformed from light energy by BR. Nature has done a very good job in optimizing the reversibility of the BR molecule. Isolated purple membranes are stable and reversible for many years when exposed to air and sunlight; surprisingly these properties persist also after embedding the purple membrane fragments into polymers. Photochromic materials should also have a large shift of the absorption spectra between educt and product state, otherwise residual absorption would reduce the operation characteristics. In most cases it is desired that the product state is thermally stable. Only photochemical activation should convert it back to the initial educt state without any side products. The photochromic shift observed in BR between B_{570} and M_{410} is large enough (160nm), but the second demand, the lack of a thermal relaxation rate is not fulfilled. The thermal relaxation depends on the time needed to pump one proton over the cell membrane and is for a suspension of wildtype BR_{WT} 10 ms. For the biological function of BR as a light-driven proton pump nature optimized the energy conversion rate during evolution and this resulted in short cycle times. However, for an application of BR as a photochromic material longer M-lifetimes are desired in order to get a purely photocontrolled system which is activated and deactivated by light-induced photochemistry only. Since no thermal reconversion from the above mentioned P/Q branch to the B_{570} state appears, this kind of photochemistry can be used for long term optical storage (Popp et al., 1993).

CHEMICAL AND GENETIC VARIATION OF THE PROPERTIES OF BACTERIORHODOPSIN

Possibilities to modify the properties of the BR-molecule are the alteration of the physiochemical conditions of the environment, the chemical replacement of retinal with retinal analogues and finally the modification of the coding gene. All these variations result mainly in a change of colour and/or kinetics of the different photointermediates.

Physiochemical methods include lowering of temperature and thereby stabilizing intermediates such as M_{410} at 230 K (Becher et al., 1978) or K_{590} at 80 K (Birge et al., 1989). The kinetics of the photocycle and especially the lifetime of the M-intermediate can be further influenced by drying out the films in order to lower the humidity, by pH changes or by less protic solvents (Oesterhelt et al., 1991).

Replacing the retinal chromophore by synthetic retinal analogues offers a wide range of possible modifications, *e.g.*, the analogue phenylretinal (Kölling et al., 1984) prevents photochemistry for steric reasons, but may be used as an optical switch in the ps-range.

However, probably the best way to control and change the wild type properties is genetic modification of the primary structure of the polypeptides. This genetic approach can be exemplified by the variant D96N. Since the functionally important role of the amino acid Asp 96 in the proton transport is known, the operation of bacteriorhodopsin can be altered by replacing this amino acid by an inert one. In the case of D96N an exchange of Asp by Asn leads to the loss of the proton donor and acceptor capabilities in BR (Miller and Oesterhelt, 1990). As a consequence the lifetime becomes dependent on the extramolecular proton availability since reprotonation of the Schiff base in the $M^{II} \rightarrow N$ transition now requires diffusion processes of protons from the outer medium. Lifetimes of up to 100 seconds for the D96N variant in suspension have been reached. This process can be externally controlled and further slowed down by higher pH-values and the exchange of water by less protic solvents like glycerol.

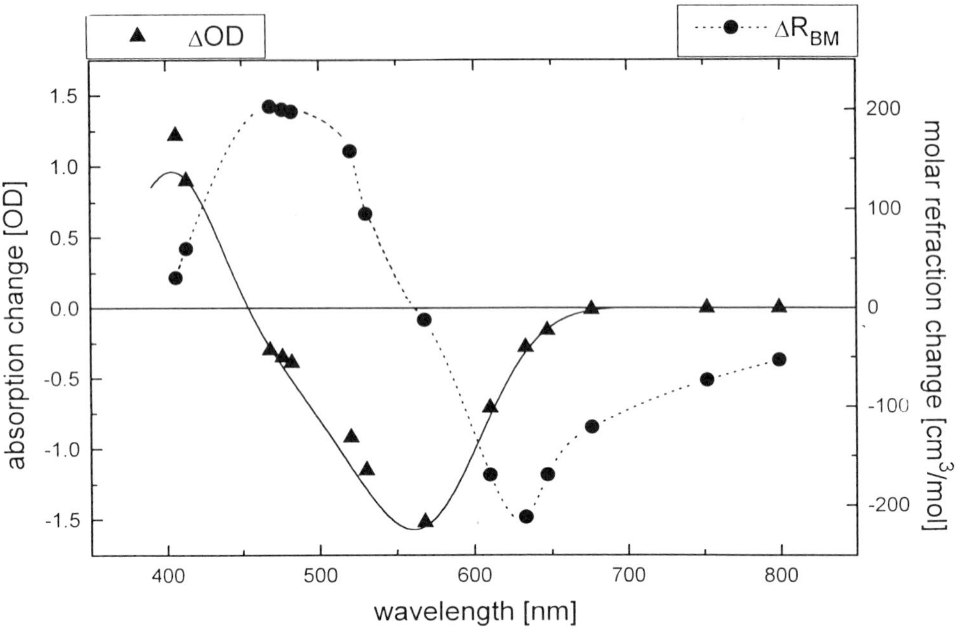

Figure 3. Spectral dependence of light-induced changes of absorption and molar refraction. The triangles stand for light-induced (λ = 568 nm, I_{568} = 20 mW/cm^2) absorption changes of a BR film containing the variant D96N with an initial optical density of iOD(568) = 3.6. The solid line fits the difference spectrum and can be obtained by a simple two level model consisting only of a population distribution of the molecules in the B_{568} and M_{410} state. The resulting spectral dependence of the difference in the molar refractions of B_{568} - and M_{410} - state is given as filled circles and is outlined by an eye-guiding dashed line.

Once a halobacterial strain carrying the modified bacterio-opsin gene has been constructed, the mutated bacteriorhodopsin can be produced by conventional biotechnological methods in virtually unlimited amounts.

Several BR-variants with potential technical importance have been generated. They can be divided into two classes: materials for optical information processing and materials for optical information storage. The first class requires fast dynamics and a high light sensitivity whereas for the second class the lifetime of the photoproduct is the most important feature (Popp et al., 1993). The BR films used can be prepared by embedding the molecules in the purple membrane fragments into polymeric matrices or by drying or centrifuging them onto high-quality optical glass substrates. No loss in the photochemical activity of BR is observed during these manipulations. The thickness of the films can be controlled by suitable metal spacers and is 20 µm.

LIGHT-INDUCED CHANGE OF ABSORPTION AND REFRACTIVE-INDEX IN BR FILMS

The photochromic characteristics of BR films under steady state illumination can be illustrated by the spectral dependence of the light-induced absorption changes. Illumination of a BR film containing the variant D96N with the krypton ion laser wavelength 568 nm and an intensity of 20 mW/cm^2 leads to the difference spectra shown in fig. 3. The absorption changes $\Delta a(\lambda)$ expressed in terms of optical densities for a sample with an initial optical density of 3.6 are marked as triangles at 14 discrete laser wavelengths. The reduced absorption around 570 nm can be explained by the lowered population distribution of the B_{570} intermediate, whereas the greater absorption measured at about 410 nm is due to the rising M_{410} photointermediate, which is the longest living photoproduct and the dominating one in the steady state approximation. In the case of the variant D96N this resulting population distribution can be sufficiently explained by a two-level model consisting only of the B_{570}- and the M_{410}-state. The theoretical relationship obtained for this two-level model is marked as a solid line in figure 3 and fits the difference spectrum well. Population of the O_{640}-intermediate, as known from the wild type system, is not observed with this variant (Hampp et al., 1992a), consequently a higher bleaching efficiency (up to 95 %) at lower light intensities can be obtained with D96N.

Immediately coupled to the light-induced absorption changes $\Delta a(\lambda, I_A)$,

$$\Delta a(\lambda, I_A) = [\sum_{i=B,M,N,O} \epsilon_i(\lambda) \cdot \Delta C_i(I_A)] \cdot d \tag{1}$$

where I_A stands for actinic light intensity, λ for the wavelength, $\Delta C_i(I_A)$ for the concentration of molecules in the intermediate i and d for the thickness of the film, are the changes in the refractive index $\Delta n(\lambda, I_A)$*$R_i(\lambda)$ is the molar refraction of the i.th intermediate and $n(\lambda)$ is the refractive index of the bulk sample.

$$\Delta n(\lambda, I_A) = \frac{(n(\lambda)^2+2)^2}{6000\, n(\lambda)} \cdot \sum_{i=B,M,N,O} R_i(\lambda) \cdot \Delta C_i(I_A) \tag{2}$$

It was shown that the relationship between absorption and dispersion can be explained and predicted by the Kramers Kronig relations and that the index of refraction $n(\lambda)$ changes in maximum about 0.005 at 633 nm simultaneously with an absorption change of Δa of 1.5 [OD] at 568 nm (Zeisel and Hampp, 1992). For the variant D96N the difference in the molar refraction $\Delta R_{BM}(\lambda) = R_B(\lambda) - R_M(\lambda)$ of the B_{568} and the M_{410}-intermediate can be calculated with the help of eq. 2 assuming identical changes of concentration $\Delta C_B(\lambda) = - \Delta C_M(\lambda)$. This change of the molecular refraction parameter is outlined in fig. 3 as filled circles with a dashed, eye-guiding line. The index change normalized per BR molecule gives a value of $\Delta R_{BM} = -3.5 \cdot 10^{-22}$ cm^3 molec.$^{-1}$, this is seven times higher than one obtained with synthetic photochromic fulgides (Kirkby et al., 1986) and near a postulated limit for organic photoisomerization processes of about 10^{-21} cm^3 molec.$^{-1}$ (Tomlinson and Chandross, 1980).

THE PROPERTIES OF BR FILMS IN DYNAMIC HOLOGRAPHY

The intensity-dependent absorption changes of BR can be used for information recording. Spatially varying intensity distributions result in related population distributions of the different intermediates in the BR-film. Since each BR-intermediate has its particular absorption spectrum and refractive index, this leads to a spatial modulation of both parameters. Each BR-molecule has dimensions of a few percent of a wavelength only. Even in the dense packing of BR in the purple membrane sheets the BR-molecules react with light almost individually, resulting in very high spatial resolution of more than 5000 lines/mm. This means that the wavelength limits the recording process, not the material. Such high resolutions are required for example in holography.

Two coherent laser beams form an interference pattern in their common overlapping region. In the case of two overlapping plane waves with spatially constant intensity distribution, sinusoidally varying intensity distributions are created. This means periodic sequences of bright and dark regions interact with the BR-molecules inside the BR-film and

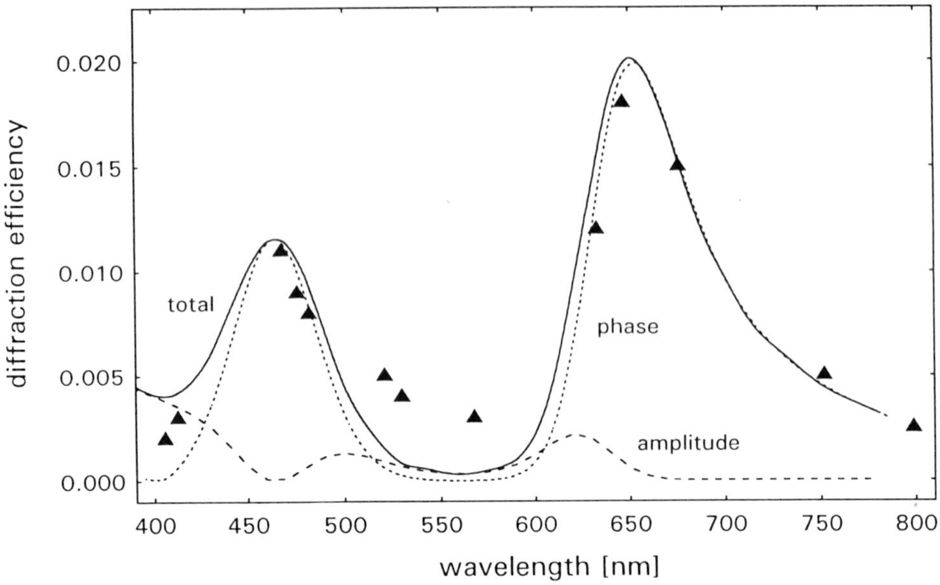

Figure 4. Spectral dependence of the diffraction efficiency of a BR film containing the variant D96N. The contributions of phase and amplitude are given by the dashed lines. The experimentally derived values are outlined as filled triangles.

result in an intensity-dependent bleaching of the BR-material. For recording of so-called B-type holograms the used wavelengths have to be chosen inside the main absorption band of the B_{568} state. The decay of the hologram is determined by the lifetime of the M_{410}-intermediate, but can be photochemically accelerated by flashing with blue light. In M-type recording a pump beam is needed to achieve a high initial population of the M_{410}-photointermediate. The information is recorded with a wavelength inside the main absorption band of M_{410}. The pump beam acts in this configuration as the read-out beam and is therefore constructive with respect to the hologram formation. With this technique and the use of the variant D96N a purely photocontrolled and fast (ms) system (B $\leftrightarrow$ M) is achieved. In both

Table 2. Holographic properties of bacteriorhodopsin films

Holographic properties of bacteriorhodopsin films	
resolution	≥ 5000 lines/mm
optical density	1 - 20 [OD_{570}]
bleaching	90 - 100 % (*e.g.* BR_{D96N})
index of refraction	1.47
refraction index change	10^{-3} - 10^{-2}
diffraction efficiency	1 - 7 % (typ. 2 -3 %)
light sensitivity	5 mJ/cm^2 (B-type) 30 mJ/cm^2 (M-type)
polarization recording	possible (chromophore anisotropy 20:1)
reversibility	≥ 10^6 cycles
thickness	10 - 500 μm (typ. 20 μm)
speed	msec - sec
aperture	unlimited commercially available up to 50 * 50 mm^2

cases the modulation of the absorption coefficient $\alpha_1 = (2.303\ \Delta a)/(2d)$ and the refractive index $n_1 = \Delta n/2$ leads to a periodically changing photochemical grating with the resulting period $\Lambda = \lambda/(2 \sin \theta/2)$, where θ is the angle between the recording beams, which can be probed by an additional reading beam (reconstruction beam). Two geometrically different cases have to be distinguished, incoming beams from the same or from opposite directions. The former denotes the transmission, the later the reflection hologram. Both hologram types can be reconstructed, if the Bragg condition is matched. The strength of the diffracted beam can usually be calculated with the help of the 'coupled wave theory' derived by Kogelnick (Kogelnick, 1969). The diffraction efficiency $\eta(\lambda_R, I_A)$ depends for a mixed phase and amplitude grating on the modulation parameters n_1 and α_1 as well as on the average absorption of the sample α_0. The resulting diffraction efficiency for the transmission type hologram is given by

$$\eta = (\sin^2 P + \sinh^2 A)\cdot\exp(-2D);$$

$$P = \frac{\pi n_1(\lambda_R, I_A)\cdot d}{\lambda_R \cos\theta_R} \qquad A = \frac{\alpha_1(\lambda_R, I_A)\cdot d}{2\cos\theta_R} \qquad D = \frac{\alpha_0(\lambda_R, I_A)\cdot d}{\cos\theta_R} \tag{3}$$

θ_R is the Bragg angle and λ_R the wavelength of the reconstructing beam. P denotes the phase, A the amplitude and D the damping part of Kogelnick's formalism.

The spectral dependence of achievable diffraction efficiencies with BR films is shown in fig.4. The main contribution to the diffraction efficiency of the material arises through the phase contribution, which is with regards to optical densities at the central wavelength 568 nm of more than two especially in the red region of the spectrum the dominating part. The experimentally determined maximum values were found at 647 nm and at 468 nm, which is in good agreement with theoretical values, although rigid assumptions such as infinite plane wavefronts, have been made. Inside the main absorption bands of M_{410} and B_{570} higher contributions from the amplitude part can be expected, but their absolute height is still strongly damped by the average absorption α_0 of the BR film during holographic recording and read-out. The residual difference between experimental data and Kogelnick's theory in this part of the spectrum can be explained by the photochemical activity of the reconstructing beam. Table 2 summarizes the main holographic properties of BR films in dynamic recording.

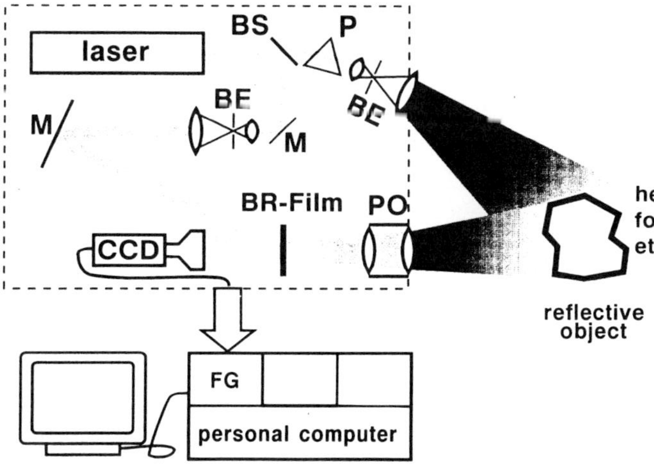

Figure 5. Experimental setup for real-time interferometry with bacteriorhodopsin films using reflection type holograms for recording and diffuse reflecting objects under evaluation. In this scheme no Bragg limitations have to be considered due to the homodyne setup with the wavelength 647 nm.

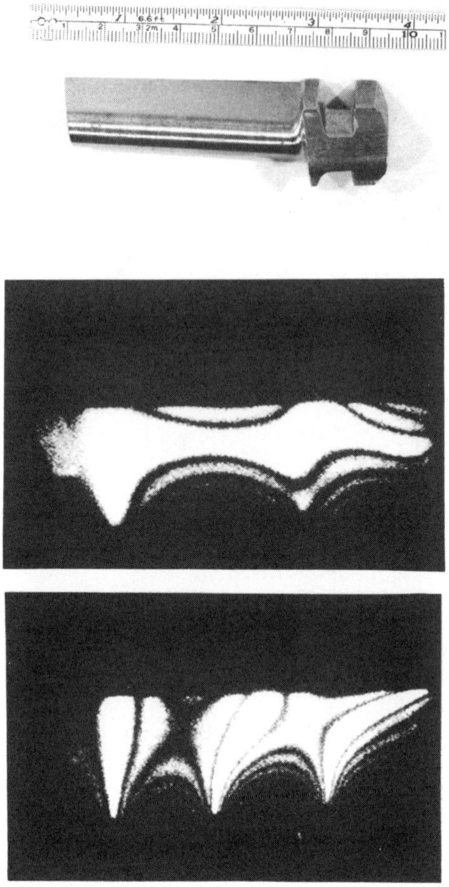

a

b

c

Figure 6. Real time interferograms of a turbine plate (real object size in a) obtained with the time-averaging technique using BR films as dynamic recording material. The plates were excited from the rear side by suitable piezo ceramics and their resonance frequencies have been determined to be 70 kHz in b) and 243 kHz in c), respectively.

APPLICATIONS OF BACTERIORHODOPSIN FILMS

The photochromic (reversible colour changes) and photoelectric (charge separation processes) properties of the BR molecule offer a variety of applications in spatial light modulation (Thoma et al., 1991; Birge et al., 1990) optical phase conjugation (Korchemskaya et al., 1990; Zeisel and Hampp, 1994), as artificial photoreceptor devices (Miyasaka and Koyama, 1992) and in the construction of optoelectronic neural networks (Haronian, 1991). Two selected examples from holographic applications, real-time interferometry and real-time pattern recognition, will be described in order to give an idea of possible technical applications of BR-films. In an holographic film the intensity distribution illuminating the material is recorded. The resulting hologram is a diffractive element which generates the original object wave in amplitude and phase when illuminated with the same reference beam as employed during hologram recording. The same happens in BR-films but since they are reversible recording materials these holograms can be erased at any time.

Applications in Holographic Interferometry

Real-time interferometry is a nondestructive technique for monitoring and evaluating objects under applied stress, heating or deformation [see for example (Schumann et al.,

1985)]. One possible homodyne setup with BR films as dynamic holographic recording media is outlined in fig. 5. The laser beam is expanded, spatially filtered and directed via a prism to the diffuse reflecting surface of the test object. Together with the reflected light the reference beam, coupled out by a beamsplitter (BS) forms a dynamic reflection grating due to opposite sides of object and reference beam. After closing the shutter (Sh) the reference beam acts as the reconstructing beam, the decaying B-type hologram can be detected with a CCD camera. Real-time interferograms can now be obtained by recording two or more holograms in a time interval $\Delta t = t_2 - t_1$. After recording a reference hologram of the object without applied stress at t_1, alterations in the object's structure due to stress or heat lead to the recording of different holograms at t_2. The read-out of these interfering holograms leads to the reconstruction of the original object, its surface covered with interference fringes. Each fringe spacing indicates a depth difference of $\lambda/2$, where λ is the wavelength of the used laser wavelength. From the location and density of the interference fringes mechanical or thermal expansion coefficients as well as local defects of a material can be detected and analyzed. Due to the use of only one wavelength (most favourable is 647 nm due to the high refractive index changes) no Bragg limitations arise and the complete angular spectrum can be reconstructed.

Figure 6 shows different interferograms recorded with D96N films and the time-averaging technique (Powell and Stetson, 1965; Renner and Hampp, 1993) applied to the homodyne setup. The test object, its original size is illustrated in fig. 6a, is a turbine plate, whose resonance frequencies have been excited from the rear side by a suitable piezo ceramic, which has direct mechanical contact with the object. Holographic recording of periodically vibrating objects means a cyclic refreshing of different interfering holograms, that have maximum diffraction efficiency in the turn around points of the oscillation due to the longer exposure time. Since BR films are reversible recording media, the interferograms in figure 6b and 6c shown could be recorded on a time scale of a few seconds by turning the piezo from 70 kHz to 243 kHz, respectively. The finite M_{410} lifetime enables the use of BR films in the described setup as a dynamic short term storage device. With the mutant D96N observation times of the interferograms of several minutes without an additional refreshing (new exposure) of the interferogram have been attained. The size of the test object in this application can be of up to half a meter, without using additional amplifying detector elements.

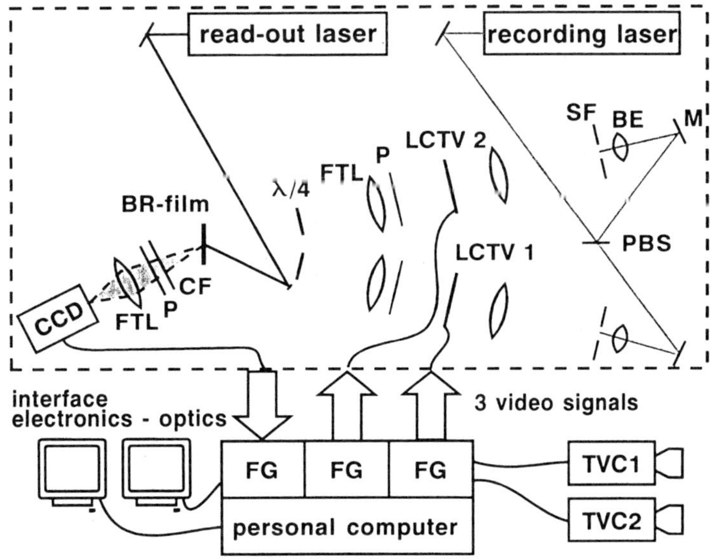

Figure 7. Experimental setup for real-time pattern recognition using a dual axis joint transform correlator. The resulting correlation signal is obtained at video frame rate by reconstructing the Fourier hologram of the two compared images with a plane wave front, the diffracted beam is detected after re-Fourier transformation with a CCD camera.

Figure 8. Photos from a real-time correlation sequence. The banknotes seen in the upper half are presented to the input video camera TVC1, which is connected to LCTV1. In the other arm (LCTV2) the three 'heads' from German banknotes are displayed as search criteria. Only when the presented banknote (upper half) matches one out of the "search set" (lower half) does an an identification spot appear (output of the correlator) below the corresponding search criterion. The photos are made from a TV screen, where the two input scenes and the output signal are displayed as a summed video signal.

Applications in Pattern Recognition

Fourier holography can be utilized to implement the mathematical convolution of two images optically (Vander Lugt, 1964; Pichon and Huignard, 1981). In this holographic technique two Fourier patterns from images that have to be compared for similarity are overlapped in the plane of the sample. The degree of similarity can be obtained by a second read-out beam that matches the Bragg condition.

The setup used, a dual-axis joint transform correlator (DAJTC) for real-time optical information processing with BR films (Thoma and Hampp, 1992), is outlined schematically in fig. 7. The two input images obtained from different frame grabbers (FG) containing arbitrary television scenes (TVC1 and TVC2) are transmitted to the DAJTC as two standard video signals and modulated onto the recording laser beams by means of a spatial light modulator, i.e. liquid crystal TV projection screens (LCTV). The two coherent, expanded (BE) and spatially filtered (SF) laser beams are Fourier transformed by suitable lenses (FTL) and overlapped to form M-type holograms in the plane of the BR film in all positions, where Fourier components common to both input patterns occur. The circular polarization of the reading beam ($\lambda/4$) together with the linear polarization of the reconstructing beam (P) enables, subsequent to retransformation and rejection of the recording light by color filters (CF), signal detection of the correlation with straylight suppression (Hampp et al., 1992b). An example for the operation of the real-time processing capabilities of the DAJTC system with D96N films is shown in fig. 8. The images to be compared consist of different German banknotes (Zeisel and Hampp, 1994b). One image contains the cental part of three selected fixed banknotes, the isolated heads serve as search criteria (illustrated in the lower parts of the summed video signal in fig.8) and are transferred via one frame grabber and LCTV1 to the DAJTC system. The second image (upper parts) contains several banknotes for comparison, which are presented sequentally in real-time to the recording TV camera, which feeds the second frame grabber and LCTV2. Whenever one part of the second image matches one part of the first image, the correlator system responds with a bright spot at the corresponding location of the first image, representing a high degree of similarity. The output video signal is recorded with a CCD camera and via the third frame grabber displayed in the summed video signal of figure 8 at the bottom and the corresponding location. Banknotes not matching the search criteria are ignored by the optical processor.

CONCLUSIONS

The two selected examples show the broad range of possible technical applications and illustrate the specific demands on the biological material in optical information processing. BR films can be used as dynamic short term storage devices in real time interferometry or as an optical processor unit in real time pattern recognition. The main holographic parameters are the light sensitivity (mJ/cm^2), the response time (ms), the resolution (5000 lines/mm), the signal to noise ratio of the correlator system (45dB) and a proper wavelength in the red region of the spectrum, where cheap laser diodes are available. The BR wildtype system, which was optimized during the course of evolution, cannot fulfil all these specific demands. Genetic engineering enables us to manipulate slightly some of the molecular properties in order to enhance their properties for nonbiological applications without disturbing their naturally balanced optimization. The new gene technological approach not only opens the field for a variety of promising technical applications, but also leads to an increase in our knowledge of the design principles of nature.

Acknowledgements

The authors thank M. Wolperdinger and R. Schmid for experimental contributions. Valuable comments of R. Thoma, D. Oesterhelt and C. Bräuchle are gratefully acknowledged. This research was supported by the Bundesministerium für Forschung und Technologie (FKZ 0319231B)

REFERENCES

Becher, B., Tokunaga, F., Ebrey, T.G., 1978, *Biochemistry* 17:2293.
Birge, R.R., 1990, *Annu. Rev. Phys. Chem.* 41:683.

Birge, R.R., Cooper, T.M., Lawrence, A.F., Masthay, M.B., Vasilakis, C., 1989, *J. Am. Chem. Soc.* 111:4063.

Birge, R.R., Fleitz, P.A., Gross, R.B., Izgi, J.C., Lawrence, A.F., Stuart, J.A., Tallent, J.R., 1990, *IEEE EMBS* 12:1788.

Dobler, J., Zinth, W., Kaiser, W., Oesterhelt, D., 1988, *Chem. Phys. Lett.* 144:215.

Fischer, U., Oesterhelt, D., 1979, *Biophys. J.* 28:211.

Hampp, N., Bräuchle, C., Oesterhelt, D., 1990, *Biophys. J.* 58:83.

Hampp, N., Popp, A., Bräuchle, C., Oesterhelt, D., 1992a, *J. Phys. Chem.* 96:4679.

Hampp, N., Thoma, R., Bräuchle, C., Oesterhelt, D., 1992b, *Appl. Opt.* 31:1158.

Haronian, D., Lewis, A., 1991, *Appl. Opt.* 30:597

Heberle, J., Dencher, N.A., 1990, *FEBS Lett.* 277:277.

Kirkby, C.J.G., Cush, R., Bennion, I., 1986, *in* Optical Bistability III (Eds. H.M. Gibbs, G. Mandel, W. Perghambarian and S.D. Smith), Springer Verlag, Berlin, p.165.

Kogelnick, H., 1969, *Bell. Sys. Tech. J.* 48:2909.

Kölling, E., Gärtner, W., Oesterhelt, D., Ernst, L., 1984, *Angew. Chemie. Int. Ed.* 23:81.

Korchemskaya, E.Y., Soskin, M.S., Taranenko, V.B., 1990, *Sov. J. Quant. Electron.* 20:381

Kouyama, T., Kinosita, K., Ikegami, A., 1988, *Adv. Biophys.* 24:123.

Mathies, R.A., Cruz, C.H.B., Pollard, W.T., Shank, C.V., 1988, *Science* 240:777.

Mathies, R.A., Lin, S.W., Ames, J.B., Pollard, W.T., 1991, *Annu. Rev. Biophys. Chem.* 20:491.

Miller, A., Oesterhelt, D., 1990, *Biochim. Biophys. Acta* 1020:57.

Miyasaka, T., Koyama, K., Itoh, I., 1992, *Science* 255:342

Oesterhelt, D., Bräuchle, C., Hampp, N., 1991, *Q. Rev. Biophys.* 24:425.

Oesterhelt, D., Stoeckenius, W., 1971, *Nature* 233:149-152.

Oesterhelt, D., Tittor, J., Bamberg, E., 1992, *J. of Bioenerg. and Biomemb.* 24:181.

Pichon, L., Huignard, J.P., 1981, *Opt. Commun.* 36:277.

Popp, A., Wolperdinger, M., Oesterhelt, D., Bräuchle, C., Hampp, N., 1993, *Biophys. J.* 64:1449.

Powell, R.L., Stetson, K.A., 1965, *J. Opt. Soc. Am.* 55:1593.

Renner, T., Hampp, N., 1993, *Opt. Commun.* 96:142.

Schumann, W., Züricher, J.P., Cuche, D., 1985, *in* Holography and Deformation Analysis Springer Verlag, Berlin.

Thoma, R., Hampp, N., 1992, *Opt. Lett.* 17:1158.

Thoma, R., Hampp, N., Bräuchle, C., Oesterhelt, D., 1991, *Opt. Lett.* 16:651.

Tittor, J., Oesterhelt, D., 1990, *FEBS Lett.* 263:269.

Tomlinson, W.J., Chandross, E.A., 1980, *in* Advances in Photochemistry (John Wileys and sons, New York) 12:210.

Vander Lugt, A.E., 1964, *IEEE Trans. Inf. Theory* 10:139.

Varo, G., Lanyi, J.K., 1990, *Biochemistry* 29:6858.

Xie, A., 1990, *Biophys. J.* 58:1127.

Zeisel, D., Hampp, N., 1992, *J. Phys. Chem.* 96:7788-7792.

Zeisel, D., Hampp, N., 1994, *Opt. Lett.* 19 in press

Zeisel, D., Hampp, N., 1994b, *IEEE Eng. Med. Bio.* 13:67.

12

NEW DEVICES MADE FROM COMBINING SILICON MICROFABRICATION AND CONDUCTING POLYMERS

Elisabeth Smela, Olle Inganäs, and Ingemar Lundström

Linköping Institute of Technology
Department Physics and Measurement Technology
S-581 83 Linköping, Sweden

INTRODUCTION

Conducting Plastics

Conducting polymers have been developed primarily for macroscopic use in batteries, electromagnetic shielding, coatings, displays, and solar cells (commercial applications have been discussed by Schoch and Saunders, 1992, and Studt, 1991). Chief among their attributes is their high conductivity in the doped state, which approaches that of copper (for an introduction, see Kaner and MacDiarmid, 1988; Reynolds, 1988; or Kanatzidis, 1990). This is due to the pattern of alternating single and double bonds, or conjugation, that gives them their other commonly used name, "conjugated polymers". (See Figure 1.) Dopants added to the material donate (or remove) electrons from the chain, delocalizing the electron cloud in the immediate vicinity and distorting the bond lengths. This results in the formation of bond alternation domain walls (called solitons or polarons and bipolarons, depending on whether the material has a degenerate ground state) which have energy levels within the polymer's band gap. At high doping levels, they can link together to form energy bands through which the electrons can travel. Charge is thought to hop between polymer chains. Because conductivity is higher along a chain than between chains, stretching a sheet of conducting polymer, which orients the chains, results in conductivity anisotropies as high as 100-1000 (e.g. Schimmel et al., 1991).

Overview of Conducting Polymer Properties

Conjugated polymers are of interest not only because of their high conductivity when doped. Unlike conventional solid state materials, their conductivity can be *changed* by orders of magnitude, and it can be done controllably and reversibly by changing the doping level. This is because, like silicon, conjugated polymers are semiconductors. But unlike silicon,

Molecular Manufacturing, edited by C. Nicolini
Plenum Press, New York, 1996

the dopants are easily inserted and removed from the spaces they occupy between polymer chains.

The doping level in conducting polymers is extremely high in comparison with silicon. Whereas doping levels in silicon are parts per million to parts per thousand, in conjugated polymers roughly one dopant ion is incorporated per every four monomer units. The ability of the polymers to store or disperse such a large ion (or charge) density is what has made them so attractive for use in batteries, but can also be exploited for sensing or control of chemical environments.

The presence of a bandgap in these materials allows electron-hole recombination, thus enabling the use of conjugated polymers in light emitting diodes (Clery, 1994). New advances are being made at a phenomenal rate in tailoring the bandgap by synthesis of new polymers and in increasing the light output by improving the quantum efficiencies (e.g. Grem et al., 1992; Berggren et al., 1994).

In the doped form, there are electronic states within the bandgap to which electrons and holes can move. This means that when the material is switched between doped (sub-bandgap states created) and undoped (no sub-bandgap states) forms, the electrons' transition energies are altered, changing the optical properties of the material. Because the optical absorption energy is changed, the color of the material can be electrically controlled. This is known as the electrochromic effect, and is being developed for use in thin film displays and smart windows (e.g. Mastragostino et al., 1993; Akhtar et al., 1988; Gustafsson et al., 1994).

In order to dope conducting polymers, ions must be physically inserted into the material. This has consequences for the volume that the polymer will occupy. As a result of the transport there is a volume change in the polymer of up to several percent. This change is due to bond length modifications along the polymer chain during doping (Murthy et al., 1987), differences in the interactions between polymer chains, and physical separation of the chains due to dopant insertion or extraction. Depending on the ion and the polymer, doping can cause either swelling or shrinking (Baughman et al., 1991). This makes actuators based on converting chemical to mechanical energy possible (for chemomechanical properties of conducting polymer gels see Yoshino et al., 1989a, b; Pei and Inganäs, 1993a; of non-conducting polymers see Osada, 1987), and this may be electrically controlled[1]. *

Conducting polymers can be doped by a wide variety of chemical species. Exposure to certain compounds can therefore change their electrical properties, which has been exploited in making chemical sensors based on changes in resistance (Pearce et al., 1993; Bartlett et al., 1989). Since it also changes their volume, smart actuators have been demonstrated (Pei and Inganäs, 1993b).

Conducting Polymers, Formation and Switching of Films

Before going on to describe how conducting polymers can be integrated with conventional silicon microfabrication, a brief review of the formation and processing of conducting polymer films will be presented. The considerations raised in this section will have an impact on device design and realizability.

There are two ways to make conducting polymers, chemically and electrochemically, but after deposition the operation of both types of films is similar. The most common method

---

* 1. The swelling and shrinking have been studied using the bending beam method by Pei and Inganäs (1992a, b, 1993a, c). Ion transport has also been investigated using other techniques, including absorption spectroscopy by Li and Qian (1989); quartz crystal microbalance by Naoi et al. (1991), Baker et al. (1991), and Orata and Buttry (1987); and a fluorescence probe by Krishna et al. (1991)

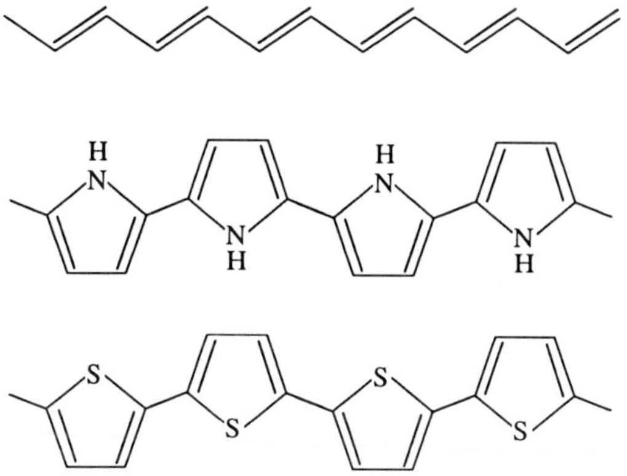

polyacetylene

polypyrrole

polythiophene

Figure 1. Three commonly used conducting polymers.

is chemical polymerization, which must usually be followed by a doping step. By adding substituents to the monomers, the chemically crosslinked polymers can be made melt processable, which makes moulding and extrusion possible. Addition of substituents has also resulted in dispersions that can be applied as coatings. Another approach is to begin with processable precursor polymers, chemically converting these polymers to obtain the desired but unwieldy long-chain polymer in the final step. In electrochemical polymerization, one begins with a solution of the monomers and grows a film directly onto an electrode; films made this way are usually automatically p-doped.

Electrochemical Growth. A polymer film is grown electrochemically onto an electrode, known as the working electrode, in an electrolyte solution containing monomers of the molecule. The working electrode is made of a material such as gold which will not oxidize or otherwise react in the voltage range in which the growth takes place: most conducting polymers cannot be grown on metals like aluminum. Since gold or platinum are commonly used and easily deposited, the microfabrication of suitable electrodes is straightforward, although with gold precautions may have to be taken in the presence of electronic circuitry because of its extremely high mobility and its poisonous effect on integrated circuits. Counter and reference electrodes are also placed in the bath. The reference electrode is typically a calomel or Ag/AgCl electrode; the counter electrode can be a Pt foil, Au wire, or indium/tin-oxide (ITO) covered glass plate, its shape depending on the geometry of the electrochemical cell. During growth, the voltage applied between the working and counter electrodes maintains the chosen voltage between the working and reference electrodes. The voltage can be constant (potentiostatic growth), time varying (potentiodynamic), or adjusted to maintain a constant current (galvanostatic); the potentials applied will affect the film qualities somewhat (e.g. Otero and de Larreta, 1993; West et al., 1993; Bilger and Heinz, 1993; Fujii et al., 1993).

There are a variety of electrolytes and solvents that can be used in the bath. The solvent must be able to dissolve the salt that one wishes to use. If the solvent is organic (typically acetonitrile or propylene carbonate are used) then the choice of salts is more limited than if the solvent is water. Of course, the solvent/electrolyte solution must also be able to dissolve the monomers, putting another constraint on the choice. Sometimes several formulas must be tried before a suitable combination can be found for growth of a particular monomer. The first step in film growth is thought to be the formation of monomer radicals, which react with each other to form short oligomer chains (Street, 1986; Kankare and

Lukkari, 1993; Lukkari et al., 1993; Otero and Rodriguez, 1993). After some time these precipitate onto the working electrode where they continue to grow. The polymer segments on the electrode are positively charged and attract anions from the solution to maintain charge neutrality. Approximately one anion for each three or four monomer units is incorporated in a typical polymer such as polypyrrole. Thus, films deposited this way are automatically doped and in the conducting state. They are also automatically patterned if the electrode has been patterned because the polymer grows onto the electrode.

Electrochemical Switching. The extraction and injection of electrons and the associated dopant transport are the cause of the conductivity, color, and volume changes in conducting polymers, so control over the movement of the electrons and the ionic species is critical. Although doping can be done chemically by exposure to a vapor, for most applications one needs electrical control over the process, so it is electrochemical doping that will be used (for general discussion see Scrosati, 1988; Heinze and Dietrich, 1989; also Heinze, 1991).

For cycling the films, either an electrolyte bath without monomers (which may be the same or different from the growth solution if the films were deposited electrochemically) or a polymer electrolyte is used. The sequence of ion movements during cycling is not completely straightforward, but depends on the salts involved (Baughman et al., 1991)[1]. For example, if a negative potential is put on the working electrode, either the anions can migrate out of the film, or the cations can move in, or both. Therefore, the doping/undoping process can be represented in different ways:

$$P^{+}(\text{bulky anion}^{-}) + \text{mobile cation}^{+} + e^{-} \longleftrightarrow P^{o}(\text{salt})$$

$$P^{+}(\text{mobile anion}^{-}) + \text{cation}^{+} + e^{-} \longleftrightarrow P^{o} + \text{salt}$$

Movement of both ionic species in a device is detrimental because it complicates the behavior of the polymer and can cause long term changes. Salt draining, in which the ions complex with each other and diffuse out of the polymer, is one such negative side effect. Therefore, it is desirable to use large, bulky anions during polymer growth that become permanently trapped in the polymer matrix. During operation of these films, only the cations shuttle in and out.

Microfabrication

A brief overview of some conventional microfabrication techniques will be helpful to illustrate the procedures that can be used to make micron-sized conducting polymer structures (for an introductory text, see for example Streetman, 1980). Microfabrication of silicon integrated circuits involves the vertical placement and lateral patterning of multiple conducting and insulating layers.

In the silicon itself, doping is used to obtain regions with excess charge carriers, either electrons (n-type) or holes (p-type), and it is permanent. The entire wafer is usually lightly doped either n or p-type when it is grown, but to allow control over charge transport in the final device, additional dopants must be put down in selected areas. These additional dopants are added to the surface of the wafer (within the first 1 μm). In general, differently doped regions must be placed adjacent to one another[2] *because patterning is confined to the plane of the wafer surface. Silicon became the most popular semiconductor for microfabrication

* 2. By controlling the depth of the doped regions and by counterdoping with the opposite dopant type, some vertical placement of the areas can be accomplished.

because it forms an oxide that serves as an insulating layer and that grows even from doped regions. Above these conducting and insulating layers that can be formed using the wafer, additional layers are deposited such as polycrystalline silicon, metals, and other insulators. Deposition is normally done by thermal evaporation, sputtering, or chemical vapor deposition.

The oxide and the deposited layers are patterned by removing areas that are unwanted, normally by etching. Photolithography is used to define the pattern and is very similar to making prints in photography. A layer of photoresist (akin to the emulsion in the paper) is put down over the layer to be patterned. This is normally done by spin coating: a drop of resist is put in the center of a wafer that is then spun at high speed to make a uniform coat. The resist is exposed to light through a mask (which plays the role of the negative). The pattern is developed, leaving some areas covered with resist and others bare. The wafer is then exposed to a chemical environment that etches the uncovered material, and the resist is removed. Circuits are built up layer by layer by multiple planar doping, oxide growth, or deposition steps followed by patterning.

Micromachining, which uses the techniques and expertise acquired over the decades with microfabrication, is aimed toward making miniature moving components on the surface of a wafer.[3 *] In this new discipline, the layers are patterned to achieve mechanical rather than electrical structures. The key to micromachining is the use of sacrificial layers which support and separate other layers during fabrication, but are removed to leave free standing parts in the final device. Anisotropic etchants, which etch different crystal planes at different speeds, are also important. New techniques for deposition and patterning have also been developed, including multi-step electroplating of metals inside patterns defined in photoresist, which allows one to build complex three dimensional shapes (Frazier and Allen, 1993; Engelmann et al., 1993).

CONDUCTING POLYMERS AND MICROFABRICATION

Because they are insulators, the use of conventional polymers in microfabrication has been limited, though clearly important. Photoresist is the most commonly encountered polymer in a clean room, but plays no role in the finished product. Polyimide is starting to be used as a component in devices, serving as an electrically and thermally insulating layer. In addition to ordinary insulating polymers, there are a number of specialty polymers like piezoelectric polymers and polymer gels. These will undoubtedly also make a contribution to microfabrication, but the number of new applications is not anticipated to be large.

Conducting polymers, by contrast, have a number of interesting properties that make them attractive for use in microfabrication. Although most of the development work on conducting polymers has been done with conventional fabrication in mind, there has also been work combining conducting polymers with microfabrication. Therefore, we will begin this section with an overview of the devices in which conducting polymers have already appeared. We will follow that with a description of our recent work on using conducting polymers as electromechanical elements and will propose several new devices that could be realized almost immediately.

---

* 3. There has been an explosion of publications in this field, and new journals such as the Journal of Microelectromechanical Systems and the Journal of Micromechanics and Microengineering have been created. For a review of the techniques, see Delapierre, 1989.

Applications to Date

The work that has been done combining conducting polymers with microfabrication has mainly been in the areas of sensors and electronic devices. This work was considerably advanced by the advent of soluble polymers, such as poly(3-alkylthiophene), and precursors that allowed thin films to be formed and patterned (e.g. Tomozawa et al., 1987). Schottky diodes have been made by contacting the conjugated polymer with one ohmic and one rectifying contact. Field effect transistors have also been studied (e.g. Paul et al., 1985; Thakeray et al., 1985; Tsumura et al., 1986; Burroughes et al., 1988; Assadi et al., 1988, 1992) and provide a way to add and remove charges from the conjugated polymer by injection and depletion rather than by chemical doping. However, these devices suffer from low mobility, in the range of $\mu = 10^{-5}$ to 10^{-2} cm^2/V-sec in comparison with $\mu = 10^3$ cm^2/V-sec in silicon. In addition, their behavior is unstable and depends on their electrical and thermal history (Inganäs et al., 1991). Although organic electronic elements cannot compete with conventional ones, they may find a niche in which it is not high performance that is required, but low cost, easy processability, and mechanical flexibility (Garnier et al., 1990).

Schottky diodes made from conducting polymers have been found to be electroluminescent, leading to a surge in research on polymer light emitting diodes (LEDs) (Buroughes, 1990; see also Burroughes et al., 1988). Although these LEDs have been macroscopic in size, the fabrication techniques are similar to those used for the microfabricated Schottky diodes and they could readily be miniaturized. Like the other electronic devices, LEDs have been helped by the development of solution processable conducting polymers that can be spun or cast to form thin films. The emitted light is polarized in oriented films, so the luminescence is caused by intrachain excitations, presumably by polarons. The mechanism can be described by using a band model: electrons are injected, directly or indirectly, from the high work function rectifying contact (often calcium) into the upper polaron state of the p-type polymer where they can radiatively recombine with holes in lower polaron states injected from a low work function ohmic contact (aluminum or ITO). One of these contacts should be transparent. The barrier heights between the polymer and the contacts can be changed with the metals. The energy gap can be controlled by tailoring the polymer: different colors can be obtained by changing the conjugation length along the polymer backbone. Quantum efficiencies of 0.1% to 4% have been reported (Braun and Heeger, 1991; Greenham et al., 1993; Berggren et al., 1994). Completely flexible LEDs with these efficiencies have also been demonstrated (Gustafsson et al., 1992). Conducting polymer LEDs are on their way to commercialization.

Sensors have been a growing application of conjugated polymers, including gas, ion, and bio-sensors (for reviews, see Zotti, 1992; Bidan, 1992; for an example, see Assadi et al., 1994). Their operation is based on changes in the oxidation state of either the polymer or the sensed molecule. These changes can be detected electrically by a change in resistance, a modification of the work function (potentiometric), or current generation (amperometric). The polymers can be functionalized to facilitate electron transfer directly from redox centers of some biomolecules (molecular wiring) or they can be used to entrap electroactive enzymes as counterions. An electronic nose with electrochemically grown conducting polymer sensors is already being commercialized (Coghlan, 1994; for a description of the sensor see Pearce et al., 1993).

New Applications

Although some of the designs that will be presented may seem quite futuristic, it should be emphasized that with present technology prototypes of most of them could be built in the laboratory today. Because these are new materials for microfabrication, experience in using them is lacking and there are still a number of technological hurdles. The most important of these problems will be discussed in the last section. Nevertheless, enough is known about handling conducting polymers because their various properties have been used to make a number of macroscopic devices, and microfabrication is well established, so new microfabricated devices akin to those suggested below could be demonstrated very shortly.

A distinction can be made between applications that require switching the polymer between conducting and insulating forms and those that operate continuously with the polymer in only one of the two states. In the former case, the design will be complicated by the requirement of having the conjugated polymer in contact with an electrolyte, either a liquid or an ion-conducting polymer, and with providing a counter electrode (and perhaps also a reference electrode). The poor room-temperature ionic conductivities of polymer electrolytes available today hinder their application, but progress in synthesizing new materials will inevitably make these materials more attractive in the future. In addition to needing increased ionic conductivity, the polymer electrolytes will also need to be patternable for some applications. For non-moving components, like electrochromic elements, this will not be an issue and the ability to spin-coat a film over the completed device will be sufficient. In that configuration, the counter electrode, which may be an aluminum film, can be evaporated over the electrolyte. In other cases it may be more advantageous to have the counter electrode patterned on the chip alongside any electrical contacts to the conducting polymer. For the present, however, it will be liquid environments in which most switchable devices like microactuators will operate. In some situations, operation in a liquid is required: handling of biological fluids is one example. The low voltages (less than 1 V, typically) needed to drive the transition from the doped to the undoped state makes conducting polymer actuators compatible with delicate environments of that kind.

Electromechanical Devices. Conducting polymers show great promise as electromechanical materials. The potential application of conducting polymers as an electromechanical material in microactuators has been evaluated conceptually (Baughmann et al., 1991). It was pointed out that a requirement of such materials is that they provide high stresses, substantial work and power densities, and large dimensional changes without requiring high voltages or high temperatures for operation. In these respects, conducting polymers compared very favorably with piezoelectric and magnetorestrictive materials, as well as with shape memory alloys. In comparison with polymer gels they have superior mechanical properties and higher conductivities. In the work by Baughman et al., several device designs were suggested that exploited the electromechanical properties of conducting polymers.

The change in volume between doped and undoped states can be on the order of 10%, giving a linear change $\Delta L/L$ of several percent. In addition, conducting polymers operate at extremely low voltages compared with other solid state materials. Although their mechanical properties have not been optimized, they have high strength and can deliver a sizeable stress (force/unit area). The ultimate tensile strength of polyaniline, for example, is 50 to 350 MPa, depending on factors such as the dopant molecule, doping level, and method by which the films or fibers were produced (Baughmann et al., 1991). For polythiophene, the tensile strength is in the same range, approximately 80 MPa. Stress is related to strain by Hook's law: $\sigma = \varepsilon E$, where the strain $\varepsilon = \Delta L/L$ (fractional change in dimension) and E is Young's modulus (a measure of the stiffness of the material). The maximum stress for an actuator is the smaller of the strain under zero load times the Young's modulus for the contracted state, and the limiting stress before failure, which can be taken as approximately 50% of the ultimate tensile stress. For polyaniline, $\Delta L/L$ is 0.005 to 0.01 and the Young's modulus is 5 to 9 GPa, so the stresses that can be produced are tens to hundreds of MPa. In a comparison of artificial actuator materials, these values compared very favorably with those for muscle and other contractile materials (Hunter and Lafontaine, 1992). For example, muscle only achieves a stress of 0.1 MPa; piezoelectric ceramics, shape-memory alloys, and magnetorestrictive materials have stress figures in the same range as conducting polymers. However, another factor to consider in applying these materials is cycle time. One of the drawbacks of conducting polymers is that, because they rely on diffusion, they are relatively slow. Because thin films respond more quickly than thicker ones, this is less of a problem for micromechanical actuators. Switching times of approximately 100 msec have been reported for polyaniline and polypyrrole films 500 Å to 3 μm thick (Baughmann et al., 1991) but piezoelectric ceramics are ten times faster, and muscle is 5 times faster (Hunter and Lafontaine, 1992).

Bilayer Microactuators. A bilayer between an active and a passive layer is one general, basic structure from which to build microactuating components. The simplicity of constructing bending beams stands in contrast with the difficulties encountered in making actuators out of silicon or other inorganic materials. The principle behind the bending beam's operation is a change in the volume of one of the layers, which results in bending of the bilayer if it is free to move. A relationship to calculate the curvature change in a bilayer is (for derivation see Pei and Inganäs, 1992b):

$$\frac{1}{R} - \frac{1}{R_o} = \frac{6\alpha_{(t)}}{\dfrac{(E'_1 h_1^2 - E'_2 h_2^2)^2}{E'_1 E'_2 h_1 h_2 (h_1 + h_2)} + 4(h_1 + h_2)} \tag{1}$$

where R_o is the radius of curvature before bending (usually infinite), R is the radius of the bent bilayer, the subscripts 1 and 2 denote the layer, $E' = E / (1-\nu)$ where E is the Young's modulus and ν is the Poisson's ratio, h is the layer thickness, and α is the lateral linear strain. Not only does the extent of bending in the strip depend on the thicknesses of the layers, but also the rate if the active layer is a conducting polymer: although it takes longer for thicker strips to reach their maximum curvature, because the change is controlled by diffusion, they curve into a tighter arc.

Some numbers that are of interest for metal/conducting polymer bilayers are (Baughmann et al., 1991; Sperling, 1986; James and Lord, 1992):

	Young's modulus	Poisson's ratio
gold	$7.44 - 8.0 \times 10^{10}$ Pa	0.42 - 0.44
polyaniline	$5 - 9 \times 10^{9}$ Pa	
plastics	$0.5 - 3.5 \times 10^{9}$	0.2 - 0.4

Using values of 5×10^9 for the Young's modulus and 0.3 for the Poisson's ratio for the conducting polymer layer gives us $E_1' = 7 \times 10^9$, while $E_2' = 1.3 \times 10^{11}$. For a volume expansion of 10%, α is approximately 0.03. To give an idea of the magnitude of the bending radius, if the two layers are of equal thickness, $h_1 = h_2 = h$, equation 1 gives us R Å 140·h. So if the layers are each 4000 Å thick, the radius is approximately 55 μm but if the thickness is reduced to 500 Å, the radius decreases to 7 μm. If we choose to make the sum of the layer thicknesses constant, $h_1 + h_2 =$ 4000 Å, we obtain the curve in Figure 2 showing that for maximum bending the best ratio is in the neighborhood of 4:1 polymer:gold.

To begin the discussion of microfabricated artificial muscles, I will describe a concrete example, microfingers that were made in our laboratory (for a full description see Smela et al., 1993a, b). Because these were actually fabricated, and because many of the fabrication issues will be similar in the devices proposed below, I will treat these in some detail.

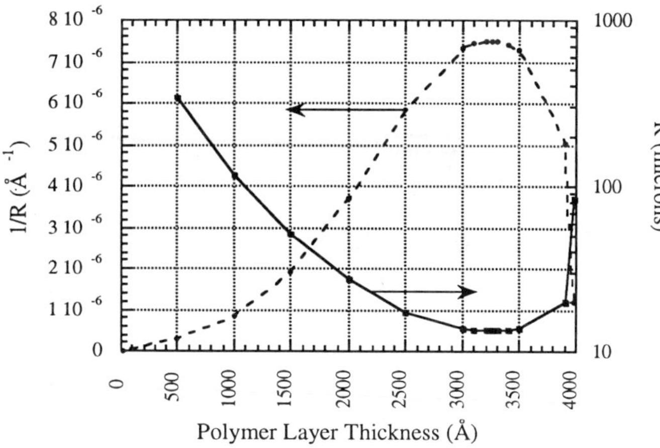

Figure 2. Radius of curvature as a function of polymer layer thickness for a constant bilayer thickness $h_1 + h_2 = 4000$ Å. For maximum bending the best ratio is approximately 4:1 polymer:gold.

Minifingers. Studies of macroscopic bilayers (areas of 0.5 x 2 cm) of polyethylene and polypyrrole had demonstrated that bilayer structures could convert electrochemical energy into mechanical energy (Pei and Inganäs, 1992a, b, 1993a, c). We were interested in using microfabrication methods to make conducting polymer bilayers that were anchored to a silicon wafer. We began with finger-like strips on the millimeter scale made of polypyrrole (PPy) and gold. Although the method used to make these fingers was primitive, it was a starting point for miniaturization and eventual combination with more sophisticated techniques.

On a silicon wafer, photoresist was patterned photolithographically into large rectangular areas. A very thin chromium layer (to promote adhesion) and a gold layer of approximately 3000 Å were vacuum evaporated through a copper shadow mask so that some of the gold would deposit on the photoresist and some over the SiO_2. The arrangement is shown in Figure 3. Polypyrrole was electrochemically polymerized at a constant 0.7 V in a solution of 0.1 M pyrrole and 0.1 M sodium dodecylbenzene sulfonate (DBS-Na) in water. DBS is a bulky anion that becomes immobilized in the polymer. The gold film was the working electrode, a second gold covered wafer, parallel to the first, was the counter electrode, and an Ag/AgCl electrode was used as a reference. Because the polymer grew over the whole area of the gold electrode, and nowhere else, it was automatically patterned using this technique. The PPy layer was grown to a thickness of approximately 4500 Å. The sacrificial layer (photoresist) was then rinsed away using acetone to leave the fingers connected to the wafer at one end but free to move at the other.

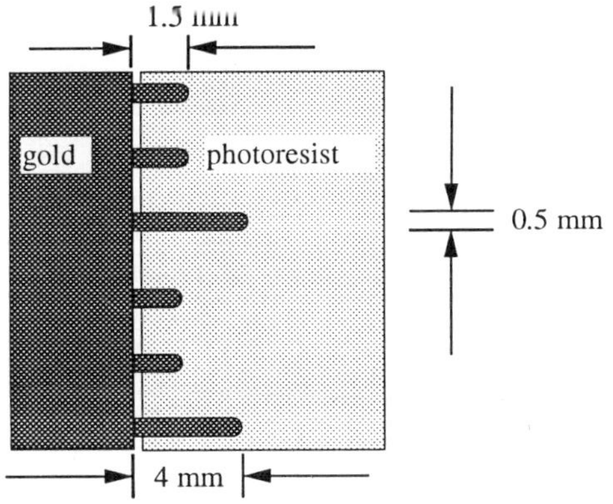

Figure 3. The gold film for the "minifingers" was evaporated in the pattern shown through a shadow mask, part of it onto the photoresist sacrificial layer and part directly onto the silicon wafer surface.

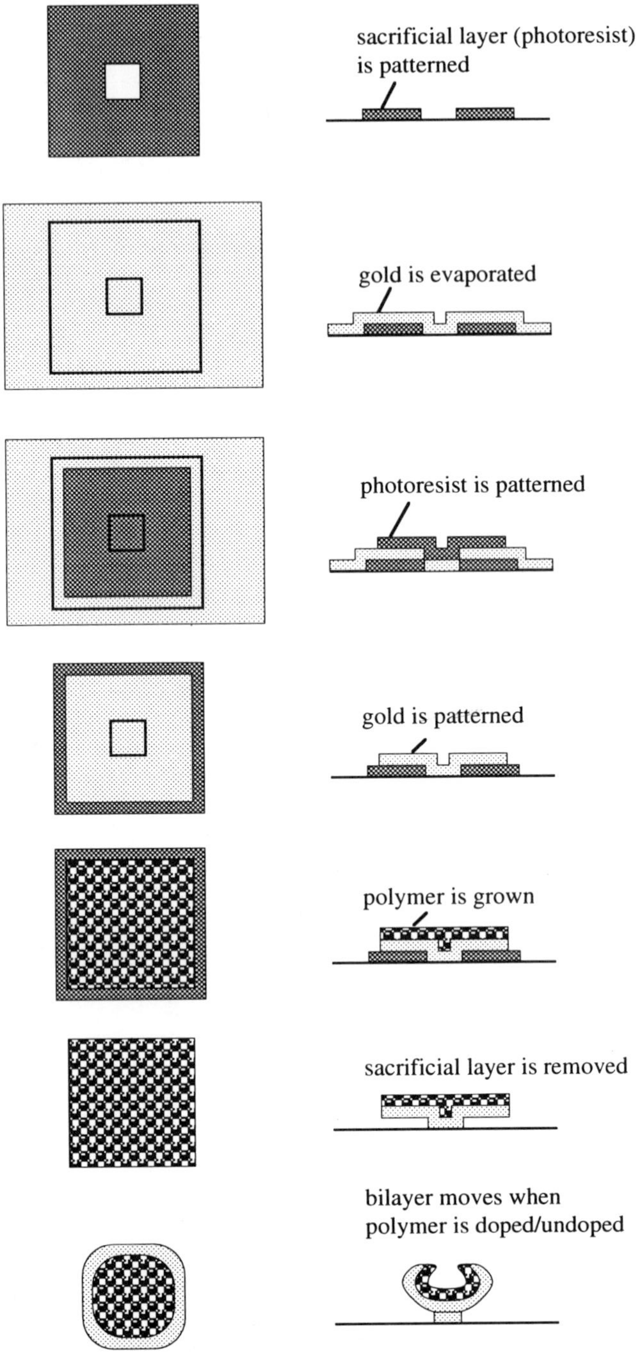

Figure 4. Processing steps for the microstructures. Note that dimensions are not to scale! (Vertical dimension is exaggerated for illustration.) Standard photolithography was used for patterning. Only two masks were needed.

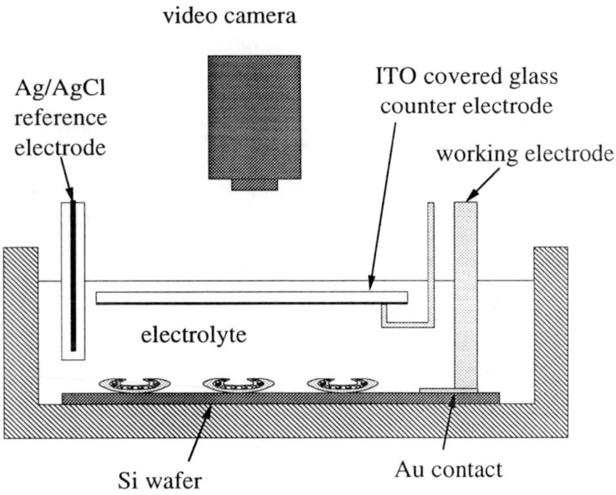

Figure 5. Illustration of the three-electrode electrochemical cell used to grow and operate the microstructures. A camera was positioned to look down from above through the transparent indium/tin-oxide counter electrode (contacted by a gold wire) and the liquid electrolyte.

The same electrolyte solution (without pyrrole) and reference electrode were used for the bending tests, but a transparent indium tin oxide (ITO) coated glass plate was used as the counter electrode so that the motion could be seen. The potential between the working and reference electrodes was a square wave between -0.85 and +0.3 V. When positive voltages were applied to the working electrode, the volume of PPy contracted as sodium was expelled, leaving the DBS in the film so the polymer acquired a positive charge, becoming p-doped. For negative potentials the PPy film expanded. These fingers had a wide range of motion: at +0.3 V they curled or twisted up and at -0.85 V they relaxed, but they did not flatten out completely. Some asymmetry caused the longer strips to spiral; it was difficult to grow the polymer to a homogeneous thickness and to control all the processing steps to achieve perfect uniformity. Switching the voltage every 5 seconds allowed the fingers to curl and straighten to their maximum extent; when using a faster rate of 1 or 2 seconds they did not go through the complete range of motion. The fingers were surprisingly robust: even with a length of several millimeters, the beams, which had a thickness of less than 1 μm, did not snap off when they were floating freely in a swirling liquid environment or under gentle rinsing.

The PPy was only weakly attached to the gold layer, sticking through Van der Waals forces. As a result, the PPy delaminated after a few thousand cycles. Improving polymer-metal adhesion is one of the major challenges of making successful microactuators from a combination of organic and inorganic materials. We have taken some steps to improve the adhesion by using an intermediate layer of thiol modified pyrrole monomers. This work will be discussed in more detail below.

Microfingers. The next step in our work was to make micron-sized structures. The smallness of the features did not permit the use of a shadow mask, so the gold had to be patterned photolithographically. Two masks were necessary. The first mask was the pattern for the photoresist sacrificial layer. The second was for the gold layer and contained thousands of structures, the smallest of which were 10 x 10 μm. The structures included "fingers", "tubes", and "paddles", and were designed to explore possibilities and pinpoint problem areas. After the sacrificial layer was patterned, Cr and Au were evaporated onto the wafer. A second layer of resist was patterned over the Au, aligned with the clearly visible resist pattern underneath, and the Au and Cr were chemically etched. In order to remove the resist from the top of the gold, but retain it underneath the fingers, the wafer was flood-ex-

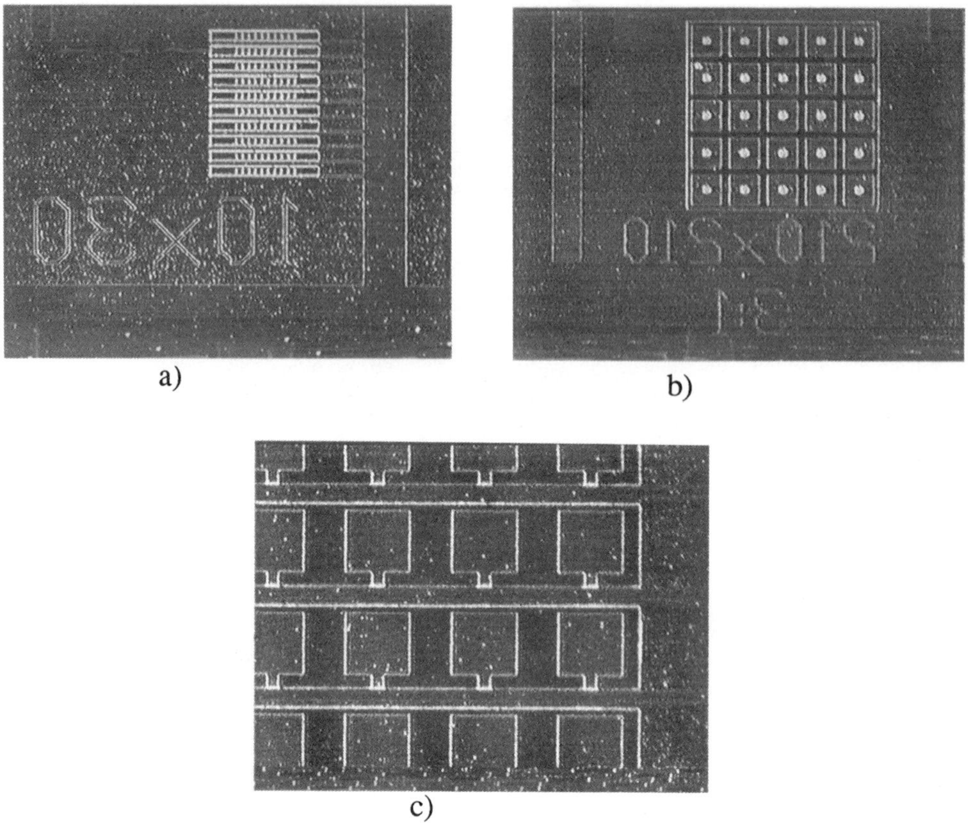

Figure 6. Photographs taken with a video camera of the microstructures before the polymer was grown and before the sacrificial layer had been removed. a) 100 element array of 10 x 30 μm fingers (smallest vertical white rectangles). Horizontal rectangles are the sacrificial photoresist. Size can be read in mirror image writing below. Gold pathways making electrical contact to the microfingers can also be seen. b) 210 x 210 μm center-attached squares, as in Figure 4. Electrical connection to these structures is made through the Si wafer; the closest gold wire is seen on the left. c) Close-up of 150 x 150 μm paddles lying over horizontal photoresist rectangles and attached at the base to the wafer.

posed and developed again. Polypyrrole was then grown onto the gold and the sacrificial layer of resist removed. The processing steps are illustrated in Figure 4.

Care had to be taken during processing to avoid over-baking the sacrificial layer. It was baked a total of three times: first it was soft-baked for patterning, then it was exposed to high temperatures during vacuum evaporation of the gold, and finally it was heated during the soft-baking of the second resist layer. If the sacrificial resist was overheated or spent too much time in vacuum, it became impossible to remove. Therefore, minimum baking and short times in the evaporation chamber were required. Another problem was that the large variation in feature sizes and the large resistive drop across the ITO counter electrode used during growth meant that the voltage required to grow PPy was higher on some parts of the wafer than on others, and growth was uneven. Delamination was also a more serious problem as the dimensions decreased. Thus, this conceptually simple step actually involved several new complications.

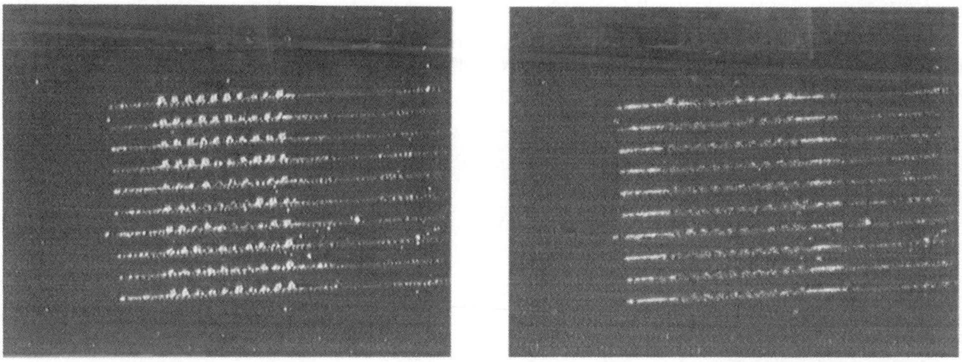

a)

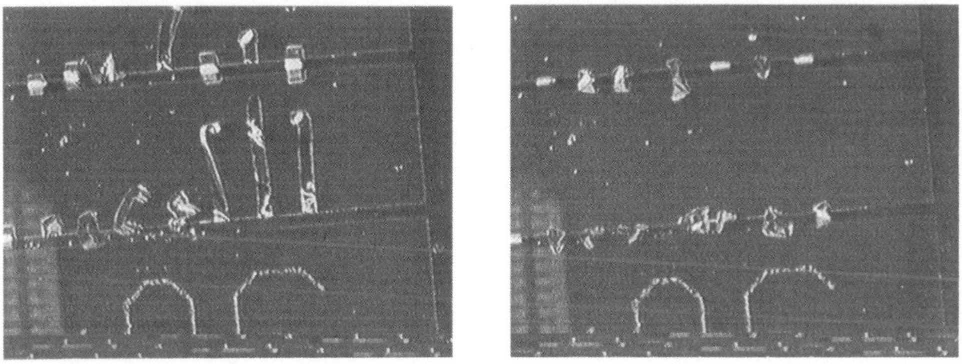

b)

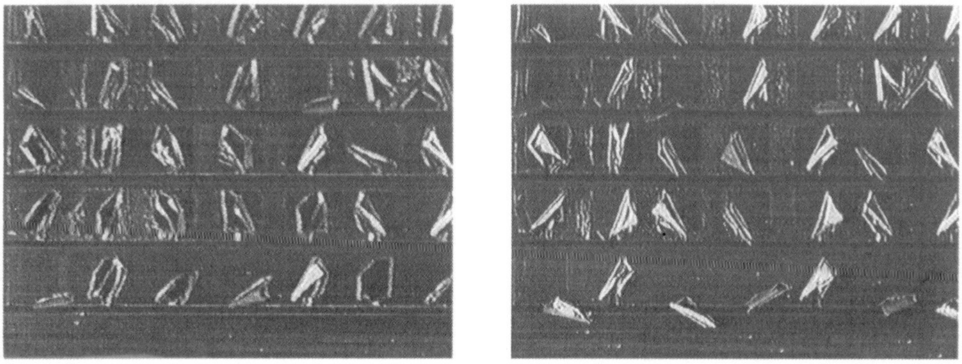

c)

Figure 7. Photographs of the microstructures during operation. Left side, undoped (-1.0 V) state and right side, doped (+0.5 V) state. a) 100 element array of 10 x 10 μm fingers; all elements in this array worked. b) 30 x 300 μm fingers; some extended and rolled up but others did not have this full range of motion. c) 150 x 150 μm paddles; since they were soft, they crumpled.

Several successful wafers were recently produced. The chromium layer was 5 Å thick and gold was 150 Å. Polypyrrole was grown to a thickness of between 1000 and 5000 Å at 0.8 V in the aqueous pyrrole/DBS solution described above. The electrochemical cell is illustrated schematically in Figure 5. Figure 6 shows photographs taken with a video camera through a light microscope of some of the structures before the PPy was grown and before the sacrificial layer was removed. Either acetone or ethanol could be used to remove the resist, but not commercial resist remover because it damaged the PPy. The microstructures were operated using the same configuration and solutions as the minifingers, but the voltage was switched between +0.5 and -1.0 V. Peak currents were 20 mA when using a square wave, average currents were 5 mA. Expansion and contraction took between 1 and 10 seconds, which is similar to the time it took for the minifingers since the PPy layers were approximately the same thickness. Larger structures such as the 1000x1000 μm tubes took longer than smaller ones like the 10x10 μm fingers, and where the PPy was thicker the response was also longer, as expected. The photographs in Figure 7 compare several structures in the doped and undoped states.

In order to observe their behavior in the presence of small particles, silica beads 100 μm in diameter were sprinkled over the structures on the chip. Tubes, paddles, and center-attached squares curled up and were able to wrap up or grab the particles, open to free them,

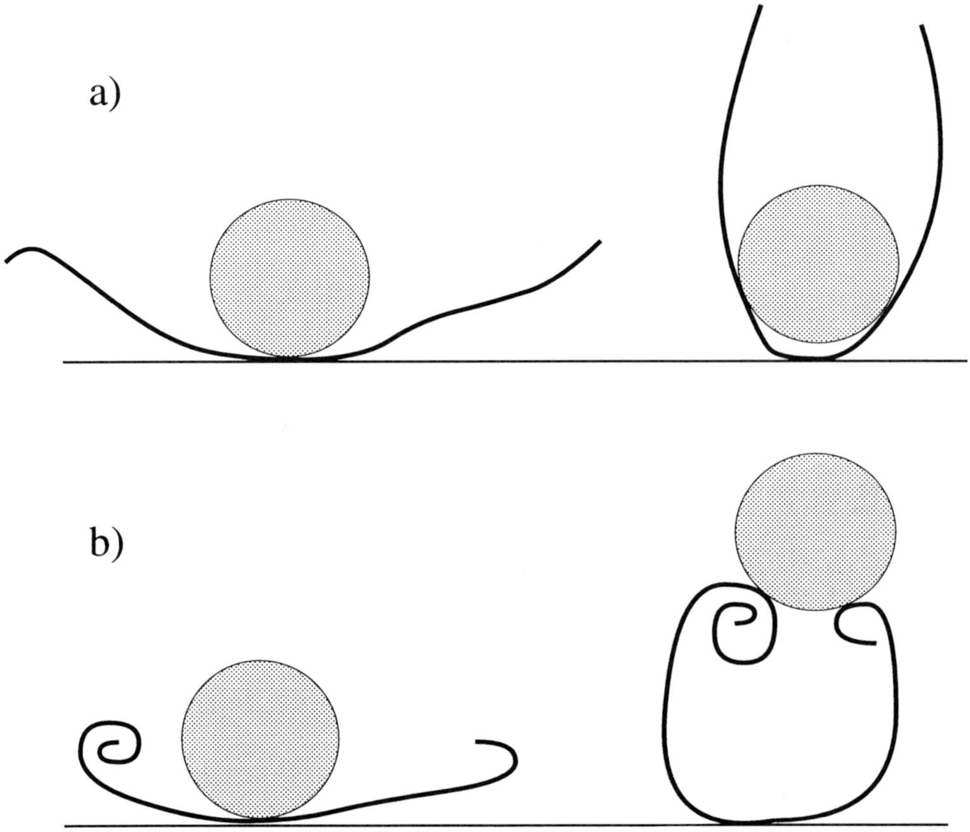

Figure 8. Particles of 100 μm diameter were sprinkled over the chip. Center attached squares were seen to a) grab the particles in some cases and b) pop them up and balance them overhead.

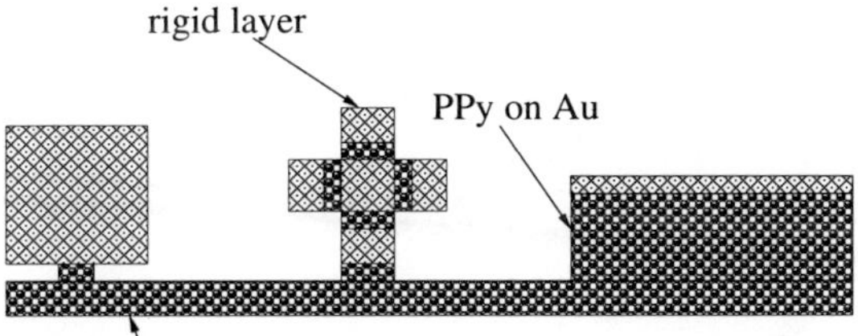

Figure 9. The next step in the bilayer microactuator work is to add a rigid layer to parts of the structures. This would allow one to make a stiff paddle bending only at the hinge (compare Figure 7c), self-closing boxes, and tubes that roll up properly.

and wrap around them again, as the voltage was switched. One of the squares was observed to alternately catch a sphere and balance it overhead as illustrated in Figure 8.

Present Work. In the immediate future, rigid elements will be incorporated into the bilayers. For example, by adding a third layer on which PPy will not grow, we hope to turn the soft, crumply "paddle" into a real paddle: the rigid layer will be patterned to go over the large square area, and the PPy will be grown only on the narrow hinge. It should also be possible to make self-closing boxes with stiff sides in this next step. By using a rigid support bar on the ends of the tubes, the corners will be prevented from curling. These ideas are illustrated in Figure 9.

Another natural development is to incorporate a polymer electrolyte into the design so that these bilayers can operate in air, "dry", rather than being restricted to liquids. This also requires careful consideration of counter (and optionally, reference) electrode placement. Finally, although in the present generation of devices all the muscles were electrically connected, so that all were synchronously operated, in future implementations separate connections to array rows can be made, allowing phase differences to be generated.

Nonconducting Polymer Bilayers. Independently of our work, a polymer bilayer microelectromechanical system (MEMS) has been made of nonconducting polymers by Ataka et al. (1993). A ciliary motion system was made from an array of oppositely facing sets of bilayer strips consisting of two layers of polyimide with different coefficients of thermal expansion and a microheater. It is interesting to compare our two approaches. The dimensions of their cilia were 100x500 μm and they were 6 μm thick. Aluminum was used for the sacrificial layer. The polyimide layers were spin coated and patterned by reactive ion etching using a nickel mask. Between the polyimide layers a thin chromium/gold/nickel layer was evaporated and chemically patterned to make a heater. The beams curled upward upon release. The layer with the higher coefficient of thermal expansion was on the top, so when heated the beams flattened. The 512 elements in the array were driven by two square waves, with a delay between them, at 1 to 40 Hz. Cilia facing one direction were activated together, followed by those facing the other direction. The array operated in air. Small pieces of silicon could be conveyed by the wave motion induced in the cilia, which took 16 V and 65 mA.

Proposals for Other Bilayer Devices. Based on the PPy/gold bilayer design, several closely related actuators suggest themselves. For example, a series of gates with a design similar to the rigid paddles described above could be placed in a channel that is later capped to make an enclosed conduit. The gates could be raised electrically to prevent the passage of particles carried by a liquid, or to reduce the flow of a gas or fluid. With appropriate phase differences, they could be used as pumps. Placing paddles at the opening of a set of channels, permitting a complete seal, would allow them to be used as valves to direct flow between

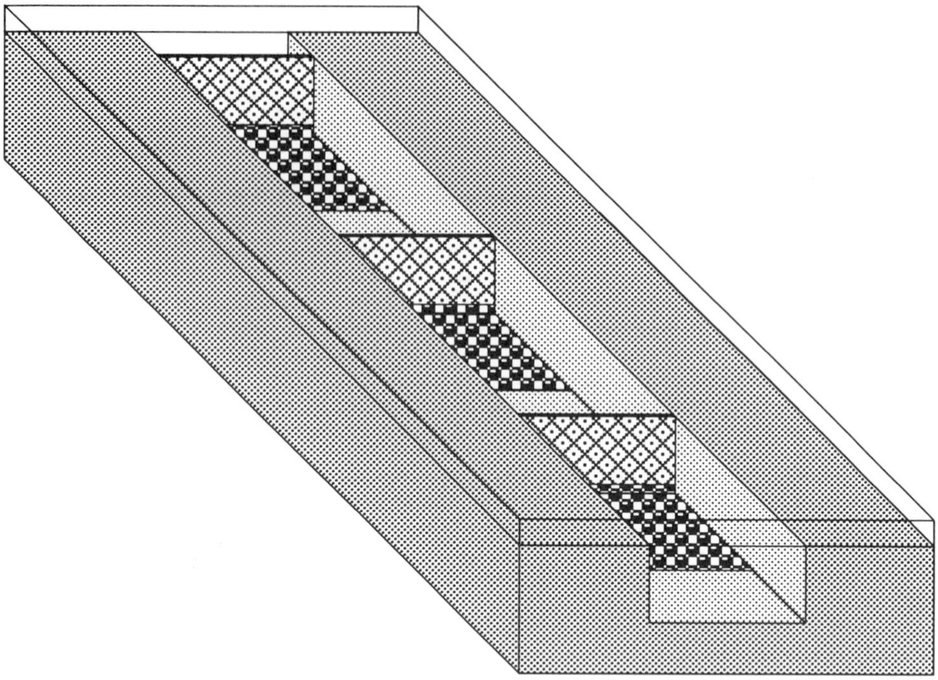

Figure 10. Bilayers set into a channel could be used as gates to control particle movement or as pumps.

channels. The gates could be chemically actuated for some applications, responding to the presence of certain gases or ions (Pei and Inganäs, 1992a, 1993b). Since the polymer is conducting while bent, the bilayer can be used to make an electrical switch. These ideas are illustrated in Figures 10-12.

Bilayers anchored at two ends could be used to hoist hinged structures off the silicon surface. Exciting work done by Pister (1992) has paved the way for this. This group has micromachined polysilicon plates several hundred micrometers on a side that were hinged by a "staple" on one edge. These were free to rotate on the hinge and lift off the surface. By using a tab and spring design, various elements snapped together to form complex 3-dimensional structures. A fraction of the structures self-assembled by gentle agitation in water, the success rate depending on the design. Electrical connection to the plates was made by doped polysilicon. Conducting polymer muscles provide an alternative to water assembly. (See Figure 13.) By controlling when each muscle contracts, parts can be assembled in a specific

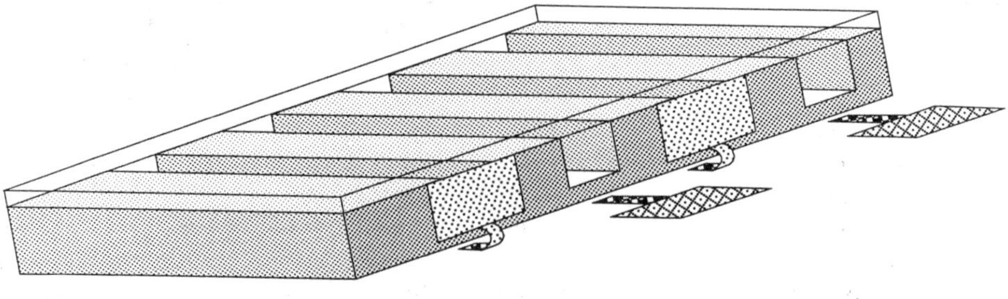

Figure 11. Bilayers in this geometry could block fluid or gas flow down selected channels, acting as valves.

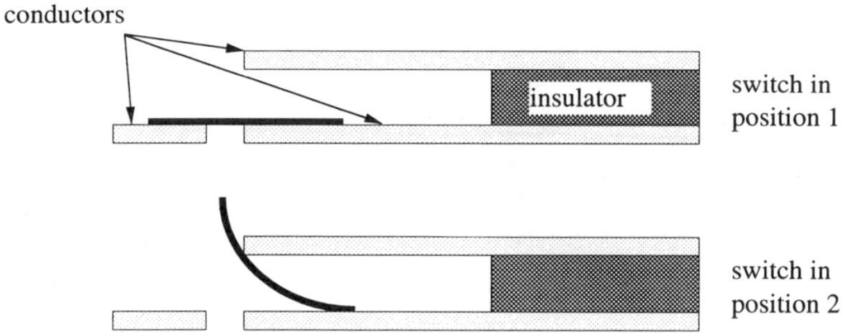

Figure 12. Bilayer electrical switch.

order. In addition, conducting polymer cables can be used as flexible electrical connections to hinged structures.

Electroplating metals inside openings in a patterned resist layer has opened up other micromachining avenues. This work began with the LIGA process (Becker et al., 1986; Guckel et al., 1990) and has since been adapted for UV light (Frazier and Allen, 1993; Engelmann et al., 1993). By plating multiple layers, one can build complex shapes with high aspect ratios. It would be possible with today's technology to build the fanciful tree shown in Figure 14, with the trees having movable branches. After the chromium/gold backbone is made, the conducting polymer could be grown electrochemically, adhering only to the top side of the horizontal layers since chromium is on the underside. If this were connected to a flexible lead and released, it might be able to swim.

Other Mechanical Devices. An example of a different sort of mechanical structure, one that doesn't use a bilayer but could probably be constructed today, is a table or mirror with hydraulic supports, suggested in Figure 15. The processing steps will be different from those needed to make bilayers. One possible processing sequence is as follows. A polymer electrolyte is deposited over gold electrodes. Photoresist is used to open holes over the gold contacts. Leaving the photoresist in place, a conducting polymer is grown electrochemically in these holes forming the table legs. An insulator is evaporated over the polymer-studded resist and patterned into small squares. Removing the resist leaves a gap between the table top and the electrolyte, allowing room for movement. The electrolyte is in contact with the lower part of each leg, allowing the conducting polymer to be doped and undoped, changing its volume. A counter electrode could be put on the wafer at the same time as the electrical connections to the conducting polymer. This device could also be operated in a liquid

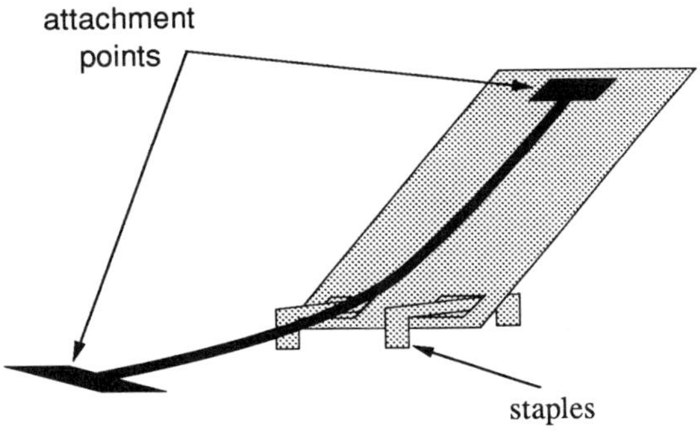

Figure 13. Bilayer micromuscles could be used to systematically pull rigid, hinged silicon elements into position with respect to other movable components.

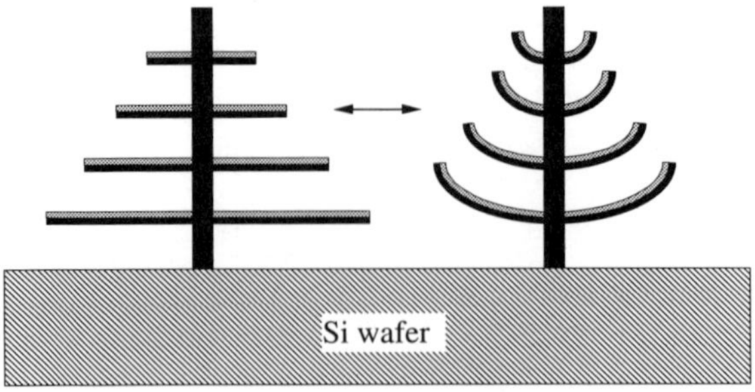

Figure 14. Using a combination of electroplated metals and electrochemically polymerized conducting polymers would allow one to construct complex 3-dimensional moving structures such as these "trees", seen from the side.

environment with external working and reference electrodes. A metal could be evaporated over the insulator to make a mirror.

An array of micromirrors has already been made using conventional micromachining materials and techniques (Younse, 1993). Each of these mirrors is mounted above a memory cell made in standard CMOS technology. A polymer layer is coated over the circuitry, and holes are opened for the legs. After that, a hinge layer and the mirror layer (an aluminum alloy) are deposited, and the polymer support layer is removed. The mirrors are hinged at two corners and their tilt of ±10° is controlled by electrostatic forces. Arrays of the digital micromirror devices (DMDs) are being developed for projection displays.

Mirrors with the polymer hydraulic supports could not compete with the DMDs for use in high speed displays. However, there are other applications where they may be advantageous. The polymer "jiggle tables" would be able to rotate in any direction by

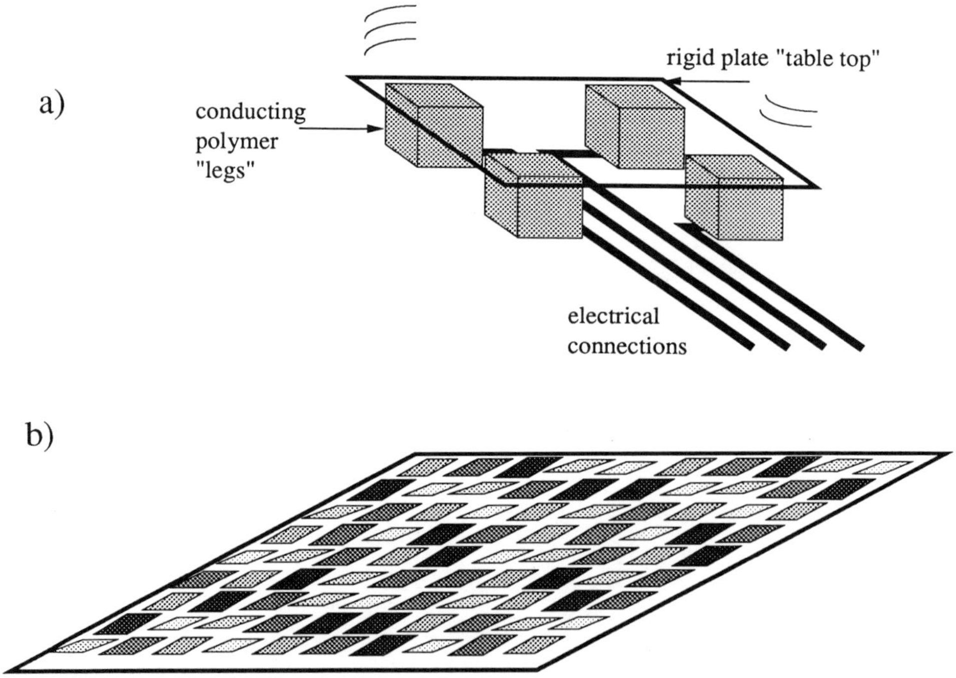

Figure 15. Conducting polymers as hydraulic electromechanical elements: in this example they are used as legs for a small table that can be tilted in any direction through a wide range of angles. a) Closeup of a single "jiggle table". b) An array of tables could be put on a wafer and controlled separately.

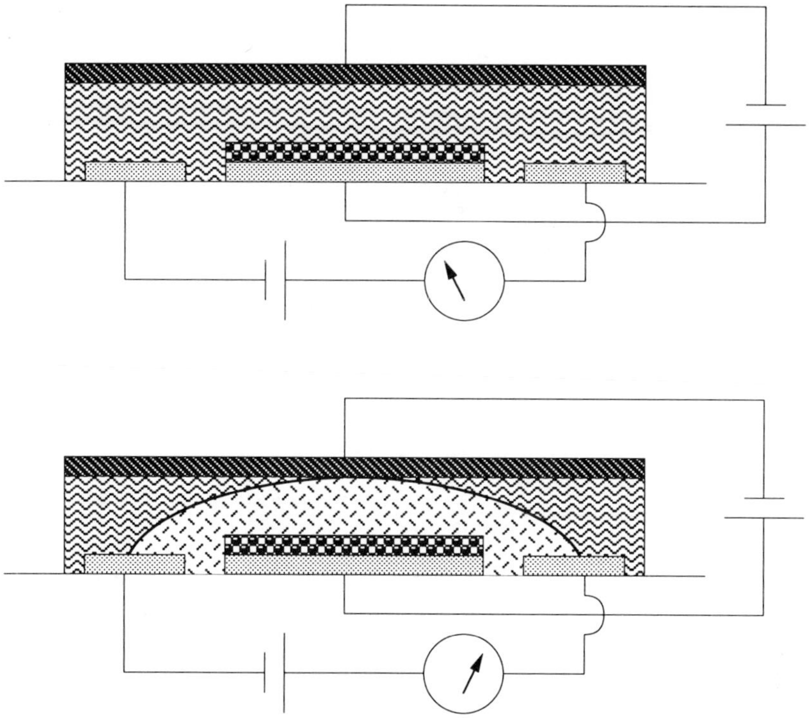

Figure 16. A conducting polymer film can be used to control the concentration of an ion species in a small volume of electrolyte.

separately addressing each leg and is not limited to tilting about a single axis. Also, the degree of tilt would be variable. Finally, polymer LEDs could be made on top of the flat surfaces, giving a light emitting surface that one could aim in any desired direction. One application for an array of jiggle tables would be a particle filter: by tilting all the tables identically one would open vias whose width was determined by the angle, allowing particles of that size but no larger to fall through. Bouncing them between that angle and 0° could facilitate moving the particles. Tactile sensing and segmented lenses may be other applications.

The ability to grow conducting polymers in micropores has generated great excitement because increased crystallinity and anisotropy has been found in these tubes (Penner and Martin, 1986; Cahalane and Labes, 1989; Martin et al., 1990; Martin, 1991; Cai et al., 1991; Granström and Inganäs, 1993). In addition, on some surfaces the conducting polymer grows on the support walls first, allowing one to grow hollow tubules. Controlled permeability membranes have been demonstrated the use of such tubules as constrictors. One may also be able to create constrictors in microfabricated channels. Insulating, hollow horizontal boxes could easily be made with metal contacts on one edge. Starting from the contact, the polymer could be grown inside the tube, whose inner walls had been appropriately treated with, for example, silane molecules that would encourage adhesion to the walls. Such constrictors could be used as very simple pumps or valves for very small fluid volumes.

Devices Based on Ion Transport. A feature of conducting polymers that may be of interest in bioengineering is controlled ion insertion and extraction. Because conducting

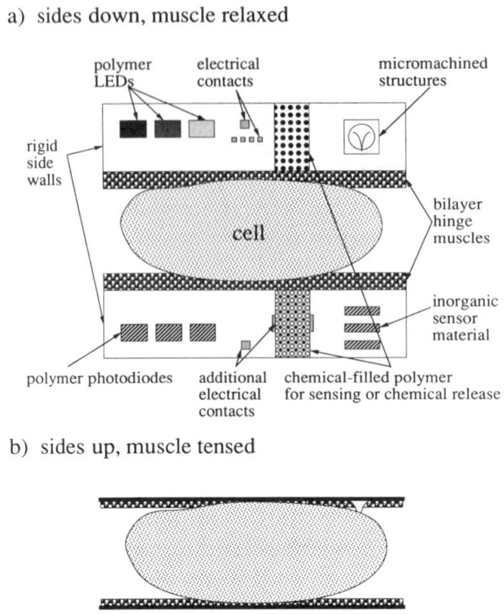

Figure 17. Top view of a "cell clinic", looking down on the device from above. The different properties of conducting polymers could be combined to make complex devices like this which can grab cells, do various measurements on them, and release them again. See text for more details. a) Various functions are put down on on the surface of rigid "flaps" using conventional, planar processing. b) Tensing the bilayer muscles raises the flaps out of the plane of the wafer surface and aligns the components with respect to each other. For example, light from the LEDs can pass through the cell and into the photodiodes.

polymers can be doped by a wide variety of ionic species, such as enzymes or hydrogen ions, they can be used to store and transfer these species to and from their surroundings. For example, a dose of some chemical could be delivered to a micro-organism.

A constant concentration of a species could be maintained in a thin film or small cavity by a 4-electrode pattern in which two electrodes are used to measure the resistivity of the environment and the third, a conducting polymer electrode, and fourth, a counter electrode, are used to adjust the ion concentration. The idea is illustrated in Figure 16. By adjusting the voltage, the ion concentration in the film could be maintained or varied at will. The area that could be controlled would depend on the size of the third electrode, which could be made quite large. Alternatively, an array of small electrodes would allow point control of the concentration and the generation of manipulatable gradients.

Combinations: the "Cell Clinic"

Even more interesting ideas result by combining several of these devices. At another level of sophistication, and a few years into the future, is the possibility of adding thin films of polysilicon or other solid state materials, such as sensing oxides, onto movable parts. If these could be mounted on the face of a rigid wall, for instance, then a variety of sensing materials, circuits, probes, etc., could be lifted off the chip and positioned in precise configurations with respect to each other.

Consider the 2-sided cell analysis chamber illustrated in Figure17. It would be the size of a cell, made for trapping a specimen, performing tests on it, and then letting it go. The two rigid flaps could be raised on bilayer hinges. Different functions could be laid out on the upper/inner surfaces of the walls, specialized for particular tasks. Conducting polymer light emitting diodes paired with polymer photodiodes could do optical measurements. These would be correctly positioned with respect to each other when the flaps were raised. Electrical probes could do impedance measurements across the cell or on the cell surface. Other conducting polymer layers with different chemical dopants, separately addressed electrically, could be used to deliver doses of those chemicals. Such layers could also be used to sense chemicals excreted by the cell since the doping level could be made to change in the presence of that species, which could be detected by a change in resistance of the polymer layer. Diverse micromachined parts or electrical circuits could also be raised into position. A silicon wafer could be completely covered with these clinics.

Other. One more thing that should not be overlooked is that conducting polymers, combined with conventional polymers, allow one to do micromachining on soft substrates. Light emitting diodes have already been made on bendable plastic, and operate despite being severely deformed (Gustafsson et al., 1992). Although the electronic circuits will not perform at the standard of their silicon counterparts, there may be a niche for flexible substrate devices. In addition to physical flexibility, the substrates can be of tremendously large area and low cost.

CHALLENGES

Some of the challenges have been touched upon already. They include providing contact to an electrolyte, preventing delamination, increasing device lifetime, and developing microfabrication techniques that are compatible with the conducting polymer.

Microfabrication Issues

There are some general questions that must be addressed when designing microfabricated structures using conducting polymers. First, will the device operate with the film in one doping level at all times, like LEDs? Or will it need to be switched, like microactuators? If so, the polymer must be in contact with an electrolyte and a counter electrode will need to be included in the design.

The process sequence must be considered. Will all the materials withstand the temperatures and chemicals to which they will be exposed? Will it be necessary to chemically modify one of the surfaces, for example to prevent corrosion, and will the surface treatment stand up to subsequent processing steps? Patterning the various polymers, and protecting them if other layers must be deposited and patterned on top of them will become important.

There is an urgent necessity to replace photoresist as a sacrificial layer. Its extreme sensitivity to drying, by heat, vacuum, or storage, makes processing very difficult. It also limits the number of layers that can be put down: the partially rigid sides are probably not achievable with this technique because of the additional baking and exposure to vacuum required in patterning another layer. In addition, the resist is very thick; the 150 Å gold layer is not continuous over the 14,000 Å step from the silicon to the resist. As a result, the PPy must grow until it is also 14,000 Å thick, and then it can continue over the fingers. The non-uniformity and discontinuity interfere with device performance: tears occur at the steps, most of the bending takes place where the gold is missing, and growth is so irregular over the pattern that some devices remain uncovered with PPy while others are tied down because the PPy has grown between and connected them. If another material is substituted for the resist, it must be removed without damaging the polymer. This rules out SiO_2 because the HF used to etch it will damage the PPy. It must also be removed fairly quickly because some of the structures are as large as 1000 x 1000 μm, which is a significant area to underetch.

Electrolyte

For a particular application, the choice between liquid and polymer electrolytes must be made. If it is to be a polymer, will it need to be patterned? What ionic conductivity is

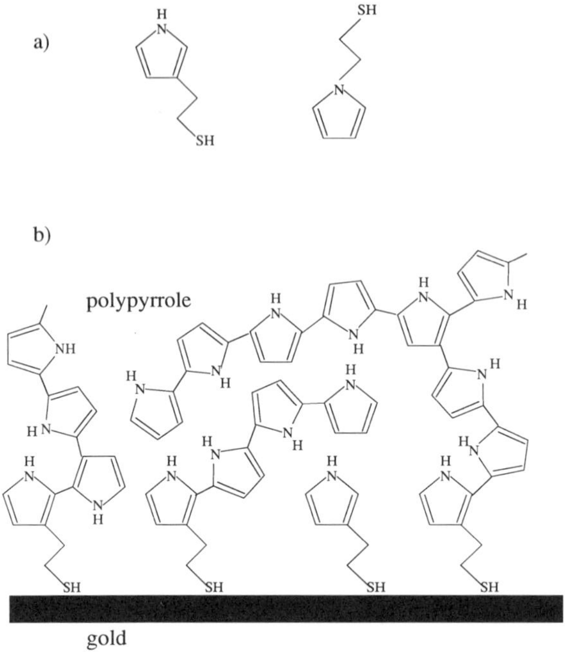

Figure 18. a) 3-(2-thioethyl)-pyrrole, left, and 1-(2-thioethyl)-pyrrole, right, were used to improve adhesion between gold and PPy. b) The pyrrole moieties are thought to be incorporated into the PPy film. The 3-TEP molecule was used in this illustration.

necessary to successfully operate the device? At the present time, the room temperature ionic conductivity of polymer electrolytes is too low for most applications. Although new materials are being synthesized and significant improvements are anticipated, operation in liquid environment will be necessary for some time. This is not a significant set-back because for some applications, like fluid handling or interaction with biological systems, it is actually required, and because the conducting polymer materials still need to be developed. Therefore, there are not many microfabricated switching conducting polymer devices ready to go on the market that are only waiting for a good polymer electrolyte. When polymer electrolytes are improved, the engineer will be able to design the devices for either liquid or dry operation depending on the application, which will be advantageous.

Molecular Engineering

Chemists will continue to play a critical role in furthering applications of conducting polymers. Molecular design enables one to improve processability and stability in the presence of oxygen and water. Rapid progress has already been made in both of these areas. It also lets one tailor and improve the various properties of conducting polymers. For example, the size of the band gap can be tuned so that polymer LEDs could emit light of all colors. The mechanical properties could be optimized for artificial muscles, or their switching speed increased.

Another area for which new molecules will be needed is surface modification. Control over the polymer/solid interface is critical. Surface treatment to prevent corrosion of materials exposed in liquid environments will become more important (Wrighton, 1986). Monolayer coatings may be necessary to stabilize polymer/metal interfaces, preventing oxide formation in the presence of the oxygen found in conducting polymers. Oxidation of aluminum in contact with PPy has been found to occur, making the interface unstable and creating high resistance contacts (Bantikassegn et al., 1993). In the work on conducting polymer tubules, it is thought that the character of the pore surfaces determines that growth rate on the pore walls and results in hollow tubes rather than solid cylinders. Treatment of SiO_2 with tailored silanes may be necessary to provide anchoring sites for initiation of polymer growth and for controlling the hydrophobicity, making a favorable chemical environment for oligomers to attach.

One of the biggest problems now faced with the microfingers is delamination. Only Van der Waals forces hold an electrochemically grown PPy film onto a gold electrode. We have found an improvement in adhesion by treating the gold surfaces with thiol modified pyrrole monomers before PPy film growth (Smela et al., in preparation). The thioethylpyrrole monomers are illustrated in Figure 18a. A short carbon chain connects an S-H group to the pyrrole in either the 1 or 3 position. The shortness of the chain ensures that molecules can conduct (long alkyl chains form insulating layers on gold). Before growth of the PPy film, the gold surface is exposed to the monomers in either the vapor phase or in solution. The advantage of vapor treatment is that solvents are avoided, which is important to the photoresist sacrificial layer that is present at this point in the processing. The PPy is then grown onto the gold electrode and is believed to incorporate the pyrrole moieties of the thiol monomers, as shown in Figure 18b. By synthesizing a separate monomer for surface treatment, the properties of the conducting polymer itself are unaltered. This allows separate tailoring of the surface interaction and the polymer properties, which is advantageous since it allows simultaneous optimization; trying to tailor the polymer to solve every problem would result in serious compromises.

CONCLUSIONS

Conducting polymers will be used as active components in microfabricated devices in a variety of ways, exploiting such properties as volume change, conductivity variation, ion storage, light emission, and optical property modulation. They have already been applied in the areas of electronic devices, in which they will have a limited use, and sensors, which is an area in which they could make a large contribution. As electromechanical elements, conducting polymers free micromechanics from the plane of the wafer and provide an easy way to create motion. As ion stores, they open up new possibilities that could not be achieved with conventional materials. Their low cost as light emitting and detecting layers will allow a new range of products to be introduced. In conclusion, the combination of conducting polymers and microfabrication will not only provide new solutions for old problems, but will also yield completely new, previously unimaginable devices

REFERENCES

Akhtar, M., Weakliem, H. A., Paiste, R. M., and Gaughan, K., 1988, *Synth. Met.* 26:203

Assadi, A., Spetz, A., Willander, M., Svensson, C., Lundström, I., and Inganäs, O., 1994, *Sensors & Actuators B*, 20:71.

Assadi, A., Svensson, C., Willander, M., and Inganäs, O., 1988, *Appl. Phys. Lett.*, 53:195.

Assadi, A., Svensson, C., Willander, M., and Inganäs, O., 1992, *J. Appl. Phys.*, 72:2900.

Ataka, M., Omodaka, A., and Fujita, H., The 7th International Conference on Solid-State Sensors and Actuators, Transducers '93, Yokohama, Japan, June 7-10, 1993, Digest of Technical Papers, p. 38.

Baker, C. K., Qiu, Y.J., and Reynolds, J. R., 1991, *J. Phys. Chem.*, 95:4446.

Bantikassegn, W., Dannetun, P., Inganäs, O., and Salaneck, W. R., 1993, *Thin Solid Films*, 224:232.

Bartlett, P. N., Archer, P. B. M., and Ling-Chung, S. K., 1989a, *Sensors & Actuators*, 19:125
Bartlett, P. N., Archer, P. B. M., and Ling-Chung, S. K., 1989b, *Sensors & Actuators*, 19:141.
Bartlett, P. N., Archer, P. B. M., and Ling-Chung, S. K., 1989c, *Sensors & Actuators*, 20:287.
Baughman, R.H., Shacklette, L. W., Elsenbaumer, R. L., Plichta, E. J., and Becht, C., 1991, in Molecular Electronics, (Ed. Lazarev P.I.), Kluwer Academic Publishers, Dordrecht, pp. 267-89.
Becker, E.W., Ehrfeld, W., Hagmann, P., Maner, A., and Munchmeyer, D., 1986, *Microelectronic Eng.*, 4:35.
Berggren, M., Gustafsson, G., Inganäs, O., Andersson, M. R., Hjertberg, T., and Wennerström, O., 1994, *Appl. Phys. Lett.*, submitted.
Berggren, M., Gustafsson, G., Inganäs, O., Andersson, M. R., Wennerström, O., and Hjertberg, T., 1994, *Adv. Mat.*, 6:488.
Bidan, G., 1992, *Sensors & Actuators B*, 6:45.
Bilger, R., and Heinz, J., 1993, *Synth. Met.*, 55-57:1424
Braun, D., and Heeger, A. J., 1991, *Appl. Phys. Lett.*, 58:1982.
Burroughes, J. H., Bradley, D. D. C., Brown, A. R., Marks, R. N., Mackay, K., Friend, R. H., Burns, P. L., and Holmes, A. B., 1990, *Nature*, 347:39.
Burroughes, J. H., Jones, C. A., and Friend, R. H., 1988, *Nature*, 335:137.
Cahalane, W., and Labes, M. M., 1989, Chem. Mat., 1:519.
Cai, Z., Lei, J., Liang, W., Menon, V., and Martin, C. R., 1991, *Chem. Mater.*, 3:960.
Clery, D., 1994, *Science* 263:1700.
Coghlan, A., *New Scientist*, 1994, Feb.5:20.
Delapierre, G., 1989, *Sensors & Actuators*, 17:123.
Engelmann, G., Ehrmann, O., Leutenbauer, R., Schmitz, H., and Reichl, H., 1993, MME'93 Workshop Digest, September 7-8 , Neuchatel, p. 85.
Frazier, A.B., and Allen, M. G., 1993, *J. Microelectromech. Syst.,* 2:87.
Fujii, M., Arii, K., and Yoshini, K., 1993, *Synth. Met.*, 55-57:1159.
Garnier, F., Horowitz, G., Peng, X., and Fichou, D., 1990, *Adv. Mater.* 2:592.
Granström, M., and Inganäs, O., 1993, *Synth. Met.* 55-57:460.
Greenham, N. C., Moratti, S. C., Bradley, D. D. C., Friend, R. H., and Holmes, A. B., 1993, *Nature* 365:628.
Grem, G., Leditzky, G., Ullrich, B., and Leising, G., 1992, *Adv. Mat.* 4:36.
Guckel, H., Christenson, T. R., Skrobis, K. J., Denton, D. D., Choi, B., Lovell, E. G., Lee, J. W., Bajikar, S. S., and Chapman, T. W., Technical Digest, IEEE Solid-State Sensor and Actuator Workshop, Hilton Head, June 4-7 1990, p. 118.
Gustafsson, G., Cao, Y., Treacy, G. M., Klavetter, F., Colaneri, M., and Heeger, A. J., 1992, *Nature* 357:477.
Gustafsson, J. C., Inganäs, O., and Anderson, A. M., 1994, *Synth. Met.* 62:17.
Heinze, J., 1991,*Synth. Met.* 41-43:2805.
Heinze, J., and Dietrich, M., 1989, *Mat. Sci. Forum* 42:63.
Hunter, I.W., and Lafontaine, S., Technical Digest of the IEEE Solid State Sensor and Actuator Workshop, Hiton Head, June 22-25 1992, p. 178.
Inganäs, O., Gustafsson, G., and Svensson, C., 1991, *Synth. Met.* 41-43:1095.
James, A. M., and Lord, M. P., Macmillan's Chemical and Physical Data, Macmillan Press Ltd., London, 1992, pp. 20, 22.
Kanatzidis, M. G., 1990, *Chem & Eng. News*, Dec. 3:36.
Kaner, R. B., and MacDiarmid, A.G., 1988, *Sci. Amer.* 2:60-65.
Kankare, J., and Lukkari, J., 1993, *Synth. Met.* 55-57:1305.
Krishna, V., Ho, Y.-H., Basak, S., and Rajeshwar, K., 1991, *J. Am. Chem. Soc.* 113:3325.
Li, Y., and Qian, R., 1989, *Synth. Met.* 28:C127.
Lukkari, J., Heikkila, L., Alanko, M., and Kankare, J., 1993, *Synth. Met.* 55-57:1311.
Martin, C. R., 1991, *Adv. Mat.* 3:457.
Martin, C. R., Van Dyke, L. S., Cai, Z., and Liang, W., 1990, *J. Am. Chem. Soc.* 112:8976.
Mastragostino, M., Arbizzani, C., Bongini, A., Barbarella, G., and Zambianchi, M., 1993, *Electrochimica Acta* 38:135.
Murthy, N.S., Shacklette, L. W., and Baughman, R. H., 1987, *J. Chem. Phys.* 87:2346.
Naoi, K., Lien, M., and Smyrl, W. H., 1991, *J. Electrochem. Soc.* 138:440.
Orata, D., and Buttry, D. A., 1987, *J. Am. Chem. Soc.* 109:3574.
Osada, Y., 1987, *Adv. Polym. Sci.* 82:1.
Otero, T. F., and de Larreta, E., 1988,*Synth. Met.*. 26:79.
Otero, T. F., and Rodriguez, J., 1993, *Synth. Met.*. 55-57:1436.
Paul, E. W., Rico, A. J., and Wrighton, M. S., 1985,*J. Phys. Chem.* 89:1441.
Pearce, T. C., Gardner, J. W., Friel, S., Bartlett, P. N., and Blair, N., 1993, *Analyst*, 118:371.

Pei, Q., and Inganäs, O., 1992a, *Adv. Mat.* 4:277.
Pei, Q., and Inganäs, O., 1992b, *J. Phys. Chem.* 96:10507.
Pei, Q., and Inganäs, O., 1993a, *Synth. Met.* 55-57:3724.
Pei, Q., and Inganäs, O., 1993b, *J. Phys. Chem.* 97:6034.
Pei, Q., and Inganäs, O., 1993c, *Solid State Ionics* 60:161.
Pei, Q., and Inganäs, O., 1993d, *Synth. Met.* 57:3718.
Pei, Q., and Inganäs, O., 1993e, *Synth. Met.* 55-57:3730.
Penner, R.M., and Martin, C.R., 1986, *J. Electrochem. Soc.* 133:2206.
Pister, K.S.J., Technical Digest, IEEE Solid-State Sensor and Actuator Workshop, Hilton Head, June 22-25 1992, p. 136.
Reynolds, J.R., 1988, *Chemtech* 7:440.
Schimmel, Th., Gläser, D., Schwoerer, M., and Naarmann, H., 1991, in Conjugated Polymers, (Eds. J.L. Brédas and R. Silbey), Kluwer Academic Publishers, Dordrecht, p. 49.
Schoch, K.F., and Saunders, H.E., 1992, *IEEE Spectrum* 6:52-55.
Scrosati, B.,1988, *Prog. Solid St. Chem.* 18:1.
Smela, E., and Inganäs, O., in preparation.
Smela, E., Inganäs, O., and Lundström, I., 1993a, *J. Micromech. Microeng.* 3:203.
Smela, E., Inganäs, O., Pei, Q., and Lundström, I., 1993b, *Adv. Mat.* 5:630.
Smela, E., Zuccarello, G., Kariis, H., Uvdal, K., Inganäs, O., Liedberg, B., and Lundström, I., in preparation
Sperling, L.H., 1986, Introduction to Physical Polymer Science, John Wiley & Sons, New York, pp. 227, 229.
Street, G.B., in Handbook of Conducting Polymers, (Ed. Skotheim T. A.), Marcel Dekker, Inc., New York, 1986, 265-91.
Streetman, B. G., 1980, Solid State Electronic Devices, 2nd Ed., Prentice Hall, Englewood Cliffs.
Studt, T., 1991, *R&D Mag.* 10:94.
Thakeray, J. W., White, H. S., and Wrighton, M. S., 1985, *J. Phys. Chem.*, 89:5133.
Tomozawa, H., Braun, D., Phillips, S., Heeger, A. J., and Kroemer, H., 1987, *Synth. Met.*. 22:63.
Tsumura, A., Koezuka, H., and Ando, T., 1986, *Appl. Phys. Lett.*. 49:1210.
West, K., Jacobsen, T., Zachau-Christiansen, B., Careem, M. A., and Skaarup, S., 1993, *Synth. Met.*, 55-57:1412.
Wrighton, M. S., 1986, *Science* 231:32.
Yoshino, K., Nakao, K., Morita, S., and Onoda, M., 1989a, *Japn. J. App. Phys.* 28:2027.
Yoshino, K., Nakao, K., Onoda, M., and Sugimoto, R., 1989b, *Solid State Comm.* 70:609.
Younse, J.M., 1993, *IEEE Spectrum*, 11: 27.
Zotti, G., 1992,*Synth. Met.* 51:373

13

MOLECULAR MANUFACTURING: A NEW FRONTIER AT THE INTERFACE OF ELECTRONICS, BIOTECHNOLOGY AND MATERIAL SCIENCES

Claudio Nicolini

Institute of Biophysics
University of Genoa
Via Giotto, 2
16153 Genoa Sestri Ponente - Italy
EL.B.A. Foundation
Marciana Marina (LI), Italy

The significant development and progress made during the last decade in the field of microelectronics and biotechnology have determined important overlapping of these disciplines, leading to the emerging of a new scientific branch, Molecular Manufacturing, aiming at manipulations and assembling at atomic scale to obtain molecular constructs capable of specific biochemical and electronic activity.

The most strategic branches of Molecular Manufacturing are Scanning Tunneling Microscopy (STM) and Atomic Force Microscopy (AFM) both of which have permitted not only to render visible atomic surfaces but also to manipulate single atoms at surfaces to design pre-determined nanostructures.

Recently, a step forward in real-time instrumentations was realized. The atomic surface becomes visible through the Scanning Tunneling Microscope (STM). Also, femtosecond laser spectroscopy can follow the time evolution of chemical reaction kinetics. A major distinction of molecular nanotechnology from bulk semiconductor technology is the bottom-up approach of building devices from the atom up.

A molecular computing and fabrication methodology aims indeed to perform massively-parallel real-time computational and imaging tasks through dynamical interactions of custom molecules patterned in lattices or distributed as substrates in solution. This research focusing on the problem of organizing complex functional patterns of custom organo-metallic molecules on computing substrates, exploits the properties of chemical reaction kinetics in tailored environments to mimic biological self-organizing, ordered-bonding processes such as the neurological synthesis of complex maps from retina to cortex.

The feasibility of this approach has been demonstrated through computer simulation as well as correlation with physiological data.

Molecular Manufacturing, edited by C. Nicolini
Plenum Press, New York, 1996

By controlling respective concentrations of reagents in solution, and by depositing controlled gradients of marker-type bonding sites on substrate surface, known epitaxy and evaporation methods can guarantee that specific types of ordered patterns arise on molecular lattices. These types of patterns mimic exactly those which exist in nature to perform all of the complex signal processing and computational tasks with which we are familiar.

Such techniques will hopefully yield a fabrication and computation technology that is fundamental to the ultimate development and fabrication of self-organizing sensors and processors on the (nanometer) molecular scale.

The potential of the Scanning Tunneling Microscope (STM) as an ultrahigh-resolution instrument for surface modification was recognized almost immediately after atomically resolved imaging was demonstrated a decade ago. Since then, controlled manipulation of single atoms on surfaces has been achieved under highly ideal conditions. We must emphasize that other Scanning Probe Microscopy techniques like Atomic Force Microscopy (AFM) were developed, thus extending the applications of the new nanotechnologies. In order to take advantage of this phenomenal level of control in the emerging field of nanoelectronic device physics, we must address the critical issues related to integration of STM-based pattern generation schemes with existing and forthcoming device processing technology.

Four key areas of STM-based nanolithography on technologically relevant silicon and gallium arsenide surfaces: surface preparation, pattern generation, pattern development, and device fabrication. Since the electrical and chemical properties of structures and interfaces must be controlled on the nanometer scale, novel diagnostic, growth, and etching techniques must be employed. In particular, STM pattern generation of ultrashallow, nanometerscale oxide masks on silicon arsenic-capped, molecular beam epitaxial (MBE) GaAs and ideally terminated GaAs wafer surfaces presented, as well as the subsequent use of these masks for pattern transfer in dry etching environment. By combining the unique lateral resolution of the STM with the atomic layer control of advanced MBE in the growth direction, the integrated STM/MBE approach to nanofabrication may lead to heterojunction device structures with superior optical and electronic properties.

The field of supramolecular engineering for nanostructure fabrication represents the tools to be developed to fabricate the nanostructures. Developments are needed in the following domains:

- *Design and synthesis of molecules and supermolecules* used as partners in molecular edifices and machineries. The molecules are getting more and more complex and difficult to synthesize as the complexity of the machineries increases, making the task of synthetic chemists more and more heavy. For maximum efficiency, chemists could be coupled tightly with physicists and biologists for such a task.
- *Molecular recognition.* Supermolecules designed for highly specific recognition capabilities towards charged or neutral species have been synthesized. The skill of the labs which perform this delicate molecular engineering is now very good, so that the molecules exhibit the expected properties. These molecules will be of great help for self-assembling and *in situ* chemistry.
- *Self-assembling* by either chemical or bio-chemical molecular recognition. This method is simple and very efficient when the partner molecules have been properly designed.
- *The Langmuir-Blodgett (LB) method* and its modern variants: the alternate layers, the semi-amphiphilic method, etc. Self-assembling and LB can complement each other very successfully.
- *Organic epitaxy* on organic and inorganic substrates. While epitaxy phenomena are very well controlled in the inorganic domain (it now dominates the industry of semi-conductors), even the basic knowledge is still unexplored in the domain of

organics. The knowledge of the driving forces for epitaxy would lead to the control of the properties of interfaces.

- *Solid state chemistry in thin films. In situ* chemistry in organized molecular films (LB films, for instance) is a very powerful tool to modify molecular edifices and give them the exact required properties: redox reactions in the case of organic conductors have been shown to give rise to mixed valence if the pristine compound is properly designed, 2D polymerization has been shown to give a floating monolayer a remarkable mechanical strength if the monomer film is pre-organized in view of being polymerized along two perpendicular directions in the plane, etc. This chemistry, which exhibits the control properties of solid chemistry, but the variety of solution chemistry because of film thinness, is being developed as a tool for molecular edifice fabrication.
- *Electrochemistry on thin films* and local areas of a sample (microelectrodes). This method should also provide means of modifying the redox state of molecules once the molecular assembly is made.

The present volume considers Molecular Manufacturing in all its multiple interweaving aspects, from *de novo* design of proteins to silicon micromachining, discussing both materials (lipids, proteins, macromolecules) and methods (scanning tunneling microscopy, 2D crystallization, optical information processing, microfabrication, nuclear magnetic resonance) used for attainment of the objectives.

The role which Molecular Manufacturing occupies presently in the R&D efforts of major laboratories, mainly in Japan, USA and since 1994 also in Italy with the launching of

Figure 1. First meeting of the Board of Trustees of the Elba Project, September 1st, 1991. From left to right: Prof. K.G. Skryabin, Director of the Center "Bioengineering", Moscow; Prof. L. Berlinguer, Rector of the University of Siena; Prof. A. Ruberti, Minister of Scientific and Technological Research of Italy; Prof. C. Nicolini, Director of the Institute of Biophysics, Genoa, and Director General of the Elba Project; Prof. S. Emelyanov, Director of the International Research Institute of Management Sciences (IRIMS), Moscow.

a Strategic Project by the National Research Council, is undoubtedly to be traced back to the overall importance given to Bioelectronics in general.

Bioelectronics was suggested as an extremely important field of innovative R&D efforts to pursue already in 1982 in a Report made by the Office of Technology Assesment of the Congress of the United States of America. As early as in 1988, Italy introduced the first National Research Program directly related to bioelectronics and since that time is one of the acknowledged pioneers in this field, achieving in its research during the last few years several goals set by the original American Technology Assesment Report, namely temperature stability of proteins (1-5) and biological self-assembly (6) processes.

One of the milestones in the development of the international cooperation in the fields relevant for bioelectronics was set up by the signing of an international agreement between Italy and Russia in December 1990, an agreement signed by Italian Minister Ruberti and Russian Vice-Minister Bortnik, for cooperation in the fields of Biomolecular Engineering and Bioelectronics, which gave the origin to the “Elba Project” between the Russian Academy of Sciences and Polo Nazionale Bioelettronica. As the result of this project, the EL.B.A. Foundation has been established on March 8, 1994, involving Italian Ministry of Foreign Affairs, Italian Ministry of University and Scientific and Technological Research, Russian Ministry of Science and Technical Policy, leading research institutions from Italy, Russia and USA, with the participation of the Commission of European Union as an observer.

As a result of the feasibility study (7) for a new R&D program on bioelectronics carried out by Polo Nazionale Bioelettronica the Commission of the European Community (now Commission of the European Union) introduced into the Biotechnology program of its IV Framework 1994-1998 a section entitled “Interface of structural biology with electronics”, a first step towards a larger program dedicated to bioelectronics.

A strong evidence of the interest of the European Community are the two Worskhops on Bioelectronics, sponsored and organized by the Commission of the European Communities. The first Workshop was held at the end of 1991 in Brussels, the second one - at the end of November, 1993, in Frankfurt (Main), summarized in recent publication by Biosensors and Bioelectronics (8). Quite recently, in August 1995, a joint symposium on Protein Arrays and Biomolecular Systems, organized in Germany by the JRDC, University of Mainz and Max-Planck-Institutes, attracted the attention of top experts in the field for a free exchange of ideas in this rapidly progressing field.

The very first meeting dedicated solely to Bioelectronics was held, however, in Italy in Novara on April 4-7, 1987, and was sponsored by CIREF, an Association for Industrial Strategic Research, upon the initiative of three top industrial Chief Executive Officers (Figures 2 and 3): Prof. Amilcaro Collina (Montedison-Donegani), Prof. Umberto Rosa (SNIA-FIAT), and Dr. Pasquale Pistorio (SGS-Thomson).

In 1990 an important step forward has been made with the creation of Polo Nazionale Bioelettronica, a consortium with headquarters in Portoferraio, Isle of Elba, that comprised several leading Universities (Genoa, Pisa, Siena, Naples, Catania), national research institutions (INBB, National Institute of Health), industrial Consortium (CIREF), industrial companies (Technobiochip, SGS-Thomson, Elsag-Bailey Finmeccanica, Sigeda, Asse-Z, Raggio Italgene, AID, Technobiochip Manufacturing (TBM), Technostone, IT.IN., Alenia Spazio, CAEN, Association of Industrialists of Livorno), regional authorities (Region of Tuscany), and financial holding companies (Raggio di Sole Biotecnologie, BIC Massa).

The already mentioned National Research Program “Technologies for Bioelectronics” was launched by the Italian Government in 1991 and the first stage of it has been most successfully concluded in spring 1995. The results of the Program were astounding in that they not only were acknowledged by the scientific community but managed to convince the industrial companies, usually all too reluctant to invest in long-range R&D. Following the first stage of the National Program, the second stage is being split into two distinct programs,

Figure 2. Representatives of major Italian companies at the 1st CIREF Workshop on Bioelectronics, April 4-7, 1987, Novara, Italy.

Figure 3. At the beginning: Prof. C. Nicolini, Prof. E. Clementi (IBM), Prof. A. Collina (Ferruzzi-Montedison-Donegani) (from left to right), Novara, April 4-7, 1987.

"Advanced Biotechnology" and "Integrated Systems for Microelectronics and Bioelectronics". Moreover, late in 1994 the National Research Council (CNR) of Italy has also started a Strategic Project "Molecular Manufacturing" which focussed on the most advanced problems which emerged at the atomic scale in course of operation of various institutions within the National Research Program.

The convincement of the industrial participants was manifested on two major occasions: the participation of numerous leading Italian companies in the Scientific and Technological Park of Elba (9) and creation of a high-tech Technobiochip Manufacturing Company. Both events required major investment of the companies and hence, are a sign that the latter believe in what they invest.

Figure 4. Dr. Enrico Villa (middle), Director General of SGS-Thomson Microelectronics, Italy, at on of the EL.B.A. Forums in Marciana Marina (LI).

Figure 5. Neuronal and Submicronics Electronics, October 9-10, 1992. From left to right: Dr. L. Briozzo (SGS-Thomson, Italy), Prof. F. Faggin (Synaptics, USA), Prof. G. Musso (Elsag Bailey, Italy), Prof. C. Nicolini.

Although the EL.B.A. Forums were started as part of the training-by-research activity of the National Research Program, they have soon outgrown this aim, becoming a meeting and discussion place for top scientists coming from all over the world (Figure 5).

The name for these Forums derives not only from the place (Isle of Elba in Italy), where the conferences are periodically held every 6 months since 1991, but also from an acronym: **EL**ectronics and **B**iotechnology **A**dvanced.

The top level of these Forums was constantly ensured by the participation (Figure 6) of the most noted experts world-wide in various fields of bioelectronics (Capasso, Faggin, Croce, Nagayama, Karplus, Barraud, Hopfield, Aizawa, Wüthrich, Göpel, Moore, Sasabe, Skryabin and many others).

Indeed, even if Technobiochip, the organizer of the EL.B.A. Forums until now, completed successfully its obligations within the National Program both in terms of strictly R&D activity and of training by research, the success and international attention to these conferences were such that the EL.B.A. Foundation decided to continue from now on these meetings on its own, continuing

Figure 6. First Generation Bioelectronics and Molecular Manufacturing, July 14-16, 1993. Prof. A. Barraud (Centre d'Etudes Nucleaires de Saclay, France), Prof. F. Capasso (AT&T Bell Laboratories, USA).

Figure 7. The very first: Biomolecular Engineering, September 25, 1992. From left to right: Prof. C. Nicolini, Prof. C. Croce (Jefferson Cancer Institute, Philadelphia, USA), Prof. G. Indiveri (University of Genoa, Italy), Prof. K. Skryabin (Center "Bioengineering", Moscow, Russia), Prof. J. van Renswoude (University of Rome "La Sapienza", Italy).

to maintain the extremely high standard set by the National Program. In fact, in February 1995, the Board of Trustees of the EL.B.A. Foundation decided unanimously to continue this most successful series of scientific workshops, converting them into conferences of world reputation, the EL.B.A. Foundation Conferences. These conferences are intended as presentation and discussion forums in the various aspects of life and physical sciences, focussing, though, their attention on problems of genome studies, biomolecular electronics, protein engineering and microelectronics.

The EL.B.A. Foundation Conferences are organized in distinct annual series in accordance with the major lines of scientific activity, among which Molecular Manufacturing plays a key role.

REFERENCES

1. Nicolini, C., Protein Nanotechnology, present volume; the whole volume and refernces therein.
2. Nicolini, C., Erokhin, V., Antolini, F., Catasti, P., Facci, P., 1993, *Biochim. Biophys. Acta* 1158:273-278; Hampp, N., 1993, *Nature* 366:12.
3. Erokhin, V., Facci, P., Carrara, S., and Nicolini, C., 1995, *Journal of Physics*, in press; Facci, P., Erokhin,V., Tronin, A., and Nicolini, C., 1994c, *J. Phys. Chem.* 98(50):13323-13327.
4. Nicolini, C., 1995, *Thin Solid Films (LB7),* in press.
5. Antolini, F., Paddeu, S., and Nicolini, C., 1995, *Langmuir* 11:2719-2725.
6. Shen, Y., Safinya, C.R., Liang, K.S., Ruppert, A.F., and Rothschild, K.J., 1993, *Nature* 366:48-50.
7. A new C.E.C. R&D Programme in Bioelectronics - EL.B.A. (ELectronics and Biotechnology Advanced) Megaproject "Towards Biocomputation", Feasibility Study, Contract n°PSS*0551, Commission of the European Communities, Brussels, January 25th, 1993
8. Nicolini, C., 1995, *Biosensors And Bioelectronics* 10:105-127; Göpel, W., and Heiduschka, P., 1994, *Biosensors And Bioelectronics* 9:iii-xiii; Göpel, W., and Nicolini, C., Eds., 1994, *Biosensors And Bioelectronics* 9:601-760 and 1995, *Biosensors And Bioelectronics* 10:1-236.
9. The Economist, 1995, Sept. 9-15:90

INDEX